PLANT ANATOMY

The Late Dr C Russell Metcalfe, who both inspired and taught me. (DFC)

Professors Chris H. Bornman and Ray F. Evert who as teachers, mentors, and colleagues encouraged me to develop a fascination and passion to study functional plant anatomy. (CEJB)

The Late Richard A. Popham who first stimulated and encouraged my interest in plant anatomy. (DWS)

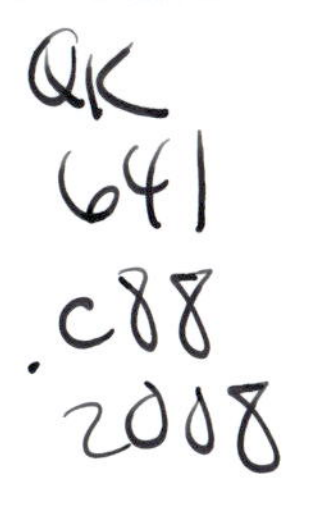

Plant Anatomy

An Applied Approach

D.F. CUTLER, C.E.J. BOTHA, D.W. STEVENSON

David F Cutler
Honorary Research Fellow
Jodrell Laboratory
Royal Botanic Gardens
Kew
Richmond, Surrey, UK

Ted Botha
Rhodes University
Department of Botany
Grahamstown
Eastern Cape Province, South Africa

Dennis Wm Stevenson
Vice President and Rupert Barneby Curator for Botanical Science
The New York Botanical Garden
Bronx, New York, USA

Blackwell
Publishing

BLACKWELL PUBLISHING
350 Main Street, Malden, MA 02148-5020, USA
9600 Garsington Road, Oxford OX4 2DQ, UK
550 Swanston Street, Carlton, Victoria 3053, Australia

First published 2008 by Blackwell Publishing Ltd

Based on the publication Applied Plant Anatomy by D.F. Cutler, published 1978 by Longman.

1 2008

Library of Congress Cataloging-in-Publication Data

Cutler, D.F. (David Frederick), 1939–
Plant anatomy : an applied approach / D.F. Cutler, C.E.J. Botha. D.W. Stevenson.
p. cm.
Includes bibliographical references and index.
ISBN 978-1-4051-2679-3 (pbk. : alk. paper)
1. Plant anatomy. I. Botha, C.E.J. (Christiaan Edward Johannes), 1946–
II. Stevenson, Dennis Wm. (Dennis William), 1942– III. Title.

QK641C867 2007
580–dc22

2007003776

A catalogue record for this title is available from the British Library.

Set in 10.5 on 13 pt Janson
by SNP Best-set Typesetter Ltd, Hong Kong
Printed and bound in Singapore
by Fabulous Printers Pte Ltd

The publisher's policy is to use permanent paper from mills that operate a sustainable forestry policy, and which has been manufactured from pulp processed using acid-free and elementary chlorine-free practices. Furthermore, the publisher ensures that the text paper and cover board used have met acceptable environmental accreditation standards.

For further information on
Blackwell Publishing, visit our website:
www.blackwellpublishing.com

Contents

Preface ix
Acknowledgements x

Introduction 1

1 Morphology and tissue systems: the integrated plant body 4
General background 4
Adaptation to aerial growth 6
The systems in detail 9

2 Meristems and meristematic growth 14
Introduction 14
Apical meristems 15
Lateral meristems 20
Practical applications and uses of meristems 21

3 The structure of xylem and phloem 28
Introduction 28
The xylem 28
The phloem 41
Structure–function relationships in primary and secondary vascular tissues 45

4 The root 48
Introduction 48
Epidermis 48
Cortex 49
Endodermis 51
Pericycle 52
Vascular system 53
Lateral roots 54

5 The stem 57
Introduction 57
Stems – cross-sectional appearance 59
Transport phloem within the axial system 64
Transport tissue – structural components 66
Concluding remarks 68

6 The leaf 70
Introduction 70
Leaf structure 74
The epidermis 76
The mesophyll 93
Strengthening systems in the leaf 102
The vascular system 103
The phloem 108
Specifics of the monocotyledonous foliage leaf 111
Secretory structures 118
Concluding remarks 119

7 Flowers, fruits and seeds 121
Introduction 121
Vascularization 121
SEM studies 123
Palynology 124
Embryology 127
Seed and fruit histology 127

8 Adaptive features 135
Introduction 135
Mechanical adaptations 135
Adaptations to habitat 137
Xerophytes 139
Mesophytes 147
Hydrophytes 150
Applications 152

9 Economic aspects of applied plant anatomy 154
Introduction 154
Identification and classification 154
Taxonomic application 155
Medicinal plants 158
Food adulterants and contaminants 159
Animal feeding habits 162
Wood: present day 163

Wood: in archaeology 165
Forensic applications 168
Palaeobotany 169
Postscript 169

10 Practical microtechnique 170
Safety considerations 170
Materials and methods 170
Microscopy 191

Appendix 1 Selected study material 195
Appendix 2 Practical exercises 203
Glossary 242
Cited references 280
Further reading 282
Index 287

Preface

Plant anatomy, the study of plant cells and tissues, has advanced considerably since the early descriptive accounts were made which consisted mainly of cataloguing what was 'out there'. Anatomical data have been applied in the better understanding of the interrelationships of plants, and in the molecular age provide confirming evidence of natural relationships of plant families in combined analyses. Plant physiologists need to know where certain processes are being carried out by plants – there are particularly interesting studies on phloem loading and the transport of synthesized materials, for example. There is a long list of applications, and these are expanded on in Chapter 1. One of us (DFC) wrote a book, *Applied Plant Anatomy* (published in 1978), aimed at reaching students who needed to know about the anatomy of plants, but found the encyclopaedic volumes daunting. This book served its purpose well, but is now very dated and long out of print.

We realized that with the passage of time, many new disciplines had been developed, and older ones expanded to a point where a much revised and updated book of this type could play an important part. Consequently, this volume was conceived, and together with the CD-ROM which takes the study of practical plant anatomy to new levels, presents a ready way for non-specialists to learn about and enjoy the subject, at their own pace and in many places, beyond the formal constraints of the laboratory.

D.F. Cutler, C.E.J. Botha, D. Wm. Stevenson

Acknowledgements

We thank The Director and Trustees of the Royal Botanic Gardens, Kew for allowing the use of photomicrographs from the first edition of *Applied Plant Anatomy:* Figs 3.1, 3.6, 3.7, 3.8, 3.12, 3.13, 3.14, 3.15, 4.2, 6.4, 6.5, 6.6, 6.25, 7.2, 7.3, 8.5, 9.2, 9.3, 9.4, 9.5, 9.6, and 9.7. Also to Dr Peter Gasson for Fig. 3.9.

Introduction

Plant anatomy is in everyday use, and remains a powerful tool that can be used to help solve baffling problems, whether this is in the classroom, or at national botanical research facilities. Many of the results may have economic value, and a good number are of increasing scientific interest. As such, the subject of plant anatomy remains alive, fascinating and very central to finding answers to many everyday structural and physiological problems. We also apply anatomy to help solve rather more academic questions of the probable relationships between families, genera and species. The incorporation of anatomical data with the findings from studies on gross morphology, pollen, cytology, physiology, chemistry or molecular biology and similar disciplines enables those making revisions of the classification of plants to produce more natural systems. The economic significance of accurate classification and hence accurate identification of plants is frequently overlooked. The plant breeder, the food grower, the ecologist and the conservationist all need accurate names for the subjects of their study. The chemists and pharmacognosists searching for new chemical substances must certainly know exactly which species or even which varieties yield valuable substances, and anatomy is important when examining relationships using molecular techniques as well. Without an accurate name and description for a plant, experiments cannot be repeated. It is impossible to say if the plants chosen for a repeat experiment are the same species as those used originally if the identity of the material is uncertain.

Plant anatomy remains a central requirement for anyone experimenting with plants. A good understanding of anatomy is essential and often overlooked by many researchers when reporting their experimental results. Misidentification of cell types and even tissues are common and difficult to correct. Our aim is to present the fundamentals of plant anatomy in a way that emphasizes their application and relevance to modern botanical research. This book is intended primarily as a reference text, for intermediate students of a first-degree course, but we hope that postgraduates will find it useful as well, as we have provided what we believe to be an understandable account of applied plant structure.

Applied anatomy is the key expression in this book. Plant anatomy is a fascinating subject, but because the tradition has been to teach it as a catalogue of cell and tissue types with only slight reference to function and development, and no mention of the day-to-day use to which this knowledge is put in many laboratories around the world, some students may be put off before they realize its interest. Textbooks have been written to suit this more usual style of teaching. These advanced texts are of excellent value for the specialist student, but can be daunting to the relative beginner. Complementary to these books are those consisting largely of illustrations. These are of great benefit to students struggling to recognize what they see under the microscope, but again have their own shortcomings in that they mainly serve to teach a set of descriptive terms, rather than the application of what is seen. This book and the associated CD-ROM, *The Virtual Plant*, concentrate on vegetative anatomy. We believe that *Plant Anatomy – An Applied Approach* will fill the niche between advanced texts and illustrated picture books, by combining core reference material with solid applied and systematic anatomy where this is relevant.

A certain amount of terminology has to be learned in order to get to grips with any subject, and here we make no excuse for using terms that are specialized in their meaning. The correct use of technical terms aids clear thought and helps to make plant anatomy as exact as possible. We define these words the first time they arise, and we have put those which we believe to be most useful into an illustrated glossary.

Far too many textbooks neglect the rich tropical flora. As such, the examples that have been chosen come from a wide range of plants from temperate to tropical environments. If you are particularly interested in the stock examples of plants used in traditional teaching, you will find many of them on the CD-ROM. Both in the book and the CD-ROM, the reader should find plants mentioned which are readily available to them to illustrate particular cells or tissues. We hope those in tropical countries will seize the opportunity to look at plants growing on their own doorsteps, instead of having to send to north temperate lands for microscope slides of unfamiliar plants! To this end, we have provided simple techniques and recipes for the preparation of non-permanent and permanent slide material as well in Chapter 10, and some examples of plants that might be studied in Appendix 1 and practicals in Appendix 2. The practical information given here is greatly expanded on in the CD-ROM, which is an essential companion to the book.

Many of us have experienced situations where budget restraints do not allow expenditure of scarce resources on expensive microscopes. Many laboratories worldwide are poorly equipped to teach plant anatomy, as fund allocation becomes more and more competitive and it becomes more difficult to justify spending money on microscopes when there is other 'must have' equipment that needs to be bought as well. It was this challenge that encouraged us to present a series of practical plant anatomy assignments in the virtual rather than the real laboratory environment. The accompanying CD-ROM achieves several things. Firstly, it allows self-paced study and exploration of plant structure. Secondly, it gives illustrated instructions on the use of the light microscope. Thirdly, it focuses attention on issues that will be encountered in the laboratory environment and, hopefully, answers more questions than it generates. Fourthly, it provides a source of reference images for instructors who need illustrations to enable them to demonstrate aspects of plant structure that they otherwise may not be able to. Finally, in its trial form it has proved to be a successful reference tool, and in this sense fulfils most of the aims which led to its creation.

The book, then, takes the reader through basic plant morphology in Chapter 1, to assist those who have little background in 'whole plant' studies; it considers the importance of micromorphology in the plants around us and looks at the challenges to which land plants are subjected. Next there is a brief account of meristems (Chapter 2). Rather than take cell and tissue types as a separate chapter, they are described in connection with the organs in which they occur, and are also illustrated in the glossary. The exception is for xylem and phloem (Chapter 3), because of their complexity and particular interest to physiologists. This is followed by chapters on root, stem and leaf. Chapters on adaptations, economic topics and techniques then follow, with appendices on selected study material and practical examples, and the glossary is the final section. An appendix of further reading completes the book.

Diagrams

Key to shading used in all diagrams throughout the book

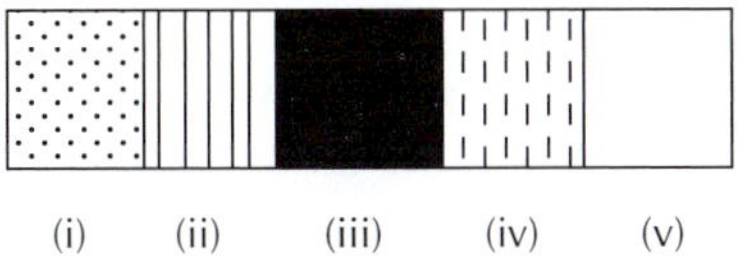

(i) (ii) (iii) (iv) (v)

(i) phloem
(ii) xylem
(iii) sclerenchyma
(iv) chlorenchyma
(v) parenchyma

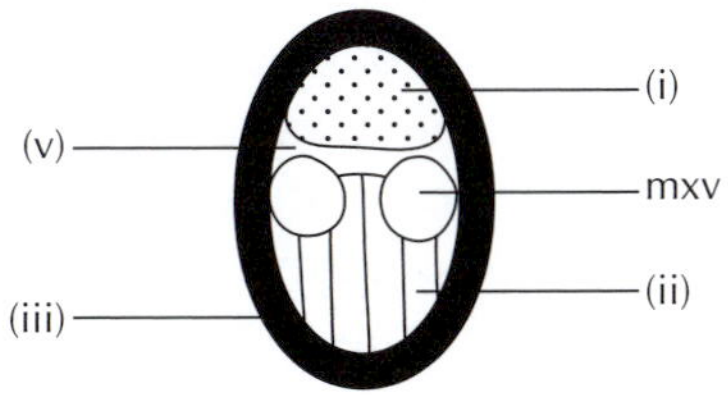

(i)
(v)
mxv
(ii)
(iii)

Example of shading in diagrams of vascular bundle T.S.
mxv, metaxylem vessel

CHAPTER 1

Morphology and tissue systems: the integrated plant body

General background

Because each organ of the plant will be discussed in detail in later chapters, this section is intended only to be a reminder of basic plant structure and arrangements of tissue systems. It is not intended to be comprehensive, and by its very nature it oversimplifies the complex and wide range of form and organization existing in the higher plants. When a specialized term is first used, it is normally defined. The glossary forms an essential part of the book, and should be consulted if the meaning of a term is not clear.

This book concentrates on the vegetative anatomy of land plants, and in particular on monocotyledons and dicotyledons (flowering plants, angiosperms, with the seeds enclosed in carpels). Some anatomical features of conifers (gymnosperms – plants with seeds but without carpels fruits, enclosing the seed) are also described. Monocotyledons (Fig. 1.1) are flowering plants that when the seed germinates start life with one seed leaf, and lack the tissues that form new (secondary) growth in thickness, the vascular cambium, and a long-lived primary root. Examples include the grasses, orchids, palms and lilies. Dicotyledons (Fig. 1.2) are also flowering plants but have two seed leaves, and like the conifers have stems that generally have the ability to grow in thickness through a formal vascular cambium, and have a long-lived primary root. Examples of dicotyledons include the bean, rose and potato families, and the conifers include such plants as pines, larches and araucarias. There are, of course, other features that distinguish the angiosperms from gymnosperms (e.g. reproductive structures and reproductive cycle).

The plant organs are shown in Figs 1.1 and 1.2. Most land plants have roots, which anchor them in the ground, or attach them to other plants (as in epiphytes). Roots also absorb water and minerals. Roots first arise in the

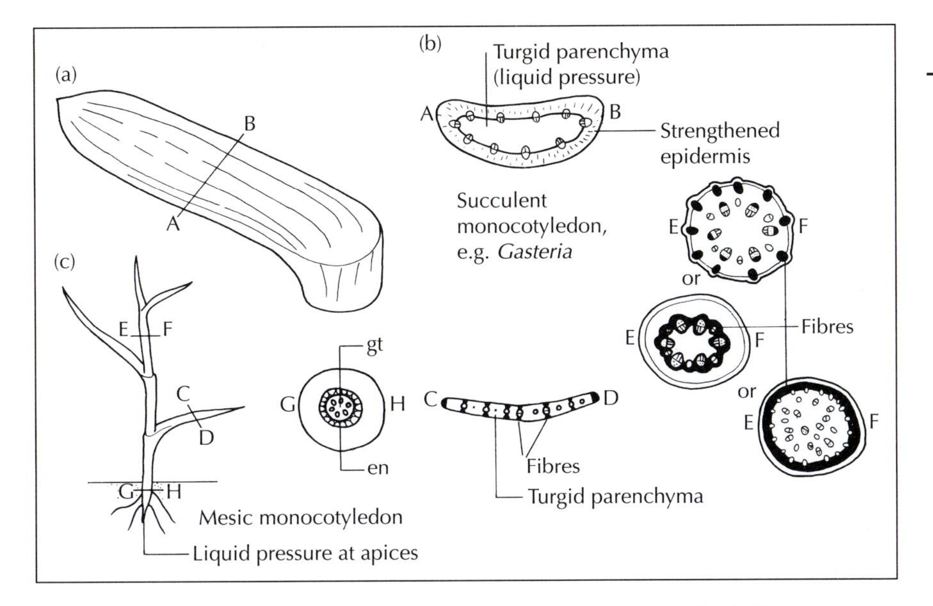

Fig. 1.1 Some mechanical systems in monocotyledons. (a) A fleshy leaf of *Gasteria*; note lack of sclerenchyma in the section (b). (c) A mesic monocotyledon, C–D shows one type of sclerenchyma arrangement in leaf TS; E–F shows three of the main types of sclerenchyma arrangements in the stem TS; G–H shows a typical root section in which most strength is concentrated in the centre. en, endodermis; gt, ground tissue, which may be lignified.

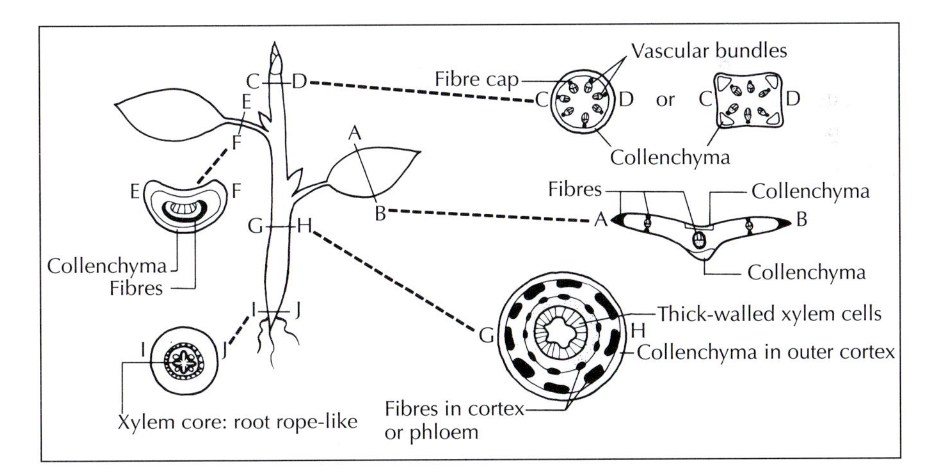

Fig. 1.2 Some mechanical systems in dicotyledons. A schematic plant with position of sections indicated. Liquid pressure occurs in turgid cells through the plant. Collenchyma is often conspicuous in actively extending regions and petioles. Sclerechyma fibres are most abundant in parts that have ceased main extension growth. Xylem elements with thick walls have some mechanical function in young plants and give a great deal of support in most secondarily thickened plants.

embryo and are there attached to the stem through a specialized region called the hypocotyl. Later in development if growth in thickness occurs, the hypocotyl becomes obscured. Many species grow additional roots, called adventitious roots, because they arise from other parts of the plant (although some roots themselves can also give rise to adventitious roots, but these do not develop from the normal sites for secondary roots). When leaves are present, they arise from the stem, either from the apical meristem (see next chapter), or from axillary bud meristems. Their particular arrangement (phyllotaxy) is usually recognizable, for example opposite one another, alternate or in an obvious spiral. Buds may be present in the axils of leaves, that is, between the leaf and the stem, close to where they join. Sometimes buds develop from other parts of the plant; these are called adventitious buds.

Adaptation to aerial growth

To understand the structure – morphology and anatomy – of land plants we have to remember that plant life started from single-celled organisms in an aquatic environment. There are still many thousands of different species of unicellular algae both in water and exposed – on tree trunks, leaves, soil and rock faces for example, in suitably moist places. Evolution of algae in the water has produced some very large, multicellular forms, for example *Laminaria* species, kelps. These large plants are fine in water, but lack the adaptations necessary for terrestrial life. They need to be bathed in water, which is a source of dissolved nutrients. Because they can absorb nutrients over most of their surface area, there is no need for a complex internal plumbing system, like the xylem (woody tissue) and phloem (cells adapted to conduct synthesized materials in the plant) in vascular bundles of land plants. They lack roots, but have holdfasts, structures adapted to anchor them to a firm substrate, but which are not absorbing organs for minerals and water, such as roots usually are. They lack a waterproof covering, a modified outer layer of epidermal cells of land plants, and rapidly desiccate if exposed to the air. Their mechanical support comes from the surrounding water, so they do not need the woody tissue (xylem) or fibres (elongate, thick-walled cells with tapered ends whose cell walls become strengthened with lignin, a hard material, at maturity; form part of the sclerenchyma) of land plants. True, they are tough and very flexible, and most can survive violent wave action. Even their reproduction depends on the release of male and female gametes into the water around them.

Some types of land plants still rely on a film of water for their male gametes to swim in to reach the female gamete and effect fertilization, for example mosses and ferns, but the higher plants like gymnosperms and angiosperms have their male gametes delivered in a protective package, the pollen grain, to a receptive female part of the cone or flower.

There is a very wide range of land habitats, and land plants show a remarkable range of shapes and sizes. This book is mostly about the anatomy of flowering plants (angiosperms), and the vast majority of these share distinct vegetative organs that are readily recognized. They are leaf, root and stem (Figs 1.1, 1.2). These organs cope with the need to obtain, transport and retain enough water to help prevent wilting, carry dissolves minerals and keep the plants cool when necessary. Most land plants contain specialized cells and tissues for mechanical support and others for movement within the plant of materials they synthesize. The tough skin (epidermis, together with a cuticle and sometimes waxy materials) prevents water loss but permits gas exchange. Small pores in the epidermis of most leaves and young stems can be opened and closed and regulated in size (see Chapter 6 for details). These are called stomata and they regulate the rate of movement of water and dissolved minerals through and out of the plant. Sometimes the epidermis is the main part of the mechanical system as well, and holds the main leaf or stem material inside under hydraulic pressure.

In many plants, the strength of the 'skin' is supplemented by tough mechanical cells arranged in mechanically appropriate areas. These are forms of sclerenchyma cells with lignified walls: fibres which are elongated cells and sclereids, which are usually relatively short; a range of types exists (see the Glossary). Collenchyma is also a supporting or mechanical tissue which occurs in young organs and in certain leaves; the walls are mainly cellulosic. Here walls are thickest in the angles between the cell walls, or in lamellar collenchyma wall thickening is found mainly on anticlinal cell walls; see below for details.

Plants submerged in water are afforded some protection from damaging ultraviolet (UV) light. Land plants need other mechanisms to prevent UV damage. The green pigment, chlorophyll, is readily damaged by UV. Since this pigment and its cohort of specialized enzymes is responsible for transforming the energy of sunlight through its action on CO_2 and H_2O into sugars, the starting point for nearly all stored organic energy on earth, it is vitally important that the UV screening methods developed are effective.

All green plants need light for photosynthesis. Plants have evolved different strategies which bring leaves into a good position for obtaining the sunlight. Some (annuals, ephemerals) put out their leaves before others neighbouring plants, complete their annual or shorter cycle and form seed for the next generation. Others retire to a dormant form (some perennials and biennials) at a time when they may be shaded by taller vegetation. Many species develop long stems or trunks and expose their leaves above the competition (some are annuals or biennials and but most are perennials). Some species do not have mechanically strong stems, but use the support provided by those which do, climbing or scrambling over them (they can be either annuals or perennials). Biennials are plants with a two-year life cycle. They build up a plant body and food reserves in the first year, and then flower and fruit in the second.

In summary, the main factors which all terrestrial plants with aerial (above-ground) stems and their associated leaves have to overcome are:

1 Mechanical, i.e. support must be provided in one way or another so that a suitable surface area with cells containing chloroplasts can be exposed to the sunlight to intercept and fix solar energy. These chlorenchyma cells may be on the surface, or just beneath translucent layers of cells. See below for more detail of the cell types that give mechanical strength. Secondary growth in thickness is another strategy that provides mechanical strength to parts both above and below ground. The growth in thickness may be relatively small in annuals, but in perennial plants it may be extensive, and requiring the use of large quantities of energy in its production. When present, the way secondary growth occurs differs between monocots and dicots.

2 Risk of excess water loss, i.e. they must be provided with protection against too much water loss from the exposed surfaces. This is generally done by a combination of a waxy outer layer and a fatty cuticle above an epidermis (the outer skin). Because water has to evaporate from some exposed surfaces so that movement of water and dissolved minerals can take place through the plant (transpiration), most leaves, and stems which retain the epidermis, have regulated pores, stomata, which can be opened and closed in response to prevailing conditions.

3 The ability to move water and minerals from the soil (transpiration) through the roots to regions where they can be combined with other materials to build the plant body, and the movement of synthesized food material from the site of synthesis to places of growth or storage and from the stores to growing cells (translocation). Of particular interest is the level of structural and physiological control of the phloem loading process. Epiphytes are attached by their roots to other plants, and obtain their water and minerals in different ways.

4 Reproduction, placement of reproductive organs enabling the pollen or gamete receptor mechanism to operate successfully, and after fertilization and spore/seed production, ensuring dispersal of the propagules.

The first three issues outlined above are dealt with by well-organized (if complex) systems in the higher plants, and will be summarized here. The fourth, reproduction, is outside the scope of this book. Secondary growth is discussed in Chapter 2, under lateral meristems.

Mechanical support systems

1 Using inflated or turgid, thin-walled cells (parenchyma): these are present in growing points, and the cortex and parenchymatous pith of many plants. They constitute the bulk of many succulent plants, for example, *Aloe, Gasteria* leaves, *Salicornia* from salt marshes and *Lithops* from desert regions. The cell wall acts as a slightly elastic container; internal liquid pressure inflates the cell so that it becomes supporting, like the air in an inflated car tyre. Its support properties depend on water pressure, so a water shortage can lead to a loss of support and wilting. Some fairly large organs can be supported by this system, but they usually rely on the additional help of devices that reduce water loss, such as a thick cuticle, and perhaps also thick outer walls to the epidermal cells, and specially modified stomata. A strong epidermis is particularly important, because it acts as the outermost boundary between the plant cells and the air. A split in the skin of a tomato, for example, rapidly leads to deformation of the fruit, or a cut in the succulent leaf of a *Crassula* or *Senecio* rapidly opens up. Not many plants rely on the turgid cell and strong epidermis principle alone.

2 Both monocotyledons and dicotyledons and have specially developed, elongated, thick-walled fibres, in suitable places, which assist in mechanical support. Alternatively, they have especially thick-walled, generally elongated parenchyma cells (also sometimes called prosenchyma); or, in those primary parts of the stem where growth in length is continuing, collenchyma cells may be present. Although there are only a few common ways in which specialized mechanical supporting cells are arranged in the stem, leaf or root, it is the variations on these themes which are of particular interest to those who have to identify small fragments of plants, or make comparative, taxonomic studies. The variations will be dealt with in detail in the chapters dealing with each organ. Obviously, to be effective the mechanical system must be economical in materials, and the cells must not be arranged in such a way as to hinder or impede the essential physiological functions of the organs.

The mechanical systems develop with the early growth of the seedling. Whilst turgid cells are the only means of support at first, collenchyma may rapidly become established, particularly in dicotyledonous plants. This tissue is concentrated in the outer part of the cortex, and is frequently associated with the midrib of the leaf blade, and the petiole.

Collenchyma is essentially the strengthening tissue of primary organs, or those undergoing their phase of growth in length. The cells making up this tissue have thickened cellulosic walls at their angles, are rich in pectin and are often found with chloroplasts in their living protoplasts.

Sometimes the only other mechanical support is provided by the wood (xylem) composed of tracheids (imperforate tracheary elements, i.e. cells with intact pit membrane q.v., between them and adjacent elements of the vascular system), as in most gymnosperms, or by the tracheids, vessels (tube-like series of vessel elements or members with perforate common end walls; vessel elements are the individual cell components of a vessel, with perforated end walls) and xylem fibres of the angiosperms. However, far more commonly there are also fibres outside the xylem (extraxylary fibres) which are arranged in strands or as a complete cylinder, such as in *Pelargonium* which can give considerable strength to herbaceous plants, and particularly in herbaceous monocotyledons in their stems and leaves. The much elongated fibres, with their cellulose and lignin walls, are not so flexible and do not stretch as readily as does collenchyma; consequently they are often found most fully developed in those parts of organs that have ceased growth in length.

Figure 1.1 shows some fibre arrangements in monocotyledon stems and leaves. In the leaf, fibres commonly strengthen the margins (e.g. *Agave*) and are found as girders or caps associated with the vascular bundles. In the stem, strands next to the epidermis can act rather like the iron or steel reinforcing rods in reinforced concrete. Together with a ribbed outline that they often confer on the stem section, they produce a rigid yet flexible system with economy of use of strengthening material.

Tubes are known to resist bending more effectively than solid rods of similar diameter; they also use much less material than the solid rod. It is not surprising then, that tubes or cylinders of fibres commonly occur in plant stems. They may be next to the surface, further into the cortex, or may occur as a few layers of cells uniting an outer ring of vascular bundles (Fig. 1.1).

The various arrangements within leaves, stems and roots will be discussed in more detail in Chapters 4–6. Mention must be made here that in some monocotyledon stems individual vascular bundles scattered throughout the stem can each be enclosed in a strong cylinder of fibres, which form a bundle sheath. Each bundle plus its sheath then acts as a reinforcing rod set in a matrix of parenchymatous cells and with a sieve cell centre so the whole unit acts as a hollow cylinder with maximum efficiency of both transport and strength.

Fibres or sclereids in dicotyledon leaves are also often related to the arrangement of the veins in the lamina and to the petiole vascular traces. These are shown in Fig. 1.2. The concentration of strength in an approximately centrally placed cylinder or strand in the petiole permits considerable torsion or twisting to take place as the leaf blade is moved by the wind, without damage occurring to the delicate conducting tissues. Primary dicotyledonous stems may have fibres in the cortex and phloem. The subterranean roots of both monocotyledons and dicotyledons have to resist different forces and stresses from those imposed on the aerial stems – tensions or pulling forces, as opposed to bending forces. The concentration of strengthening cells near the root centre gives it rope-like properties. See Chapter 4 for a further development of these themes.

The transport systems

It is not possible to present a simple, comprehensive model to demonstrate the wide range of arrangements of vascular systems that occur in vascular plants, or in either dicotyledons or monocotyledons for that matter. Dicotyledons that are composed of wholly primary tissues tend to be a little more stereotyped than monocotyledons, but even then there is a very wide range of arrangements.

The essential elements of both systems are the xylem, concerned with transport of water and dissolved salts, and the phloem, which translocates synthesized but soluble materials around the plant to places of active growth or regions of use or storage. Xylem strands and phloem strands are normally associated and together form the vascular bundles, and are often enclosed in a sheath of fibres, and in addition, in some instances, an outer sheath of parenchyma cells (the bundle sheaths). Vascular bundles make up the 'plumbing system' of primary tissues, and organs without secondary growth in thickness.

In the apex (tip) of the shoot and root, where vascular tissue is not yet developed, soluble materials and water move from cell to cell through specialized very fine strands of protoplasm (called plasmodesmata) in these relatively unspecialized zones. Not far back from these growing points, however, more formal conducting systems are needed to cope with the flow of assimilate and water. Procambial strands, strands of elongated, thin-walled cells which are the precursors of the vascular bundles, are seen first and then, further from the tips, differentiation of protophloem (first formed primary phloem) alone followed by protoxylem (first formed primary xylem) and then by the metaphloem and metaxylem (next formed phloem and xylem cells respectively). The protoxylem and metaxylem, protophloem and metaphloem together constitute the primary vascular tissues. In most dicotyledons, the newly formed strands join the previously formed vascular bundles in the stem through a leaf or branch gap, which is composed of parenchyma cells, and 'breaches' the harder tissues associated with the plumbing of the stem.

In most dicotyledons the leaf lamina (blade) has a midrib to which are connected the lateral veins. The latter form a network composed of major and minor systems. The midrib is directly connected to the petiole trace, the vascular system of the petiole or leaf stalk. This enters the stem and joins into the main stem system through a leaf trace gap as described above. In the primary stem, all vascular bundles are separate from one another except at the nodes – those parts of the stem where one or more leaves are attached. Vascular bundles in the stem may remain separate in many climbers, e.g. *Cucurbita*, *Ecballium*, but in most dicotyledons the bundles become joined into a cylinder by growth of secondary xylem and phloem from vascular cambium (a lateral meristem composed of thin-walled cells from which the secondary vascular tissues develop); it is made up of the fascicular cambium forming within the vascular bundle and the interfascicular cambium between vascular bundles.

A complex rearrangement of tissues takes place in the primary plant where the systems of the stem and root meet (hypocotyl). In the stem vascular bundles, the phloem is normally to the outer side of the xylem in the majority of plants. In the root, as seen in cross-section, the xylem is central, and may have several lobes or poles, with the phloem situated between these. After secondary growth has taken place, the hypocotyl becomes surrounded by secondary xylem and phloem, and the shoot and root anatomy become more similar. Secondary growth is discussed in Chapter 3.

Transfer cells are specialized parenchymatous cells found in various parts of the plant, but in particular, in regions where there is a physiological demand for transport, but where more normal phloem or xylem cells are not in evidence. A good example is the junction between cotyledons (first seedling leaves) and the shoot axis in seedlings. Transfer cells may also be present near the extremities of veins, or near to adventitious buds (buds developing in an unusual position, e.g. on a stem in addition to or replacing those in leaf axils, or buds on root or leaf cuttings).

Thin sections of the walls of transfer cells show them to have numerous small projections directed towards the cell lumen (the part of the cell to the inner side of and enclosed by the cell walls). These greatly increase the plasmalemma–cell wall surface interface. a site of metabolic activity concerned with the rapid, energy-mediated movement of materials between adjacent cells. The projections are so fine that conventional sections with a rotary microtome are too thick for them to be seen.

Monocotyledons are quite different from dicotyledons in their vasculature. Leaf and stem are commonly much less readily separable as distinct organs. There is no secondary growth by a true vascular cambium, so a cylinder of vascular tissue does not form. When secondary growth occurs, as in *Dracaena* and *Cordyline*, it is by means of specialized tissue, situated near to the stem surface, which forms complete, individual vascular strands and additional ground tissue.

Vascular bundles are usually arranged in the stem with the xylem pole facing towards the stem centre (but this is not invariably so). The arrangement of leaf vascular bundles is very variable. Grasses and some *Juncus* species, for example, often have one row as in Fig. 1.3. Some of the other types of arrangement are discussed in Chapter 6.

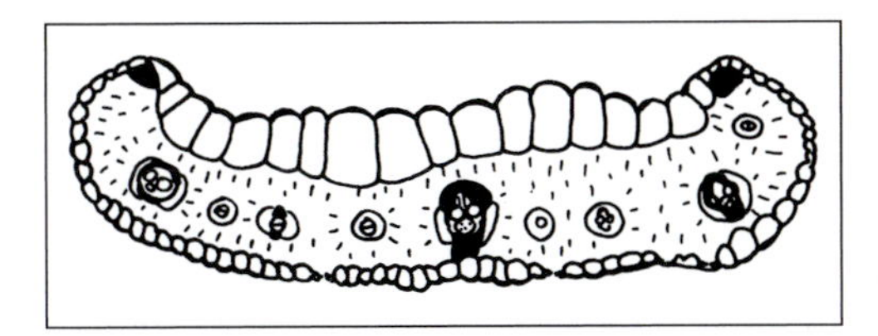

Fig. 1.3 *Juncus bufonius* leaf (TS, ×48), showing one row of vascular bundles, with the xylem poles directed towards the adaxial surface. Note the marginal sclerenchyma strands and the difference in size between adaxial and abaxial epidermal cells. Each small vascular bundle has a parenchyma sheath; in larger bundles sclerenchyma caps interrupt the parenchyma sheath.

Because there is no vascular cylinder in monocotyledons, where leaf traces (bundles) enter the stem they do not form gaps. They may join at nodes, where all the bundles at that particular level of the stem form a type of plexus, as in aloes. Sometimes, in stems with nodes, the leaf traces may continue downwards from their points of entry into the stem for a complete internode before joining the nodal plexus below (e.g. *Restio*, *Leptocarpus*, Restionaceae). In other plants without nodes (e.g. palms), the leaf traces follow a simple path curving inwards towards the stem centre, and then gradually 'move' towards the outer region of the stem lower down. These leaf traces join onto the main bundles by small, inconspicuous bridging bundles. This system is beautiful in its simplicity, but very difficult to analyse because there are so many (several hundred) vascular bundles even in the narrow portion of a stem of a small palm like *Rhapis*. As one follows the course of bundles in a palm, they are seen to spiral down the stem.

The primary root does not develop in a majority of monocotyledons. Its function is usually taken over by numerous adventitious roots that arise at an early stage, usually at the nodes, and join the stem vascular system in what frequently appears as a jumble of vascular tissue with very short elements both in the phloem and xylem.

CHAPTER 2

Meristems and meristematic growth

Introduction

Growth takes place in two stages in plants: first there is the division of cells of an undifferentiated type (simple, thin-walled parenchyma) adding to the number of cells; then there is the enlargement of some of the cells produced by these divisions.

Dividing cells of the undifferentiated type are not present throughout the plant, but are concentrated in particular places. In addition to these, certain cells in most organs remain relatively undifferentiated and may begin to divide if the appropriate conditions arise and after they have undergone a process known as dedifferentiation. Such cells give rise to adventitious roots and buds, or to the callus tissue which forms during wound healing. They are of great importance to the horticulturalist. The ability of such cells to divide is a basic requirement for the success of many forms of vegetative propagation and grafting.

Cells that divide actively to produce the primary plant body are associated together in meristems. These comprise the apical meristems at the tips of shoot and root and the tips of lateral shoots or roots. Some plants have active meristems just above and near to most nodes; these are the intercalary meristems.

When secondary growth occurs, that is, growth in thickness, lateral meristems are involved. The vascular cambium occurs in dicotyledons and gymnosperms and is the best known of the lateral meristems. Growth in thickness of stem and root causes the primary covering layer of the plant, the epidermis, to split. A secondary protective barrier between delicate tissues and the outer world is developed to replace the epidermis. It consists of layers of cork cells, derived from the specialized cork cambium or phellogen, also a lateral meristem.

In the dicotyledon leaf, cells continue to divide in various areas of the expanding lamina, some until the mature size has almost been attained, when they cease division and the products expand. Leaves in monocotyledons are different in that most have an additional basal zone of meristematic tissue that continues growth for long periods, until the mature leaf size has been reached.

Certain monocotyledons have secondary growth in stem thickness (secondary thickening meristem), although many of the larger ones do not, for example the palms. *Dracaena* (Ruscaceae) and *Cordyline* (Laxmanniaceae), *Klattia*, *Pattersonia*, *Nivenia* and *Witsenia* (Iridaceae) serve as examples in which there is a special zone of meristematic cells in the outer part of the cortex (that part of the stem to the outside of the region containing primary vascular bundles). In the cortex entire vascular bundles are formed, with new secondary ground tissue between them.

Clearly, the growing plant is exceedingly complex, containing areas that are juvenile and have actively dividing cells close to other tissues that are fully formed and mature.

Apical meristems

There are detailed differences between the meristems at the apex of shoot and root of monocotyledons, dicotyledons and gymnosperms. Three shoot apices are shown in Fig. 2.1.

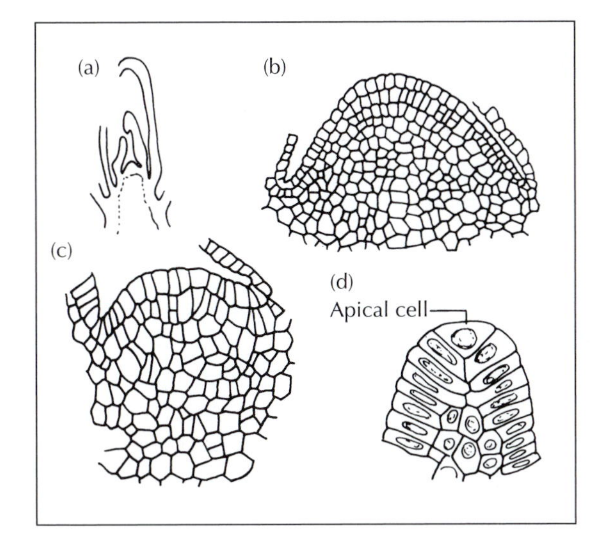

Fig. 2.1 Vegetative meristems. (a) Low-power diagram LS of *Rhododendron* apex, ×15. (b) Detail of (a), ×218. The second layer may be 'tunica' but has some periclinal division, as does that in (c), *Syringa* ×218. (d) *Equisetum* ×218 has an apical cell and not a group of meristematic cells.

Since the earliest observations, writers have attempted to classify the various layers of cells in apices that are visible in longitudinal sections. Classification of these layers is made based on the fate of cells derived from the distinct layers, or on the dominant planes of cell division apparent in the layers. For example, in the tunica and corpus theory, the (outer) tunica layers are distinguished from the inner, corpus layers because their cell divisions normally occur in the anticlinal plane only (that is, at right angles to the surface of the plant). In the corpus, divisions are both anticlinal and periclinal. Periclinal divisions are parallel with the outer surface of the plant. If a formal naming of layers must be made, the tunica–corpus system is more reliable than Hanstein's dermatogen–periblem–plerome system. In Hanstein's system, the layers are defined in relation to the tissue systems to which they are purported to give rise. It has been shown by experiment that particular layers do not consistently produce the same tissue system in the same species. As such, it is possibly better to use a topographic system and label layers L1, L2, L3, etc., and define various zones descriptively.

In the shoot apex, leaves usually arise from the tunica layers normally (Ll, or L1 and L2) and buds from the tunica and some corpus layers. The tunica produces the epidermis and usually most if not all the cortex. Usually the mature epidermis is composed of one layer of cells, but in some species cell divisions occur very early in the epidermis during the development of leaves, leading to the production of a multiple epidermis. This can be seen in Fig. 2.2, part of the apex of *Codonanthe* sp. (Gesneriaceae). Part of the mature multiple epidermis of this plant is shown in Fig. 2.2. The corpus produces the vascular system of the stem and the central ground tissue. Occasionally, the cells below the apical meristem proper may appear to be a relatively inactive zone with little or no division; this region is termed the quiescent zone, but its inactive state is not acknowledged by all, and experiments using radioactive tracers indicate that there is some cell division in these regions, albeit significantly less than the regions surrounding it. A regular rib-like arrangement of cells can also be detected in some apices below the tunica and corpus. The cells of the meristem have dense cytoplasm, lacking large vacuoles (liquid-filled spaces within the cytoplasm). Below the apical areas of active cell division, the cells begin to enlarge and vacuolate.

Attempts have been made to define different types of organization of the cell zones in both gymnosperms and angiosperms, and these types may have some significance in indicating interrelationships. They go beyond the scope of this book, but the interested reader can follow up references in the further reading section at the end of the book.

As leaf buttresses arise in sequence at the shoot apex, in the phyllotaxy (leaf arrangement on the stem) characteristic of the particular species, the procambial strands become apparent, and from them are derived the first formed phloem and then the xylem of the primary bundles. Figure 2.3

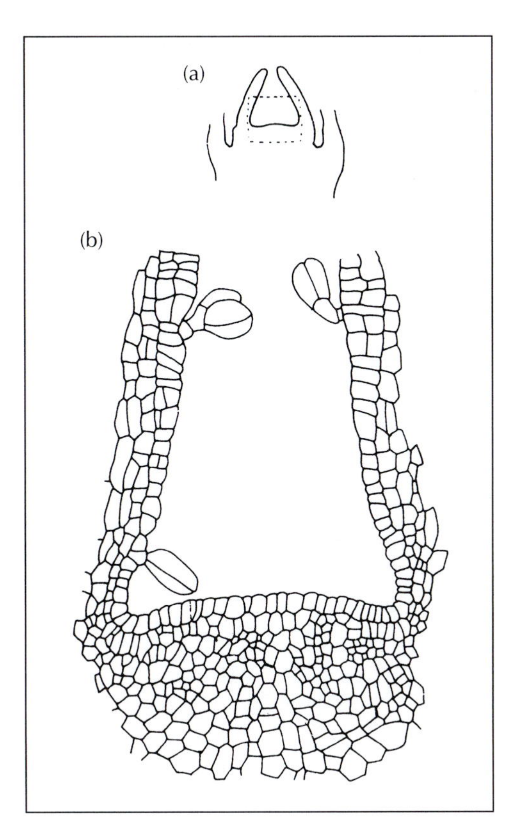

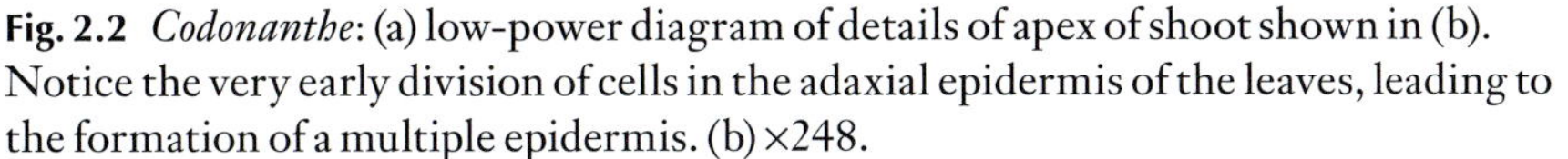

Fig. 2.2 *Codonanthe*: (a) low-power diagram of details of apex of shoot shown in (b). Notice the very early division of cells in the adaxial epidermis of the leaves, leading to the formation of a multiple epidermis. (b) ×248.

shows leaf buttresses (arrow) in *Coleus* longitudinal section. Many experiments have been conducted to try to discover the mechanisms that regulate the orderly development of these dynamic, growing apices. Control of spacing of leaf buttresses is not fully understood. Numerous experiments involving the use of mechanical devices to try and isolate one part of the apex from the rest have been carried out, and despite these painstaking experiments with growth hormones there is still a great deal to learn. It is very difficult to conduct experiments in which only one variable at a time is studied. Also, apices develop in the very enclosed, protected environment of the leaf bases that has to be substantially disturbed so that observations can be made.

The root apex is similar in many respects to the stem apex and may also have a quiescent zone, but it has one conspicuous, major difference. It has a root cap, or calyptra, frequently produced by a meristematic zone called the calyptrogen (Fig. 2.4a). The cap acts as a buffer between the soft apical

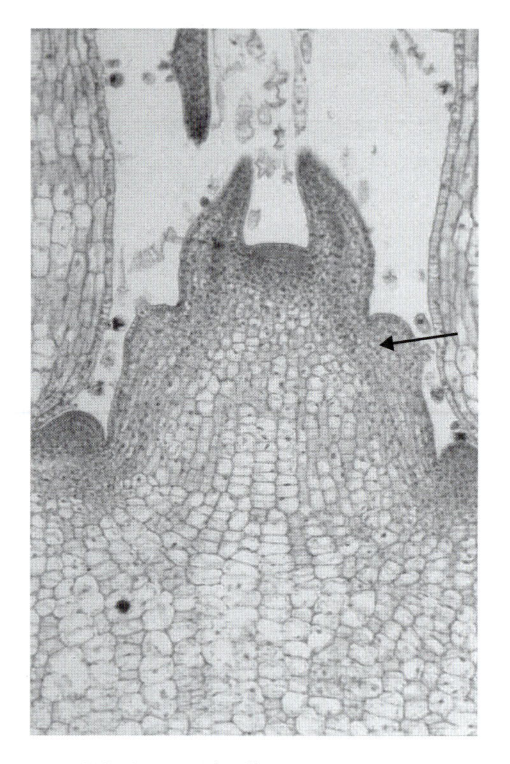

Fig. 2.3 *Coleus* shoot apex, LS. Arrow = leaf buttress. C × 100.

meristem and harsh soil particles. It wears away as growth progresses, but it is constantly renewed. It is believed to be the source of growth regulating substances that are involved in the positive geotropic response of most roots. Root caps can be seen easily on aerial roots of *Pandanus* and many epiphytic orchids and submerged roots in aquatics (*Pistia*). Besides the calyptrogen, the cell layers involved in the production of root epidermis and cortex, and the primary vascular system can be readily defined in thin, longitudinal sections suitably stained. In some roots no distinct calyptrogen is produced. In *Allium* (Fig. 2.4b, c) a column of cells develops.

Whereas the shoot apex soon produces leaves and buds exogenously (in the outermost cell layers), the root organization is quite different. Lateral roots arise endogenously, from the pericycle cells (a single or multiple layer of parenchymatous cells to the outer side of the vascular system and to the inner side of the cylinder of cells with characteristically thickened walls, the endodermis), some distance from the apex (Fig. 2.5). This deep-seated origin requires that the lateral roots grow forcibly through the endodermis and cortex to reach the exterior. The distance to be spanned between the lateral root vascular system and that of the root from which it arises is short.

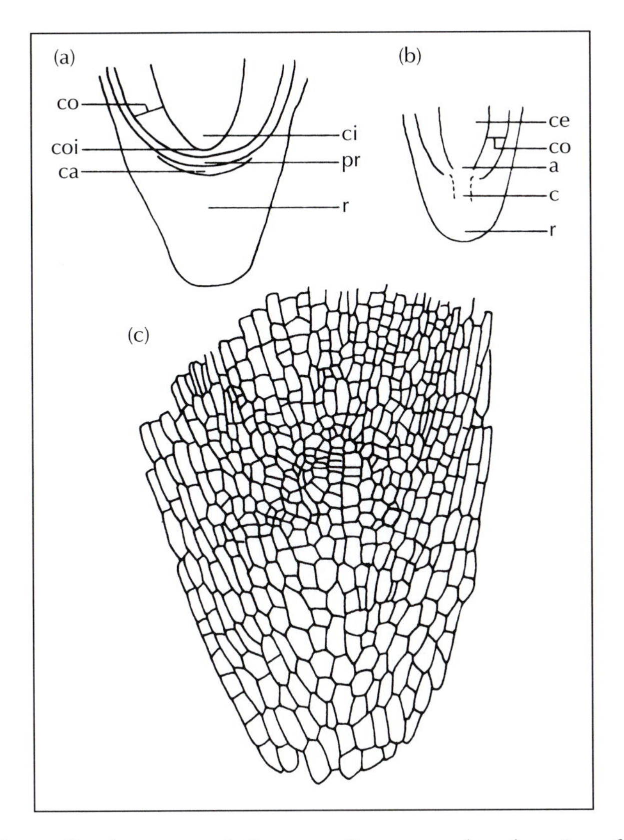

Fig. 2.4 (a) Generalized monocotyledon root, diagram to show location of various zones. Root apex, LS, in *Allium* sp.; (b) low-power diagram to show location of various cell zones in (c) (×109). a, apical meristem; c, column; ca, calyptrogen; ce, central cylinder; ci, central cylinder initials; co, cortex; coi, cortex initials; pr, protoderm initials; r, root cap.

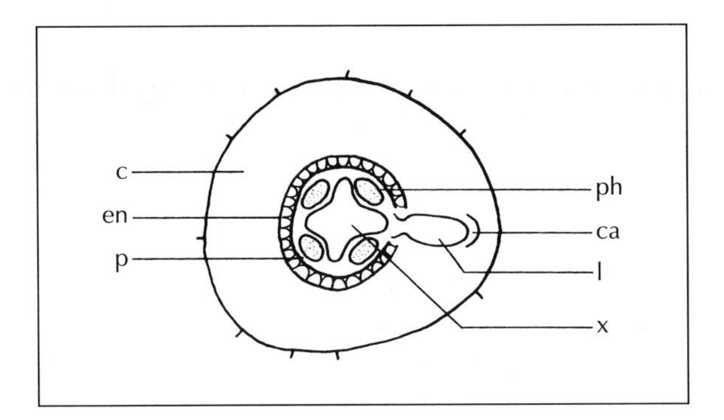

Fig. 2.5 Endogenous development of a lateral root, in TS. c, cortex; ca, small cavity ahead of developing lateral root formed by lysis of cortical cells; en, endodermis; l, lateral root; p, pericycle; ph, phloem; x, xylem.

The vascular system from a bud apex has to develop towards the main stem system and eventually becomes joined to it.

Numerous papers have been published on the apical organization of shoots and roots in many plant families. Some are comparative and aim at conclusions of taxonomic significance, but others, and probably the more useful, are concerned with the development of the particular plants under study. As mentioned before, proper developmental studies demand a high degree of competence, and are vital to an understanding of mature plant forms.

Lateral meristems

The principal lateral meristems are the vascular cambium and the cork cambium (phellogen). The vascular cambium, a feature of dicotyledons and gymnosperms and other higher plants, is described briefly below. It consists of one to several layers of thin-walled cells arranged in a cylinder, and its actual thickness is difficult to define, since those cells cut off to the outer side develop into phloem, and those on the inner side into xylem and development is gradual, so early stages of either tissue may still look like cambial initials. Normally, many more of the products of divisions of the cambial initials are xylem than phloem cells. Formation of new layers of xylem effects the displacement of the cambium itself away from the centre. Some of the cambial initials divide anticlinally to allow for the necessary increase in circumference, which grows at approximately six times that of the radius (i.e. $2\pi r$). The first formed cambial initials are axially elongated or fusiform initials. From these the elongated cells of the xylem develop. Some of these become subdivided, by one, two or more cell divisions, to form an axially oriented row of shorter cells, the ray initials. From these the rays develop (Fig. 2.6). See Chapter 3 for further information.

The phellogen can arise near the epidermis, or deeper into the cortex, as a cylinder of cells. It divides to form several layers of thin-walled, radially flattened cells to its inner side, called the phelloderm, and phellem to the outer side, the radially flattened cells of which may all become thicker-walled, and a fatty substance called suberin is incorporated into the walls. In some species, several layers of suberized cells are separated by layers of thin-walled parenchyma cells, so that alternating bands are produced (see Chapter 5 for more details). Often to the inner side of the stomata of the original epidermis, the suberized cells are not flattened, but rounded, and have air spaces between them. These are the lenticels.

In monocotyledons that have secondary thickening, the lateral meristems form in the cortex and produce both vascular bundles and ground tissue, as described above.

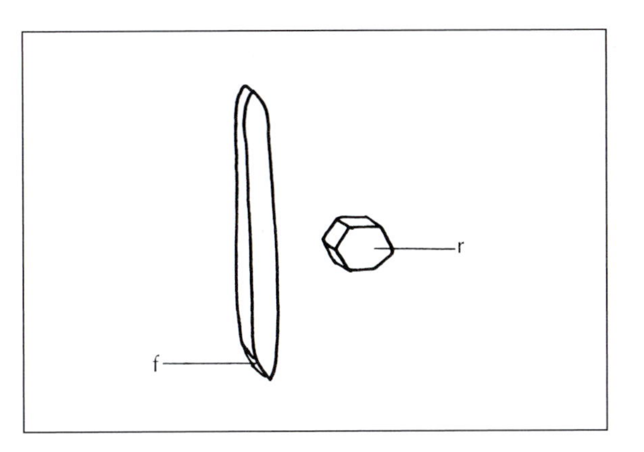

Fig. 2.6 Diagram of fusiform cambial initials (f) and ray initials (r).

Practical applications and uses of meristems

The special properties of the simple, thin-walled meristematic cells, whose anatomy and location in the plant are described above, make possible a number of horticultural techniques. Understanding the position of the meristems in the whole plant, and the delicate nature of meristematic cells, is of importance to those who wish to increase plants by vegetative propagation, increase the number of branches in a plant, make grafts, and improve the chances of good wound healing.

Apical meristems

The main practical uses for apical meristems, particularly shoot meristems, is meristem culture so that new plants can be produced vegetatively. The cells of meristems are undifferentiated parenchyma, and are in an ideal state for cell division to take place. As soon as the meristems are removed from the plant, their formal organization appears to be disrupted, and they are vulnerable to desiccation. They must be cut off carefully and transferred immediately to a nutrient medium. All stages of this technique must be aseptic so that no pathogens are introduced. If the process is successful the apex will first form a mass of callus-like tissue, similar to the orchid protocorm. Small embryonic shoots and roots form subsequently. If the tissue mass is subdivided, a number of small plantlets can be produced. It is important to have a correctly formulated growth medium often particular to the plant under culture or the tissue mass may produce only shoots or roots!

There are several circumstances where it is desirable to reproduce plants by meristem culture. For example, the required plant may be infertile, as in the case of a triploid, or it could be an F1 hybrid that would not breed true. It is also a useful method for the rapid increase of nursery stock for commercial purposes. Other vegetative methods of propagation might take several additional years before a similar number of plants could be produced. Virus diseases rarely infect apices, and meristem culture can be used to produce virus-free stock from otherwise infected plants, for example in the raspberry and the potato. Species nearing extinction may be rescued and multiplied by meristem culture. This may be the only practical approach if the breeding population is very small, or if those plants remaining are self-incompatible. Unfortunately, all of the products of an individual meristem culture will have the same genotype, so genetic diversity cannot be increased by this method of multiplication.

As a method of propagation, meristem culture would appear to have a bright future. It probably has more potential than the longer established callus culture method, whereby small portions of excised tissue (usually parenchyma) from various parts of a plant are cultured in, or on, a nutritive medium. It may take a long time to induce embryonic plants to differentiate from such a callus.

When large enough to handle, the embryonic plants are detached and grown on a sterile medium to a size at which they can be potted on, in normal potting compost.

Intercalary meristems

Intercalary meristems are also used in horticulture for propagation. In the plant one of their functions is to cause a stem that has fallen over to grow back in an upright position, for example in *Triticum*, carnation. Carnations will serve as a practical example of where an intercalary meristem is capable of producing adventitious roots. In Fig. 2.7 a carnation stem is shown cut off the plant just below a node. It is split longitudinally through the node, into the intercalary meristem zone. In horticultural practice, the split is often held open with a small piece of stick, until adventitious roots develop from the split sides.

A large number of plants quite readily form adventitious roots from the nodes, whether split off the plant or not. Considerable use is made of this property in horticulture for propagation.

Lateral meristems

Lateral meristems are also used in techniques designed to propagate plants by cuttings and in grafting, or in promoting wound healing.

The cork cambium is so specialized as to be of little value in plant propagation. It frequently plays a part in wound healing, and is, of course, employed commercially in the production of 'cork' from *Quercus suber*, the

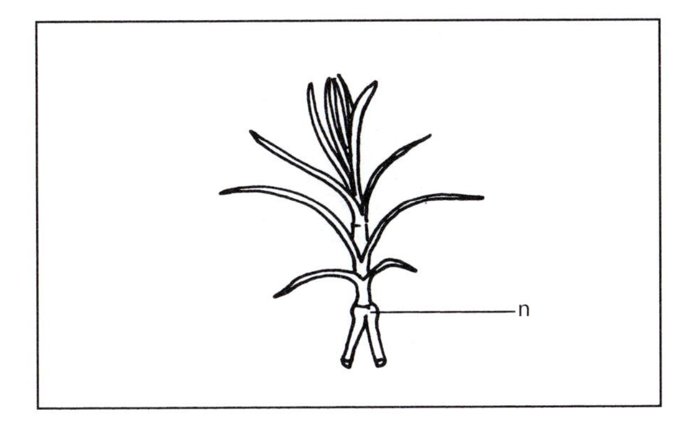

Fig. 2.7 *Dianthus* (carnation) cutting. Adventitious roots will develop from the split sides. n, node.

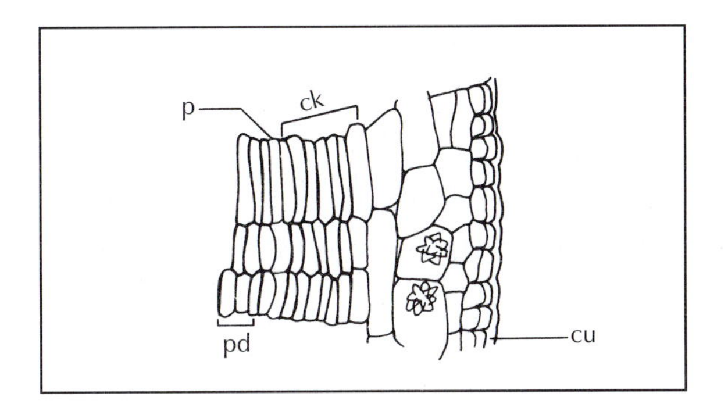

Fig. 2.8 *Ribes nigrum*, TS of sector of outer part of stem to show deep-seated cork cambium, ×218. ck, cork; cu, cuticle; p, phellogen; pd, phelloderm. Note cluster crystals in cortical cells.

cork oak, in which the cork layers are harvested at approximately ten-year intervals. Cork cambium or phellogen re-forms after the cork is carefully removed. Figure 2.8 shows a cork cambium in *Ribes nigrum*.

Of the lateral meristems, it is the vascular cambium between phloem and xylem that is most often employed by horticulturists. Its normal function in the healthy woody or herbaceous dicotyledon is to produce new cells of phloem and xylem (see Chapter 3).

If a cambium is wounded, it will normally regenerate and by influencing the developmental pathways of callus cells next to it, will assist in the healing process, so that cambial continuity is regained, and new cylinders of phloem and xylem established.

The forestry practice of removing lower branches on conifers at an early stage enables the wound to heal over (Fig. 2.9) and entire rings of sound new wood may become established. If 'snags' or broken ends of branches are left, it is some time before they are grown over, and bad knots of dead tissue and hence weak places in the harvested timber will inevitably result.

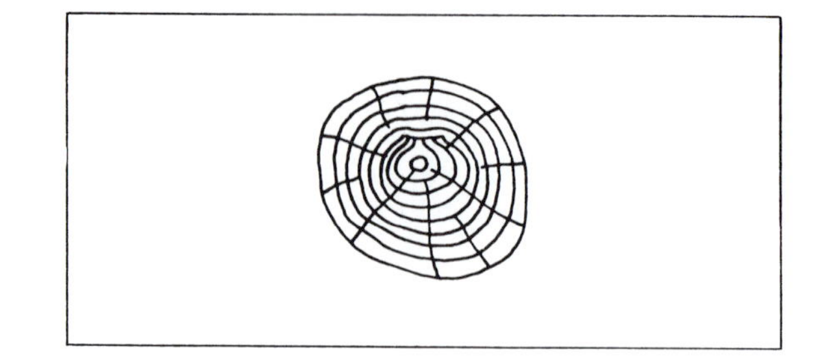

Fig. 2.9 Diagram of TS of conifer log showing resumed continuity of growth rings after lateral branch has been cut off.

The inherent ability of wounds to heal is widely used in grafting techniques. In order to establish a graft, the parts of the two plants to be united are 'wounded'. This is done by cutting the root stock and scion (short section of stem, with buds, or bud alone). The two are brought together in such a way that the cambia of stock and scion are aligned as closely as possible. When new growth is formed by callus cells, the two cambia can then quickly establish continuity by specialized differentiation of some of the callus cells and a firm, even bond is produced. No cell fusion takes place, but eventually the xylem products of the two (joined) cambia firmly bond stock and scion together. It is essential that the stock and scion should not be able to move relative to one another during the early stages, and in these stages grafting tapes are used which both give a secure bond and permit diffusion of oxygen essential to cell growth. The tape must either perish by itself in due course, or be easily removed when the graft is secure. Aftercare is very expensive for the horticulturist, and the simpler the method and the less handling involved, the better. Air gaps between stock and scion must be avoided; they can harbour pathogens, or permit the entry of water and pathogens.

Bud grafting works in much the same way, and the chip bud graft (Fig. 2.10) is becoming increasingly popular and replacing the older T-cut method. The bud cambium can be aligned more accurately by this newer method. A chip bud is removed and inserted behind the small lower lip of bark of the stock at the depth of the cambium and the bud secured by grafting tape.

The advantages of grafting are manifold. For instance the roots of some desirable species may be very weak, and vigorous roots can be grafted in their place, as in *Juniperus virginiana* where *J. glauca* stock is employed. Water melon with wilt-prone roots can be grafted onto a gourd root stock that is *Verticillicum* wilt resistant. The sizes of fruit trees, particularly apples and pears, may be regulated by careful selection of root stock vigour. Trees of relatively fixed mature size can be produced, and earlier fruiting induced.

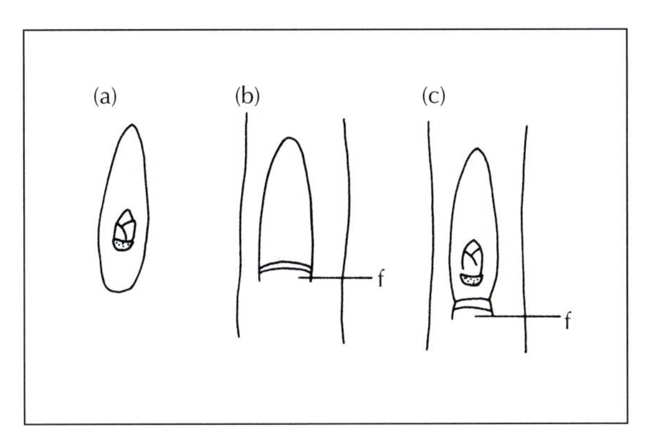

Fig. 2.10 Chip bud graft. (a) Chip with bud; (b) stock prepared; (c) chip inserted behind small flap of bark (f), ready for taping.

In the UK, the Malling Merton system provides trees with numbered rootstocks guaranteeing a mature tree with specific characteristics. Uniformity of size is essential for good husbandry. The dwarfing stocks have xylem vessel elements that are much narrower in diameter than those of the stocks producing large trees.

Bridge grafts can be used to repair trees that have been ring barked (Fig. 2.11). It is important to use twigs from the same species, because compatibility between stock and scion is essential. The twigs must be inserted so that their distal ends point away from the roots, to retain the correct polarity. In fact, the interrelationship between plants can be tested to a limited extent by their intergrafting ability. Species from the same genus will frequently unite, for example, *Prunus* species. *Solanum* species can also be grafted together. Graft hybrids between genera are much less common, for example *Laburnum/Cytisus*. Grafts between plants from different families occur rarely. However, it has been possible to demonstrate the relationship between the Cactaceae and the Madagascan endemic family Didieraceae from the production of successful interfamily grafts. In the wild, it is quite common for roots of individual trees of the same species growing closely together to become grafted together. Roots abraded by the soil, or damaged by other organisms, form callus, and are thus 'prepared' for grafting. The rapid spread of Dutch elm disease between trees in hedgerows is thought to have been due in part to root grafting between individuals.

Bud grafts are used to propagate material rapidly, for example to bring a new rose onto the market quickly. Roses, particularly hybrid teas and floribundas, are often poor performers on their own roots, and of course will not come true to type from seed. In such instances, grafting onto a healthy vigorous rootstock performs the dual function of providing a vigorous root and helps in rapid propagation.

Fig. 2.11 Twigs grafted across a damaged area of bark on a tree trunk.

If the vigour of the scion greatly exceeds that of the stock, ugly overgrowth can occur, and where there is no desire to regulate scion vigour, a stock of suitable vigour should be selected.

The callus cells produced by wounding two (or more) plants can sometimes be grown in culture, and groups of cells centrifuged together. The resulting complex of cells can be grown on, producing cytohybrid plants of the most complex type of graft imaginable, that is, except for the fusion of protoplasts of two different organisms that represents the extreme form of grafting! This latter process involves the enzymatic removal of the cell walls, leaving naked protoplasts that fuse more readily.

Monocotyledons are virtually impossible to graft, although there are a few questionable reports of such grafts in the literature. Most monocots have no secondary growth in thickness. The vascular bundles are 'closed' and produce no cambium. Some appear to have cambial division, but this may merely be the late, rather regular divisions of cell layers that take place in the central region of the bundle when it is approaching maturity. However, not enough is known about this phenomenon, and it remains questionable whether grafts can become established in monocotyledons.

The monocotyledonous bundle is often firmly enclosed by a sclerenchyma or parenchyma sheath, or both; it is termed a 'closed' bundle for this reason. So the monocotyledonous bundle lacks the meristematic cells needed to effect fusion, and the accurate positioning of bundles in a graft would, furthermore, be virtually impossible.

As mentioned earlier, secondary growth in thickness does occur in some monocotyledons, but special tissues at the periphery of the stem bring it about. These are in effect a lateral meristem and produce both new, entire vascular bundles by cell division and the new ground tissue between the bundles. *Cordyline* has the type of secondary thickening common to a number of monocotyledons. Again it is easy to see that grafting would fail because it is not possible to align enough bundles and little or no vascular continuity can be achieved.

Adventitious buds

Some plants have the ability to produce new, adventitious buds from various organs when the plant or part of the plant is placed under some unusual physiological stress. The stress may be caused by an injury or even by the separation of one organ from the rest of the plant. The development of buds in this way is usually thought to be related to the loss of a constraint, for example, the loss of some inhibitory hormone or similar chemical substance. When the apical dominance of a shoot system is removed, new adventitious buds (not related to leaf axis) may develop. This enables us to lop certain species of mature tree and get new growth. For example, *Salix* and *Platanus* will grow new branches from adventitious buds. The practice of pollarding and harvesting the pole-like young branches would not be possible if this type of recovery did not take place. Some crops such as *Quillaja* and *Cinchona* bark are obtained from coppiced trees; extracts from these are used in the preparation of medicines. Much fuel wood production results from pollarding or coppicing trees. For example, hazel (*Corylus*) coppice has been a traditional form of husbandry for centuries for forming new stems for charcoal production.

CHAPTER 3

The structure of xylem and phloem

Introduction

Xylem and phloem are the main tissues concerned with the movement of substances through the plant. Xylem transports mainly water and dissolved solutes usually in the form of minerals, and phloem translocates substances synthesized by the plant. Xylem and phloem are normally found together, and their functions are coordinated. Structure–function relationships are considered at the end of this chapter, but simply put xylem conducts water and dissolved minerals from the roots to the aerial parts and phloem conducts assimilates from the leaves to the stems and roots.

The xylem

The structure of primary xylem is dealt with in Chapters 3, 4, 5 and 6 and will not be repeated or enlarged upon here. Instead, we will concentrate on secondary xylem only, insofar as this is related to use and as an aid to classification and identification. The CD-ROM contains numerous additional images.

Secondary xylem construction

Whilst primary xylem consists of the axial cell system only, that is, xylem cells that are elongated parallel with the long axis of the organ or vascular trace in which they occur, secondary xylem, one of the products of the vascular cambium, is more complex. As we have seen in Chapter 2, the cambium is composed of two sorts of cells, the axially elongated fusiform initials, which give rise to the axial system of cells, and the short, more or less isodiametric ray initials giving rise to the radial system or rays. Figure 3.1 shows the axial and radial systems in *Alnus glutinosa* wood.

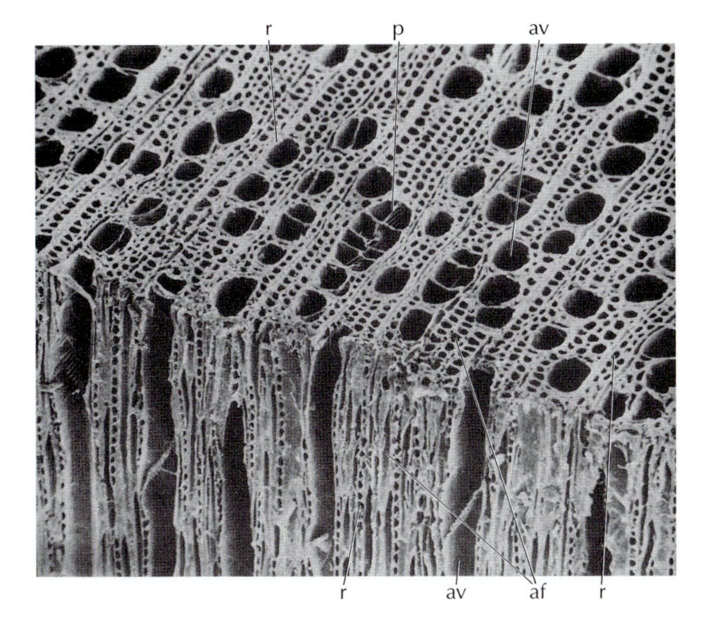

Fig. 3.1 *Alnus glutinosa*: SEM photograph of secondary xylem showing transverse and tangential longitudinal faces. af, fibres in axial system of cells; av, vessel of axial system; p, perforation plate (scalariform); r, uniseriate ray of radial system. ×100.

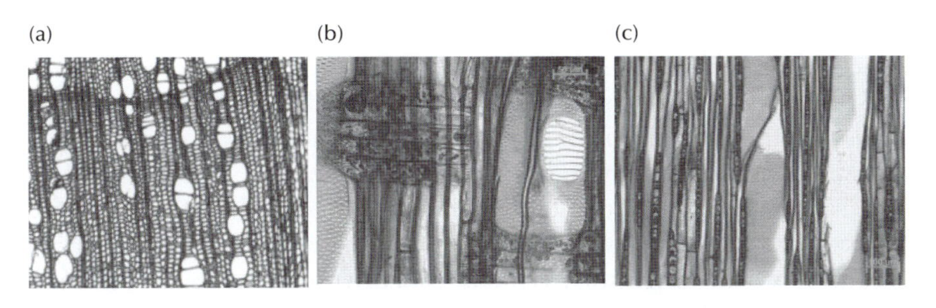

Fig. 3.2 (a) Transverse, (b) radial longitudinal and (c) tangential views of the wood of *Alnus nepalensis*. Note the narrow-diameter vessels. These have inclined compound scalariform-reticulate perforation plates. The vessels are interspersed with narrow tracheids and parenchymatous elements. Rays are of variable length and uniseriate. a, ×70; b and c, × 250.

Because both axial and radial systems are present, the study of secondary xylem can only be carried out properly by examination of three specific planes of section from a block of wood. These are the transverse section (TS) the radial longitudinal section (RLS) and the tangential longitudinal section (TLS). These expose details of both systems of cells. Figures 3.2 and 3.3 show these planes of section.

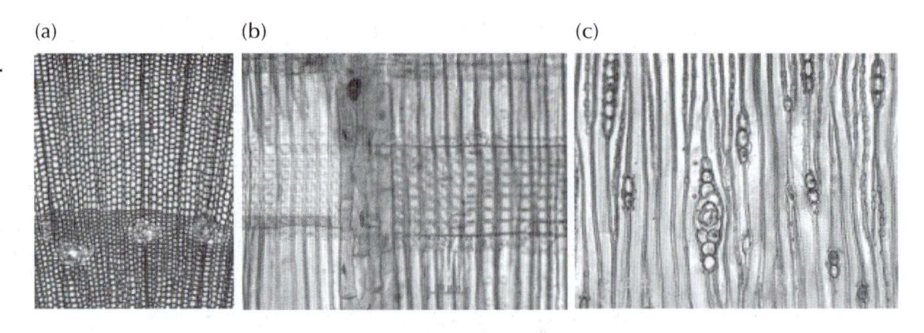

Fig. 3.3 (a) Transverse, (b) radial longitudinal and (c) tangential longitudinal views of the wood of *Pinus sylvestris*. Rays are variable, from one to many cells high and of variable width. Wider rays may be associated with resin canals. a, ×75; b and c, × 250.

The complex nature of secondary xylem will be addressed by considering first the less complex wood of gymnosperms, and next the much more varied wood found in the angiosperms.

Gymnosperm wood (conifer and Ginkgo *wood)*

In conifer woods (and gymnosperms generally), the axial water conducting system is composed largely of tracheids (imperforate cell involved in water transport, i.e. with intact pit membrane between it and adjacent cell). Tracheids are rather like elongated boxes, with a rectangular cross-section and tapering upper and lower ends. They have thickened walls. The primary wall is thin, and overlain on the side adjacent to the lumen by the secondary wall. The secondary wall is usually constructed of a series of layers, in which the microfibrils, submicroscopic, thread-like and usually cellulosic components are laid down. Microfibrils are often arranged in a spiral, wound along the long axis of the cell. The orientation of the microfibrils differs in the successive layers, each layer often reversing the direction of spiral winding of that in adjacent layers. The matrix of the wall is lignified. Tracheids communicate with one another mainly through bordered pits in the lateral (radial) walls. The size, number of rows and details of pit structure are often characteristic for given species or genera. Figure 3.3 shows sections of *Pinus* wood, and the glossary (see Pit, bordered) gives a reconstruction of the bordered pit pair between adjacent tracheids. Tracheids formed during the flush of spring growth are usually wider radially than those formed later in the growing season. It is normally easy to see the extent of thickness of a growth increment for this reason. Note the torus, a central thickening on the pit membrane, which is characteristic of tracheid pits in confers. It acts like a plug, closing against the pit aperture if there are potentially damaging changes in pressure between adjacent tracheids. In this way, the spread of air embolisms may be controlled and reduced. Thickened areas between the pits termed bars of Sanio. These are characteristic of conifer wood.

Sometimes a band of helical (spiral) thickening occurs inside the secondary wall of the tracheid. This is characteristic of *Pseudotsuga* and *Taxus* and some *Picea* spp., *Cephalotaxus* and *Torreya* in the mature trunk wood. However, many conifers have such a tertiary helical thickening on tracheid walls in twig wood, so if narrow diameter specimens are to be identified this must be borne in mind. In badly decomposed wood, helical splits may appear in tracheid walls, following the alignment of cellulose microfibrils in one of the wall layers; these can be confused with true helical thickenings.

There are no vessel elements.

Fibres are not normally found in conifer woods; if present, they run alongside the tracheids, axially. Axial parenchyma cells are rare; the cells are narrow, elongated axially, are situated alongside the tracheids, and have square end walls. Members of the Pinaceae (except *Pseudolarix*) and *Sequoia* spp. of Cupressaceae have axial resin ducts. These ducts are lined by a secretory epithelial layer of cells which remains thin-walled in *Pinus* species but becomes lignified in the other genera. The presence of resin ducts is, then, of taxonomic significance in the conifers. Many genera, like *Cedrus*, may have traumatic ducts (ducts arising from damage) which must not be confused with the regular resin ducts of Pinaceae.

The radial (ray) system in gymnosperms consists of parenchymatous cells, which are, on the whole, procumbent. Some genera have radial tracheids (ray tracheids) as well, and some Pinaceae have radial resin ducts. (Fig. 3.3c). The part of the ray cell wall bordering onto the tracheid is usually pitted. This is termed the cross-field area and the pits here are commonly characteristic for the genus or for a group of genera and can be used diagnostically. Figure 3.4 shows some such pit types. The wall pitting and thickenings of ray tracheids may also be of a characteristic form, for example, the 'dentate' ray tracheids of *Pinus* species. Rays are normally one, sometimes two cells wide, and several to many cells high.

Detailed variations between genera are too numerous for mention in this text, but useful references are given at the end of the book for further reading. There is a well-illustrated glossary of gymnosperm wood published by the International Association of Wood Anatomists.

Angiosperm wood

In dicotyledonous hardwoods there are more sorts of cell to be considered, and their appearance and distribution in the wood give rise to a great range of wood types. Although details of wall pitting still remain important in the identification of specimens, as for gymnosperms, there are many more characters that are readily observable in dicotyledonous woods than there

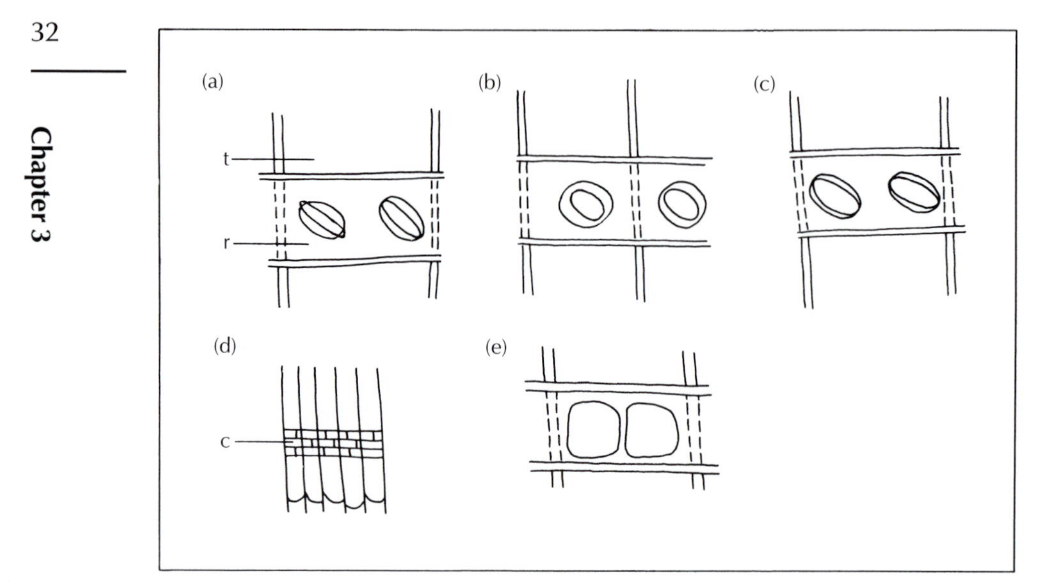

Fig. 3.4 Some types of cross-field pits in conifers. (a) Piceoid, e.g. *Picea*, *Larix*. (b) Cupressoid, found in most Cupressaceae and *Taxus*. (c) Taxodioid, e.g. Taxodiaceae, *Abies*, *Cedrus*, some *Pinus*. (d) Diagram showing location of cross-field pits (c) where ray and tracheid walls are adjacent. (e) Some *Pinus* spp. have large 'window' pits. r, ray; t, tracheid.

are in coniferous woods. Families or groups of families often exhibit quite characteristic features for study and comparison.

In the axial system of dicotyledons, tracheids are usually relatively sparse, most of the cells being vessel elements (the component cells of vessels; vessel elements are arranged in an axial file, with the terminal elements imperforate at their outer ends, but with all element to element junctions with perforations (perforation plates) in their end walls) and fibres or fibre tracheids (cells intermediate in appearance between fibres and tracheids; the pits usually intermediate in size) with varying amounts of axial parenchyma. These are all derivatives of the fusiform cambial initials. Fibres can be longer than the initial from which they were derived. Many fibres are capable of elongation by apical intrusive growth. Axial parenchyma cells are usually shorter than the initials that gave rise to them because the derivative cells often divide twice or three times, to form a chain of four or eight cells without significant enlargement. Xylem parenchyma cells normally have somewhat thickened, lignified walls. Vessel elements are extremely variable in their mature form, but particular forms are fairly consistently present in a given species. Details of wall pitting and perforation plates are illustrated in Fig. 3.5. Each of these features may be used

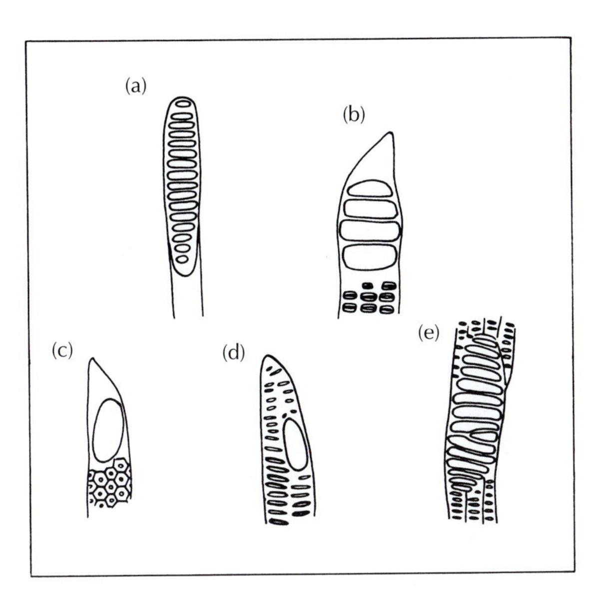

Fig. 3.5 A range of vessel element perforation plates and wall pitting. (a) *Camellia sinensis* scalariform. (b) *Liriodendron tulipfera*, scalariform; pits opposite. (c) *Sambucus nigra*, simple plate, pits alternate. (d) *Euphorbia splendens*, simple plate, pits alternate. (e) *Scirpodendron chaeri*, scalariform plate, pits opposite (from primary xylem). All ×218.

diagnostically. Tertiary spirals may occur; an example is shown in *Tilia* in Fig. 3.6.

The main cell types in the axial system of most angiosperms are, then, vessels, fibres, tracheids and parenchyma. Some species have secretory canals, lined by epithelial cells. Vessels, fibres and axial parenchyma may be arranged so that clear growth rings can be distinguished in TS, as in north and south temperate and montane species and those from other strongly seasonal regions. Lowland tropical species generally do not show growth ring boundaries. The overall density of a wood is governed not only by the sizes and proportions of the cell types present, but also by the thickness of their cell walls. Wall thickness also has a bearing on the hardness of a wood. For example balsa wood, *Ochroma lagopus*, is lighter than cork (Fig. 3.7 shows *Ochroma pyramidalis* wood in TS; note the thin-walled fibres); lignum vitae, *Guaiacum officinale*, is extremely dense and heavy (Fig. 3.8; note thick-walled fibres) and is used for making bowls, mallets and pulleys, among other things, and black ironwood, *Krugiodendron ferreum is* even more dense.

Where growth rings are apparent, vessels may grade in size, from wider in early growth, to narrower in late season growth. This produces a diffuse porous wood. In ring porous wood, vessels formed at the start of a growth

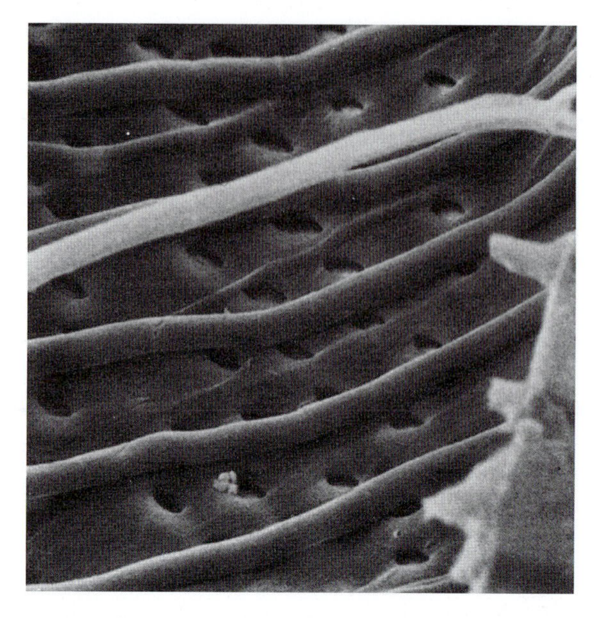

Fig. 3.6 *Tilia europaea*, LS, tertiary spirals on vessel element wall. SEM, ×3000.

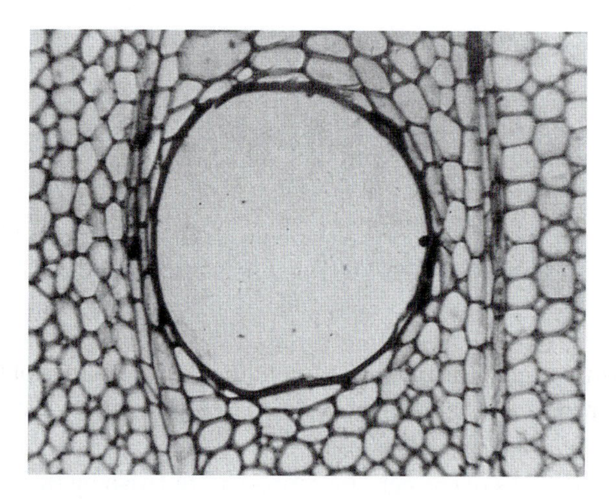

Fig. 3.7 *Ockroma pyramidalis*, wood TS; note the vessel, the thin-walled fibres and abundant parenchyma. ×200.

ring are much wider than those formed soon afterwards. Vessels as seen in TS may be solitary, in pairs or in small groups. The arrangement of these groups or multiples varies greatly. For example, there may be radial chains, oblique chains, tangential groups and so on. It is variations in vessel distribution, as seen in TS, that give the characteristic appearance of the wood. This is often further enhanced by the distribution of the axial parenchyma,

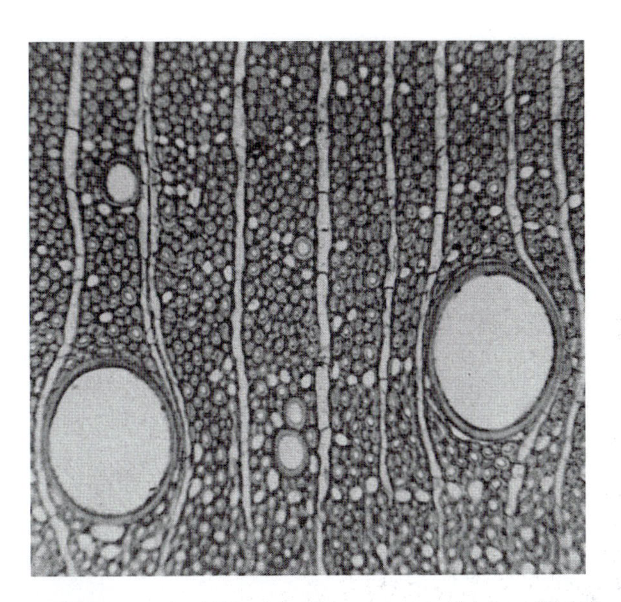

Fig. 3.8 *Guaiacum officinale* wood TS; note the numerous thick-walled fibres and the scattered parenchyma. ×200.

when present. If the strands are solitary and scattered, they might not make much impact on the appearance of the TS. If there is abundant parenchyma in strands, or groups which are either closely associated with the vessels, or clearly separated from them, a whole new range is established.

Rays are far more complex and show a wider range of variation in dicotyledons than in gymnosperms. They do not normally contain tracheids and may only sometimes contain vessels or perforated parenchyma cells. The parenchyma cells of which they are composed may exhibit a range of shapes and sizes.

The cells making up the rays are evident in all three planes of section. In TS both their frequency and width (uniseriate – one cell wide to multiseriate, many cells wide) contribute strongly to the appearance of the wood. Where there are growth rings, rays in some species may show a widening of their cells at the rings. Rays may be all uniseriate, or they may be a mixture of uniseriate and wide multiseriate, or there can be a gradation from uniseriate to wide multiseriate, without a wide gap in width. In *Castanea* and *Lithocarpus*, for example, the rays are all uniseriate. *Ulmus* and *Fagus* are examples of genera in which rays may be from one to several cells wide. *Quercus* has rays of two distinct sizes, some uniseriate and others wide and multiseriate, with no intermediates (Fig. 3.9) shows uniseriate rays.

Rays may appear to have a 'random' distribution as seen in TLS, or, in some species, they may be storied, and arranged in clear horizontal bands (relative to the long axis of the wood). This difference can be used

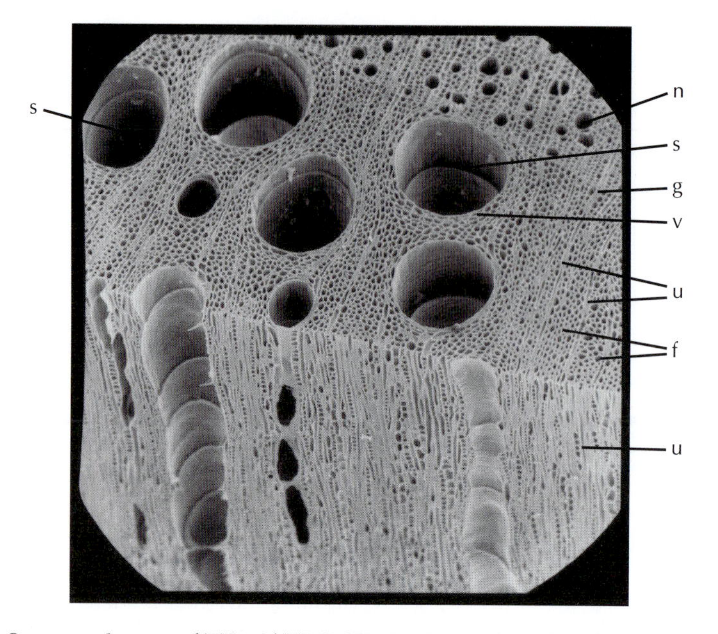

Fig. 3.9 *Quercus robur*, wood TS and TLS, SEM. Note wide spring-formed vessels (s) and narrow later-formed vessels (n). Numerous uniseriate rays can be seen (u). Small parenchyma cells of the axial system occur in more or less tangential bands among the fibres (f). A growth ring (g) is shown, as is vasicentric parenchyma (v). ×60.

diagnostically because storied rays are characteristic of some families, for example Leguminosae (= Fabaceae), or of some genera, for example *Hippophae* (Elaeagnaceae)

In radial view (RLS) ray cells appear to look like courses of bricks in a wall. In some species all the cells are of similar size and proportions (homocellular); in others distinctly recognizable differences in cell size may occur (heterocellular). Cells of any particular size or shape are usually arranged in regular 'courses' or may be in particular positions, for example at the top and bottom of a ray. A range of different ray types is shown in Fig. 3.10, and includes procumbent and upright types.

With the possible combinations of cells and rays it is easy to see how different particular woods may be from one another.

Secondary xylem – properties and uses

Secondary xylem, or wood, is put to an extremely wide variety of uses. The extensive range of species from the gymnosperms and angiosperms which are used as sources of wood is reflected in the diverse properties of the various kinds of wood.

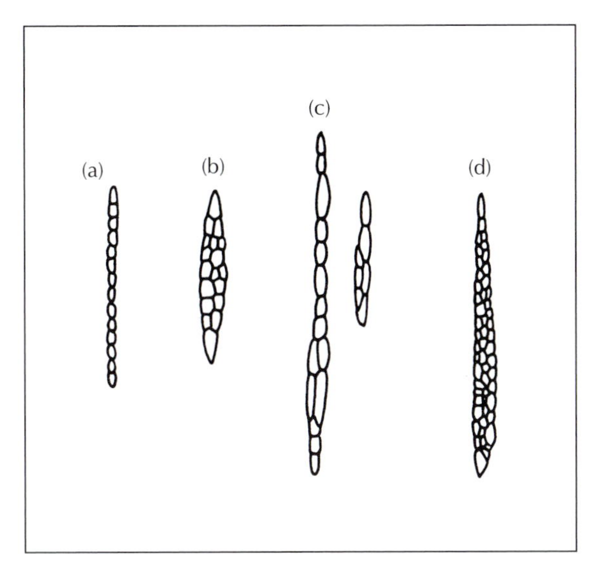

Fig. 3.10 Some ray types in TLS. (a) *Alnus glutinosa*, uniseriate, homocellular, all cells of procumbent type. (b) *Swietenia mahagoni*, multiseriate, heterocellular, with upright cells at margins and procumbent cells between. (c) *Sambucus nigra*, biseriate, with tall uniseriate portions, heterocellular. (d) *Musanga cecropioides*, multiseriate, heterocellular, procumbent and upright cells together in body of ray, upright cells at margins. All ×72.

There is archaeological evidence that our early ancestors were well aware of the best woods for burning for warmth or metal smelting, those most durable and strong for making boats or buildings and those most suited as shafts for tools or weapons. They even selected carefully for their musical instruments and decorative carvings. In our more advanced stage of technology, we make use of the different characteristics of strength, workability, durability, appearance, density and pulping potential in our selection of woods for a vast range of primary and secondary products.

Obviously, this wide range of properties is a result of the variation in the histology and fine structure of woods. In fact, there are many characters in which wood can vary, but it is not always clear what effects they have on the properties of the wood. The possible variation is so great that the set of characters shown by wood from a particular species can provide clues to the identity of the species. Sometimes the set of characters may indicate only the family or genus, but occasionally it is confined to a species. In other words, one would expect individuals of the same species to share very similar wood characters, but another closely related species might be so similar that it cannot be distinguished by wood features alone.

We shall explore the sorts of differences which occur in wood, and look at the ways in which these help in identification and in establishing the

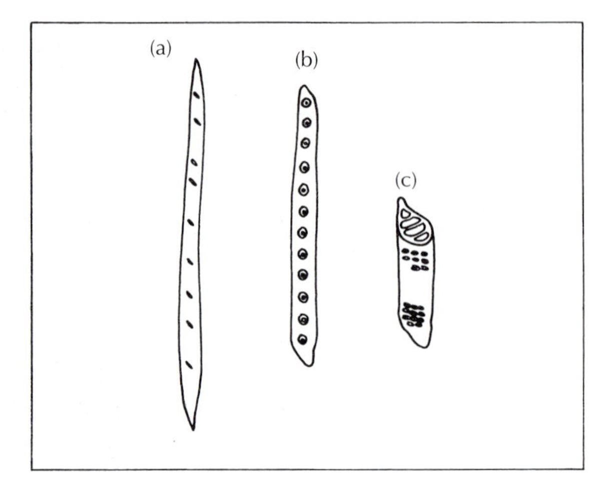

Fig. 3.11 (a) Fibre, (b) tracheid and (c) vessel, contrasted; intermediate cell types exist between each.

relationships between species, and how they affect the properties of the timber.

Evolution in secondary xylem

It is generally accepted that secondary xylem has undergone a long evolutionary history. The main trends can be seen because the various stages are often related to other 'marker' characters in flowers, fruits, etc. of the plants concerned, or in molecular markers. There are instances where habitat has seemingly reversed some of these trends in various species, but overall, their 'direction' can be fairly safely defined.

Taken in its simplest form, the evidence to hand indicates that the tracheid, a dual purpose cell combining properties of both mechanical support and water conduction, in evolving groups of plants gave rise to fibres with simple mechanical function and to perforate cells, the vessel elements, concerned with the conduction of water and dissolved salts (Fig. 3.11). This division of labour is seen as a specialization, or advance. Most angiosperms still have tracheids, as well as fibres and vessel elements.

The primitive vessel element shows much similarity to the tracheid; it is axially elongated, with oblique end walls in which are grouped perforations making up the scalariform, reticulate or otherwise compound perforation plates. The lateral walls bear bordered pits, often in an opposite arrangement. The advanced vessel element is seen as a broad short cell with large, simple transversely arranged perforation plates at both ends and alternating bordered pits on the lateral walls. Between these extremes is a variety of forms (Fig. 3.5). Vessel elements are found in axial files, and a strand of them constitutes a vessel. The terminal vessel elements in a vessel have an imperforate wall at their extremities, but a perforated end wall where in contact with the vessel element next in line.

In the monocotyledons, where xylem is entirely primary, the vessel element probably evolved first in roots, then in stems and finally in leaves. Evidence for this is found in many plants. There is no record of a species being found with vessels in the leaves only, and not in the roots.

There are some flowering plants which have been considered to be primitive because of floral characters, and which are vessel-less (e.g. *Drimys*, Winteraceae/Magnoliales, among the dicotyledons). However, recent evidence indicates that vessels have been lost in these species.

Confidence in the evolutionary sequence is such that characters of the vessel elements have often been used as an indicator of the relative phylogenetic advancement of plants. Measurements of vessel element length and width must be made on a statistically sound basis for such comparisons. It has been found that a ratio of vessel element length to tangential width produces a useful figure for application in advancement indices. There are many pitfalls in this method. Great care must be taken in ensuring that comparisons are made between plants growing in fairly similar conditions, since habitat can influence the vessel diameter. There can also be a degree of natural variability, which must be accounted for in sample size. Comparisons between the trends in families or genera are naturally more reliable than those between species. Overall trends in orders are again of more significance. So, even in the detailed measurements of fibres, tracheids and vessel elements, we see features which can be applied in evolutionary terms, or used when we are attempting to establish the possible origin of a group of plants. For example, it would be unlikely that monocotyledons with vessels in root, leaf and stem would be ancestral to those with vessels in the root only. These data are of considerable interest in phylogenetics.

The appearance of rays in the TLS is also important in systematics. From this plane of section it is easiest to see just how many cells make up the width of the rays. What may have appeared to be a wood with two ray widths from the TS could have rays with a wide, multicellular central region and elongated uniseriate tails. Cutting such rays at different heights in TS would give a misleading appearance. The overall height of a ray can be determined with certainty only from the TLS.

Some woods have their rays and fibres arranged in regular horizontal rows as viewed in TLS. This storied type of wood gives a particular 'figure' to planks, and is often of decorative value. Many of the Leguminosae are like this.

Unfortunately, it is rarely possible to say which particular anatomical features of a wood make it suitable for specific mechanical uses, although

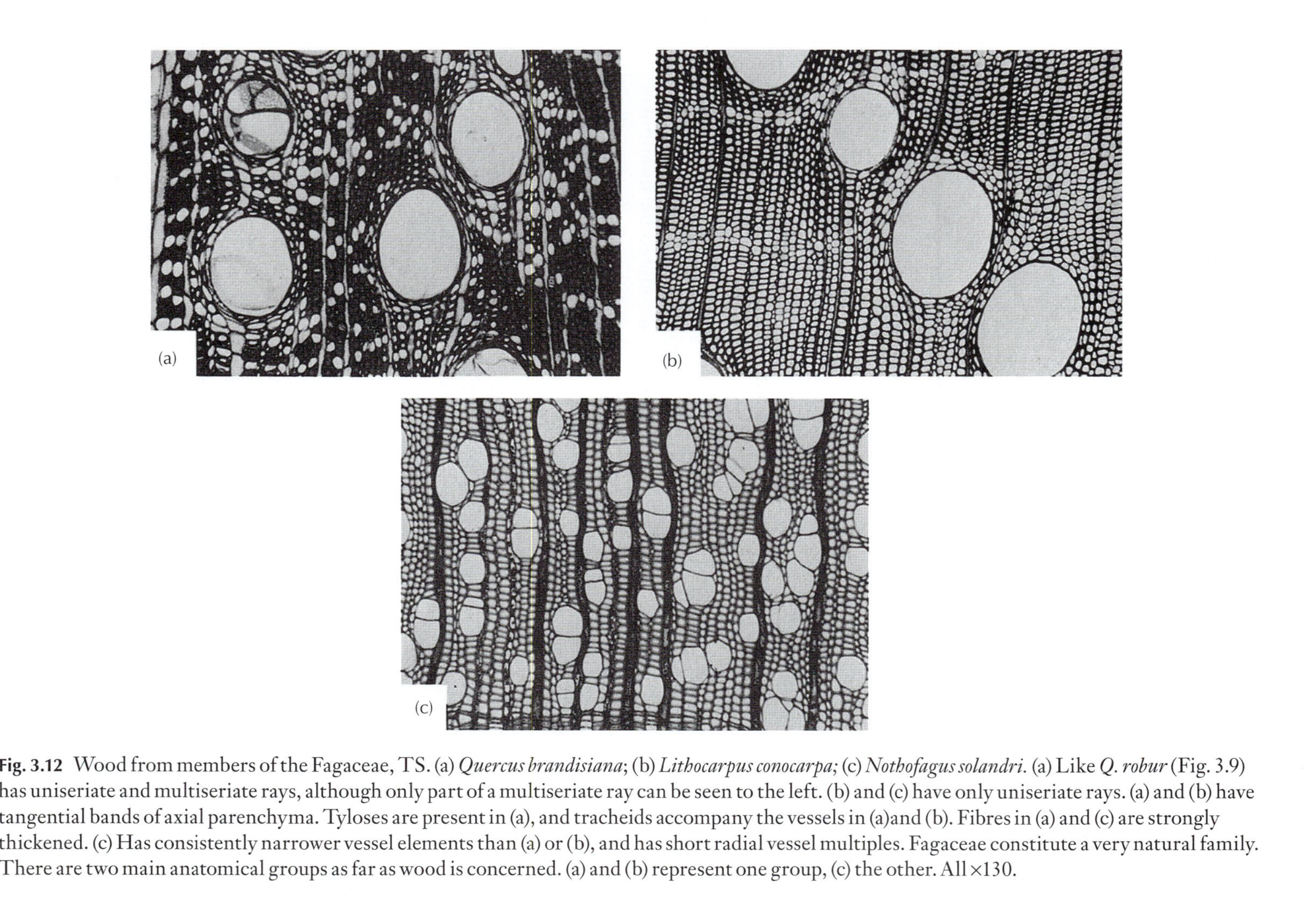

Fig. 3.12 Wood from members of the Fagaceae, TS. (a) *Quercus brandisiana*; (b) *Lithocarpus conocarpa*; (c) *Nothofagus solandri*. (a) Like *Q. robur* (Fig. 3.9) has uniseriate and multiseriate rays, although only part of a multiseriate ray can be seen to the left. (b) and (c) have only uniseriate rays. (a) and (b) have tangential bands of axial parenchyma. Tyloses are present in (a), and tracheids accompany the vessels in (a)and (b). Fibres in (a) and (c) are strongly thickened. (c) Has consistently narrower vessel elements than (a) or (b), and has short radial vessel multiples. Fagaceae constitute a very natural family. There are two main anatomical groups as far as wood is concerned. (a) and (b) represent one group, (c) the other. All ×130.

the presence of tyloses makes woods particularly useful for liquid containers. In ring porous woods, for example, it seems that the number of growth rings to the centimetre is often of relatively more importance than histological details.

Evenness of texture or straightness of grain are features which belong to certain woods. Lime, *Tilia* spp., and pear, *Pyrus* spp., for example, have properties which make them good for carving; the wood cuts well in any direction. Ash, *Fraxinus*, and hickory, *Carya*, have a straight grain and are resilient, and are chosen for axe and tool handles. Long fibres or tracheids are one of the requirements for making certain types of paper. Softwoods (conifers) without resin canals are usually preferred to hardwoods for pulping for this reason.

Light (less dense) woods with cells having moderately thickened walls are often more resilient and recover their shape better after denting than more dense woods. The cricket bat willow, *Salix alba* var *caerulea*, is such a wood. Many woods with good resistance to decay contain oils, gums or resins. Teak, *Tectona grandis*, is a good example and was extensively used in boat building. *Bulnesia sarmienti* has gums and resins which produce an incense. *Cinnamomum camphora* is the source of natural camphor.

Spruce, *Picea* sp., has good resonating qualities and is widely used in the resonating chambers of stringed instruments. Oaks, *Quercus* spp., were extensively used in buildings and boats since the time of Iron Age people. Oak can be split, using wedges, along lines of weakness formed by the broad rays, and planks or posts can be formed with simple tools.

A range of types of wood is illustrated in Figs 3.12–3.15. These are chosen to demonstrate variations in vessel, fibre and parenchyma distribution and in ray type. Several excellent general books exist on wood anatomy, and there are many volumes on woods from particular parts of the world. Some of these are listed at the end of this book. The International Association of Wood Anatomists has published an illustrated glossary of the cell types and their arrangements for angiosperm woods. Metcalfe and Chalk (*Anatomy of the Dicotyledons*, Vol. II, 1983) lists the occurrence of a range of xylem characters in various taxa.

The phloem

The primary phloem of monocotyledons and is described in Chapters 4–6. As with secondary xylem, secondary phloem has both axial and radial arrangements of cells. The same initials in the cambium which divide to form xylem to their inner side also cut off phloem cells to their outer side. Sometimes growth rings can be seen.

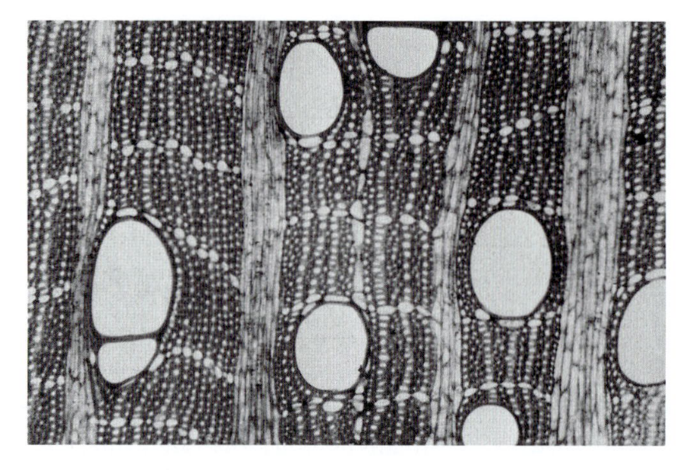

Fig. 3.13 *Platymitra siamensis*, Annonaceae, TS. Vessels diffuse, porous; rays uniseriate and multiseriate; axial parenchyma in uniseriate tangential bands; fibres thick-walled. ×130.

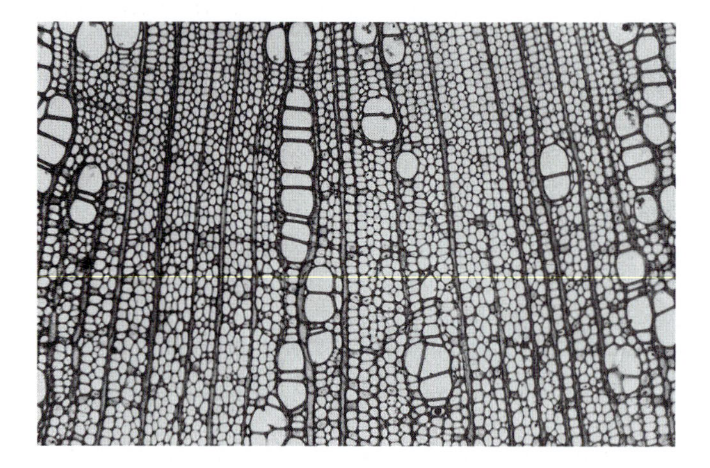

Fig. 3.14 *Carpinus betulus*, Carpinaceae. Vessels diffuse, porous, in long radial multiples. Rays uniseriate (aggregate rays also occur, but are not shown). Axial parenchyma is seen in poorly defined, interrupted tangential bands, the cells have dark contents. × 65.

Gymnosperm phloem

In gymnosperms, the axial phloem consists of sieve cells and parenchyma cells, some of which become albuminous cells (see Fig. 5.5); some gymnosperms have fibres in the phloem as well. The homocellular rays are normally uniseriate. There is often very little wall thickening but sclerification can take place. The outermost phloem layers either become compacted, or are incorporated into the 'bark' or rhytidome.

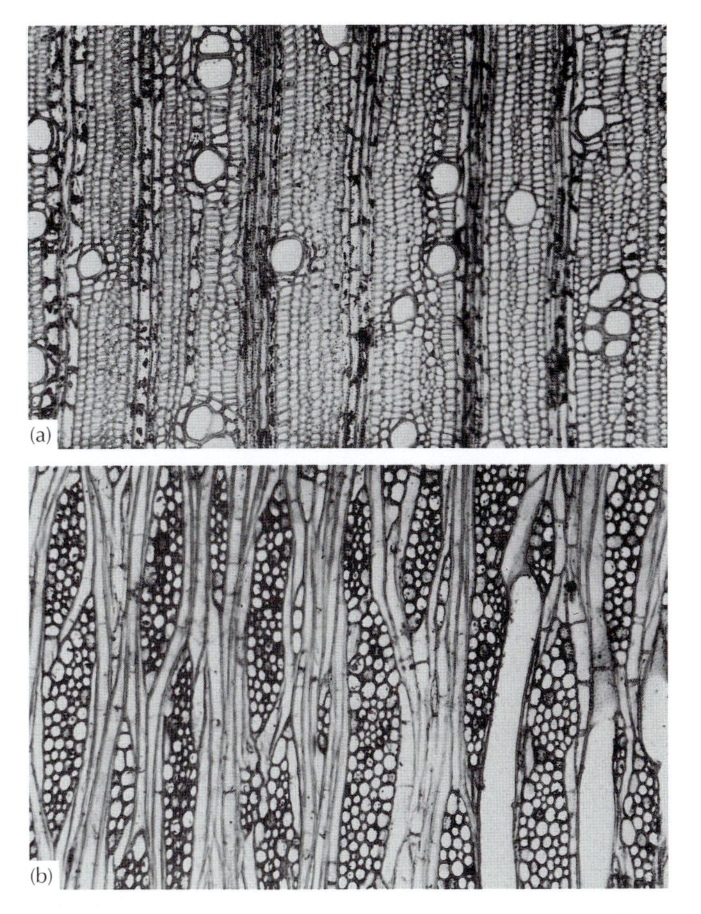

Fig. 3.15 *Laurus nobilis*, Lauraceae. Vessels diffuse porous, narrow, solitary or in small multiples, perforation plates simple. Rays uniseriate and multiseriate, heterocellular. Fibres septate. (a) TS; (b) TLS; both ×65.

Angiosperm phloem

The phloem cells of dicotyledons show evidence of evolutionary trends similar to those of the xylem. The sieve areas, which are areas of dense pitting in lateral walls of sieve cells, are a feature of the more primitive dicotyledons. Well-organized sieve plates, simple and transverse, situated at either end of the sieve tube members are considered to be advanced. Oblique, compound sieve plates also occur (Fig. 3.16); sometimes these are found in advanced genera such as *Quercus* and *Betula*, and also in lianes, for example *Vitis*, where physiological demands may call for large areas of sieve plate which are necessary for rapid, long-distance translocation of materials. Perhaps the best example of this is to be found in *Cucurbita maxima*, watermelon, where massive compound sieve plates are found in the peduncle of the fruit where high volumes of assimilate are being translocated. Even in

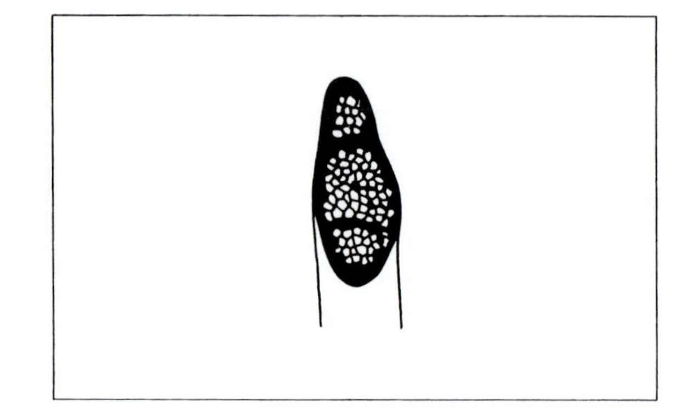

Fig. 3.16 *Aesculus pavia*, compound sieve plate. ×720.

species where sieve plates are well developed, the lateral walls of the sieve tube members often have distinct areas of pitting, called sieve areas.

Companion cells, usually much narrower than the sieve tube member to whom they are adjacent, are a feature of dicotyledon phloem. Their counterpart in the gymnosperms is thought to be the albuminous cell. Companion cells are nucleate whilst sieve tube members are not. Killing a companion cell appears to prevent the adjacent sieve element from translocating, so the function of the companion cells seems in part to be regulation of the physiological activities of the sieve elements. The axial system of secondary phloem often contains parenchyma, idioblastic cells (idioblasts, cells of a particular distinct type dispersed in another tissue), sclereids and fibres. In some species, sclereids and fibres are absent from the functioning phloem, but differentiate at a later stage. Fibres often alternate, in bands, with conducting cells, for example in *Tilia* and various Malvaceae (Fig. 3.17). Primary phloem fibres of *Linum* (flax) are of economic importance.

The rays within phloem may be either homocellular or heterocellular. In some species they remain of even width, but in others they may become wider towards their outer ends (e.g. *Tilia*). The rays may be uniseriate to multiseriate. As in secondary xylem, the secondary phloem may be storied; naturally, the storied arrangement originates from the storied cambium in these plants. Laticifers (cells containing latex) and lysigenous cavities of various kinds may occur in the phloem.

The application of phloem anatomy in taxonomy has not been of such widespread interest as might be expected. The partially sclerified, often highly crystalliferous tissue is difficult to section well. This has put off many people! When phloem tissues form part of the rhytidome they have been more intensively studied than when they are distinct from the rhytidome, and a number of valuable contributions exist on 'bark' anatomy. In

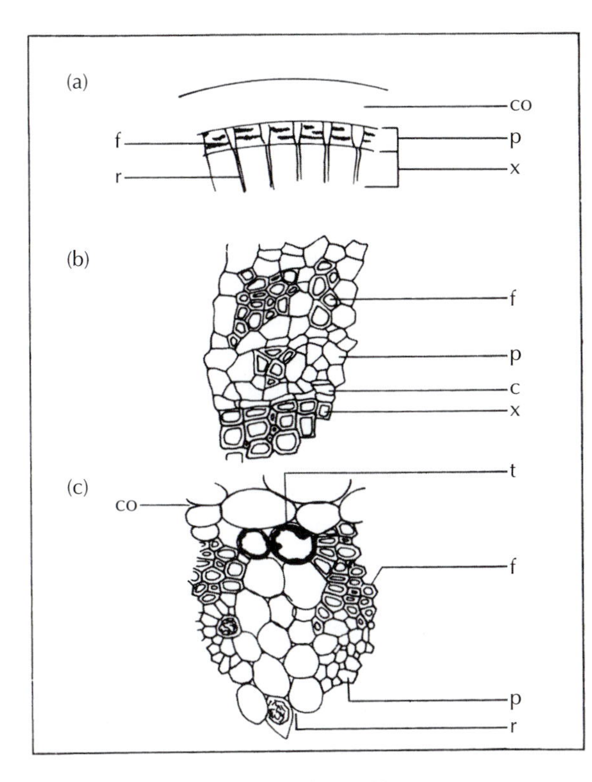

Fig. 3.17 (a) Diagram to show location of phloem fibres in *Tilia* stem TS. (b) *Malvaviscus arboreus*, ×218. (c) *Gossypium* sp., ×218. c, cambium; co, cortex; f, phloem fibres; p, functional phloem; r, ray; t, tannin; x, xylem.

root identification work, the bark of secondarily thickened roots provides data that complement those from the xylem.

The other area of applied study is in relation to diseases of the phloem. The normal structure of functioning phloem must be understood before disease symptoms can be interpreted, or the effects of chemical treatments defined.

Structure–function relationships in primary and secondary vascular tissues

The structure of the xylem and phloem in higher plants has been reviewed in a number of excellent texts. Here we propose to highlight some aspects that we consider relevant to the general student. For a study in more depth, readers are advised to refer to texts mentioned under the further reading.

The evolution of the conducting system, together with the development of the lignin synthesis pathway, must be among the important factors which contributed to the evolution of vascular plants.

Physiologically, the elements within the xylem and phloem can act independently of one another, yet the phloem relies on water provided by the xylem in order to support the driving force required for long-distance translocation of assimilated material, hence the two are almost invariably found together. Structurally, many of the elements within the xylem (with the exception of parenchymatous elements and, sometimes, fibres) are dead at maturity and have highly modified wall structure. By contrast, the phloem (with the possible exception of sclerenchymatous elements) consists of cells which contain protoplasts at maturity. Sieve elements, including the more primitive sieve cells which occur in gymnosperms, are unique, in that they either lack nuclei or contain only remnants of nuclei, which are of unknown functional or regulatory capacity.

Transport through the xylem is driven in part by root pressure and by the evapotranspirative processes which take place mostly through stomata, lenticels and possibly through cracks in cuticular layers. Transport in the phloem, on the other hand, relies on a build-up of solutes (loaded into the sieve tubes at the sources) and the subsequent attraction of solvent to this area. Increasing pressure and the resultant enhancing of flow within the sieve tubes, but away from the source, moves them to some local or distant regions of the plant (termed sinks) where the solutes are unloaded and utilized. This is termed translocation.

In the primary plant body in leaves and stems, xylem and phloem generally occur either in vascular bundles; or in roots, they are found as strands, with xylem and phloem on alternating radii. In plants that have undergone secondary thickening, the xylem and phloem in roots and stems become penetrated, by a series of secondary rays, which are usually parenchymatous. Whilst the interrelationships are easier to follow in primary vascular tissues, clearly the development of radially arranged ray tissues are of paramount importance in the regulation of solute and solvent transport, and storage of metabolized materials.

One could ask 'why do these disparate systems occur in close proximity?' Clearly, the xylem does not require any direct inputs from the phloem, but does the phloem require or obtain any input from the xylem? Examination of the leaf blade bundles in gymnosperms and angiosperms demonstrates close spatial relationships between the tissues. Functional sieve tubes may occur adjacent to tracheary elements in many monocotyledonous plants, particularly among the grasses and sedges. If not in direct contact, they are separated by a few layers only of narrow parenchyma cells. In grasses and sedges these sieve tubes are the last to differentiate and mature, and curiously, they have thick walls, which in some instances (barley and wheat) have been reported to undergo lignification. Even more curious is the lack of the identifiable companion cell–sieve tube complex, found in the early metaphloem in these plants, and commonly in all other vascular bundles in angiosperms.

Perhaps the answers to the spatial proximity of the xylem to the phloem lies in the physiological requirements for successful phloem loading at the source, the maintenance of long-distance transport and the unloading process in local and distant sinks elsewhere in the plant. Figure 6.28 illustrates the difference in size between loading, transport and unloading phloem sieve tubes and companion cells in *Nymphoides*. In the root, the mature metaphloem is about 5–10 μm in cross-section, and the companion cells are 15–40 μm in cross-section; the increase in the size of the companion cells reflects its role in the phloem unloading process.

CHAPTER 4

The root

Introduction

Primary roots have not been the subject of as many or such full studies as has been the case with stems or leaves. They do, however, show a wide range of variation which is influenced both by environment, in terms of ecological adaptation, as well as by the genotype. Compared with stems and leaves, root fragments can be difficult to identify in the primary state. Roots of monocotyledons show some variation, but generally not enough to provide reliable data for identification of unknown samples. This is not entirely because they are relatively undescribed or poorly represented in reference microscope slide collections, but partly because there is, overall, less variation.

In Chapter 1, we noted that the root is an organ which has to take strains or pulling forces. It rarely has to bend or flex, since it is usually found in a more or less solid medium. As a consequence of this, the main strengthening tissues are positioned in the central region of the root and function like a rope.

The typical primary root is bounded by an epidermis. Inside this is the cortex, which may be several- to many-layered, and this is bounded to the inner side by an endodermis. Next follows the pericycle, and then the vascular system is in the middle. Each of these parts is described in detail below, and the terminology is defined.

Epidermis

In all except aerial roots and the non-anchored roots of aquatic plants, root hairs are usually present a short distance from the growing apex. These develop from the rhizodermis or root epidermis. Often the root hairs arise centrally from the basal part of the cell; occasionally they arise from near one end – this is a useful diagnostic feature. Again, whilst many root hair bases are level with other cells in the rhizodermis, in other plants they may be bulbous and protrude; they can be sunken into the outer cortical tissues (e.g. *Stratiotes*).

A short distance further away from the apex the root hairs often die and shrivel, but in some plants the root hairs persist for a long time. To the inner side of the rhizodermis an exodermis may develop, particularly in monocots. This characteristically consists of angular cells with somewhat thickened, lignified walls.

A multiple epidermis or velamen is found, for example, in the aerial roots of epiphytes (e.g. orchids; Fig. 4.1) and aroids. Frequently, cells in this situation have specialized spiral or reticulate or irregular thickening bands and are capable of storing water absorbed from a humid atmosphere, mist or rain.

Cortex

The cortex is sufficiently variable to be used to assist in identification. Unfortunately, from that point of view, the various types of cell arrangement seem to have more ecological than systematic significance.

Two basic types of cortex are recognizable among other, less frequent, variations. These are the 'solid' and the 'lacunate' cortex (Fig. 4.1). The 'solid' cortex is composed of parenchymatous cells which are relatively compact, with intercellular spaces confined to the angles between cells. It is usual for there to be a gradual increase in the size of such cells from the outer to the inner layers, but the innermost few layers are frequently of a smaller and more compact type of cell. Such an arrangement is common in both monocotyledon and dicotyledon roots on plants which grow in well-drained soil types. The 'lacunate' cortex has a few outer layers of compactly arranged cells and the innermost layers may be similarly compact. Between these, radiating plates of cells are seen in TS, with large air spaces between them (Fig. 4.2). In TLS these may be seen as longitudinal plates, but more commonly they are arranged in a net-like pattern, thus enclosing the air cavities or lacunae. Diaphragms of stellate (literally star-like cells with radiating thin arm-like protrusions) and other cell types may traverse the lacunae, as in the stem. Most plants with this type of root cortex have their roots in soil which is periodically waterlogged, or even immersed in water. The condition can be induced in *Zea*, which normally lacks lacunae, by growing the plants in waterlogged soil.

The number of layers of cells in a cortex can vary in specimens of a given species, but within bounds. It might be possible to distinguish between species in a genus if some of them have many layers and others few, but this is not a very sound exercise.

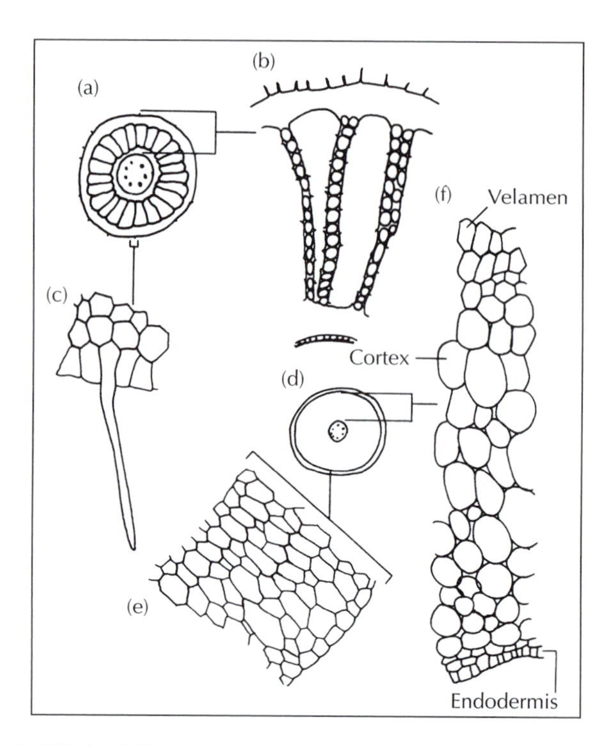

Fig. 4.1 Roots in TS. (a–c) *Juncus acutiflorus*: (a) diagram; (b) lacunate cortex, ×54; (c) root hair, ×218. (d–f) *Cattleya granulosa*: (d) diagram; (e) velamen, ×68; (f) 'solid' cortex, ×68.

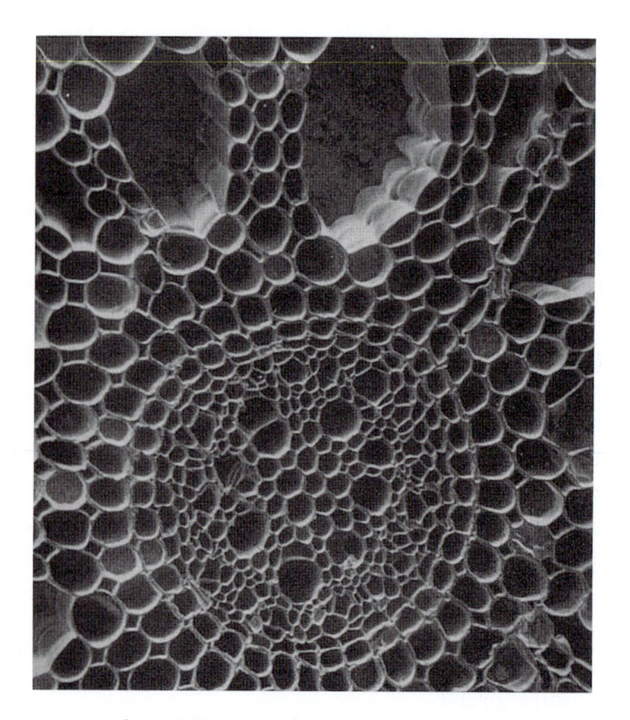

Fig. 4.2 *Stratiotes*, part of root TS, SEM photograph, note air spaces in the cortex, ×75.

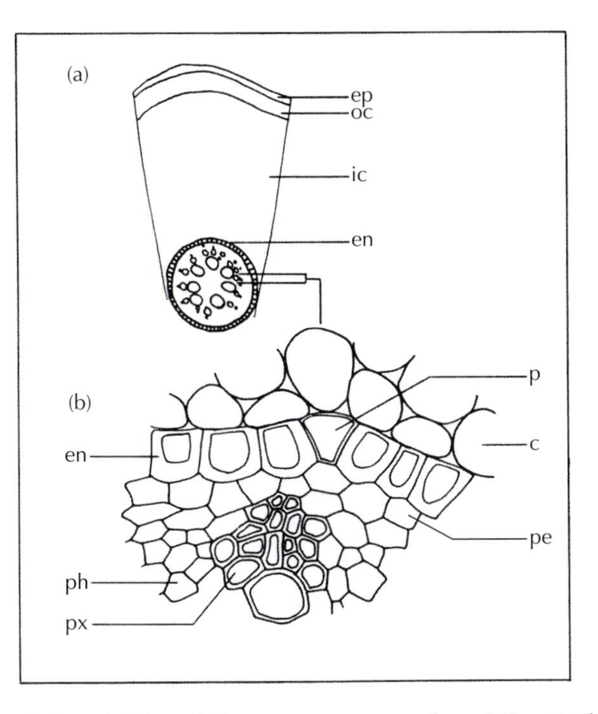

Fig. 4.3 Root endodermis *Iris* sp. (a) Low-power, sector of root TS, ×20. (b) Detail from (a), ×290. c, cortex; en, endodermis; ep, epidermis; ic, inner cortex; oc, outer cortex; p, passage cell; pe, pericycle; ph, phloem; px, protoxylem.

Sclereids, fibres, tannin cells, mucilage cells and crystal containing cells can be found scattered in the parenchymatous cortical tissue in a wide range of families. Their presence and distribution can be helpful in narrowing down possibilities for identification, but are rarely of taxonomic significance.

Endodermis

The cortex on its inner side abuts onto the endodermis. This characteristic, physiologically active tissue is frequently one layer thick but in some plants it can be two or more layered. Although the endodermis can be composed of cells with evenly thickened walls, in the majority of plants the inner and anticlinal walls are more heavily thickened with lignins and suberins than the outer periclinal walls. Consequently, in transverse section they are readily distinguished from adjacent cell layers, the so-called U-shaped thickenings making them conspicuous (Fig. 4.3).

At intervals, certain cells of the endodermis are thin-walled. These are the passage cells and are usually opposite protoxylem poles. They are supposed to afford a more ready pathway from root hairs, via the cortex to the protoxylem elements, for water and dissolved solutes. The other cells in the endodermis are supposed to restrict water flow between cortex and stele. The anticlinal walls of all endodermal cells are equipped with special suberized 'waterproof' impregnations, the Casparian strips, or bands with the cell membrane, the plasmalemma, attached to the Casparian strip. These are most easily seen in young, unthickened cells near to the root apex; they stain readily in Sudan III or IV. When the cells are plasmolysed the cytoplasm can be seen as a band because of the cell membrane being attached to the strip. Because of the range of variation in cell height and width and differences in the shape and degree of wall thickening of endodermal cells throughout the monocotyledons and dicotyledons, it is often possible to give a close description of an endodermis that is characteristic for a species or group of species. There may be many species that fit any one description, but if authenticated material is available for comparative purposes in making identifications, then a close matching of endodermis cell types must be achieved for accurate identification. As with all other minute characters, the appearance of the endodermis could never be used on its own to identify an unknown plant, but if it is simply a matter of choosing between several possible plants, then a close match on the endodermis would be quite sound evidence on which to base the identification.

Pericycle

The cells of the next layers to the inside are usually narrower than those of the endodermis, and they frequently have relatively thin walls (Fig. 4.3b). They constitute the pericycle. Very few species lack a pericycle, among them members of the southern hemisphere Centrolepidaceae. Roots lack any nodes, and lateral roots arise endogenously, that is their growing points or apices first develop in the pericycle. Division of cells in this region produces a lateral root which has to grow through the tissues of the endodermis and cortex to reach the exterior of the primary root. Because the pericycle bounds the vascular system of the root, vascular continuity between the new lateral root and the main root can soon be established once active growth has started. Some roots may have quiescent, potential laterals in the pericycle which require some hormonal stimulation or the removal of a hormonal restraint before they develop. The relatively simple nature of the pericycle and comparative lack of variation from species to species renders it of little use as a diagnostic feature.

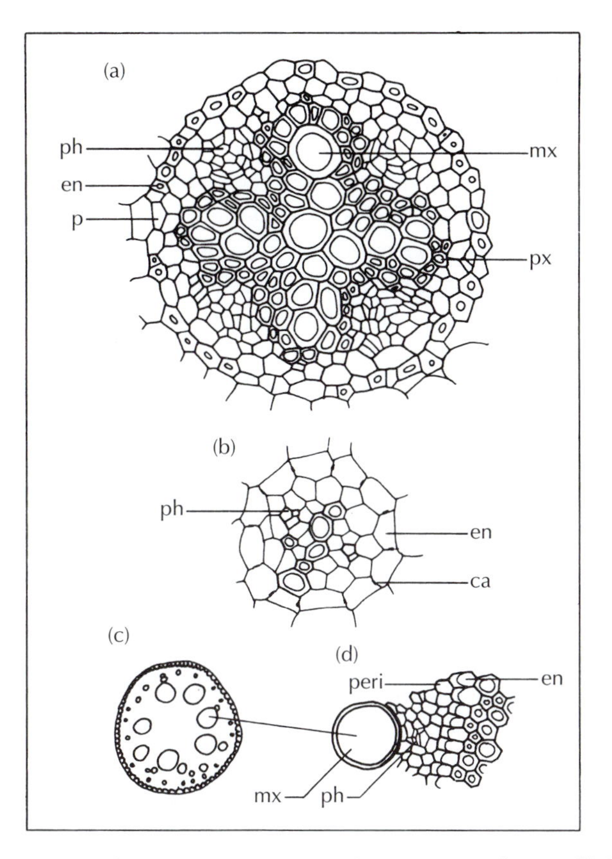

Fig. 4.4 Some root vascular systems. (a) *Ranunculus acris* tetrarch root. (b) *Echinodorus cordifolius* diarch root. (c,d) *Juncus acutiflorus* polyarch root. ca, Casparian strip; en, endodermis; mx, metaxylem; p, passage cell; peri, pericycle; ph, phloem; px, protoxylem. (a,b,d,e), ×300; (c), ×35.

Vascular system

The vascular system can take on one of several forms (Fig. 4.4). In most dicotyledons there are between two and up to about six protoxylem strands alternating with phloem strands. Large numbers of species have triarch (three-stranded) or tetrarch (four-stranded) arrangements, or a mixture of both. If above six strands are present, the roots are described as polyarch. Most monocotyledonous roots are polyarch. The metaxylem tracheids or vessel elements are normally conspicuous, and are on the same radial axis as the protoxylem poles, and to their inner side. This arrangement is characteristic of roots, and distinguishes them from stems. Root anatomy is initially exarch, with the protoxylem poles to the outer side, and endarch in stems, with the protoxylem to the inner side of the metaxylem. The transition usually occurs in the hypocotyl or the top of the primary root.

Commonly in monocotyledons there is one ring only, but there may be more or scattered metaxylem elements in the root centre. In most dicotyledons several metaxylem elements are grouped together in strands.

Although phloem is normally confined to the outer ring, occasional genera, for example *Cannomois* in Restionaceae, can have additional strands associated with the dispersed metaxylem elements.

In monocotyledons, the vessel element is thought to have had its origin in roots; in the least advanced plants, if vessels are developed at all, they are found in the root only, and not in the stem or the leaf. The next stage of advancement is for vessels to occur in the root and the stem, and in the most advanced plants vessels occur in the root, the stem and the leaf. A number of plants have shorter, wider vessel elements in the root than in the stem or leaf, bearing out the theory of the evolutionary sequence from the primitive narrow long elements to the more advanced shorter, wider elements. In supposedly primitive monocots, then, it becomes of great interest to see if vessels are present in the roots. The methods used are as described on pages 66 and 67.

Vessel and tracheid wall pittings and thickenings are similar to those of the stem wood. The phloem cells of the root also have the same range of forms as are found in the stem.

The centre of the root may be made up entirely of xylem in the dicotyledon, or in monocotyledons or some dicotyledons especially at the base of the primary root, it may contain a ground tissue composed of parenchyma with thin or thickened walls, sometimes termed pith (Figs 4.5–4.7). Sclereids could be present.

If a root is to be identified, it must be matched accurately in respect of all its tissues with authentic reference material. Descriptions and drawings are rarely full enough for one to be absolutely certain of a match. We have examined roots suspected to be the food supply of underground larvae of various insects. Only by having samples of roots from all the plants growing in the region where the larvae were found was it possible to make identifications of the chewed fragments. The division into monocotyledon or dicotyledon roots is relatively simple; it is only after that that the real problems begin!

Fortunately, secondarily thickened roots of dicotyledons (all monocotyledon roots are primary) are much simpler to identify. Secondary thickening is described in Chapters 2 and 3.

Lateral roots

In primary roots, lateral roots arise opposite the protoxylem poles. They develop from root apices which form in the pericycle. Each apex has a root cap. The new roots have to grow through the cortex to reach the outside,

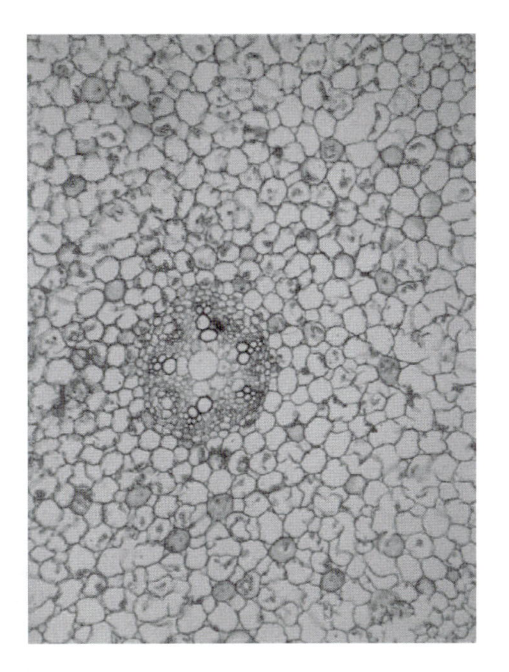

Fig. 4.5 *Ranunculus* (buttercup) root TS illustrating the relatively simple structure of a young dicotyledonous root. The xylem is tetrarch, and four strands of phloem alternate with the protoxylem. This root is just beginning to undergo limited secondary growth, with a cambial zone. ×100.

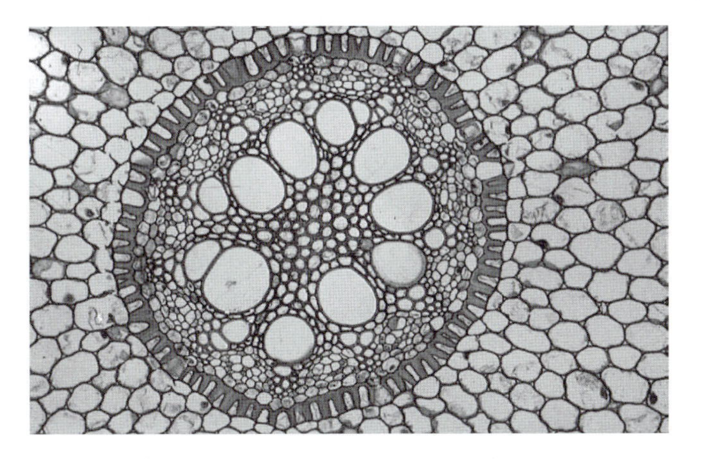

Fig. 4.6 *Iris* root TS, showing a very prominent endodermis – seen here as the layer of cells with striking thickening of the radial and inner tangential walls. The endodermis is the innermost layer of the cortex. The wall thickening forces water and other molecules to take a symplasmic route from the cortex to the stele, and vice versa, through the unthickened passage cells. ×350.

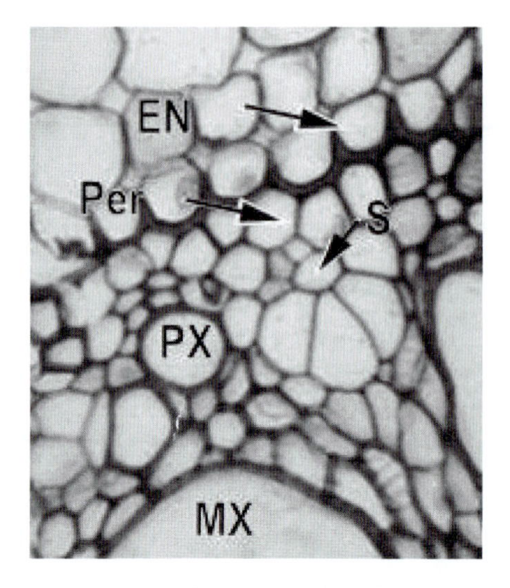

Fig. 4.7 Portion of a *Zea mays* root in this micrograph illustrates the arrangement of the separation of the cortical from the stelar tissues. *Zea*, like all primary roots, has an endodermis that forms the boundary between the cortex and the stele, and a layer immediately beneath this, the pericycle, that is the outermost layer of the stele. EN, endodermis; MX, metaxylem; Per, pericycle; PX, protoxylem; S, stele. ×500.

with enzymes dissolving ahead of the growth the material that bonds cells together (middle lamella), or by physically forcing through the cortex and epidermis. This type of development is described as endogenous. Because the origin of lateral roots is close to the central vascular material, new vascular connections can be made readily. Because the lateral roots arise opposite the protoxylem poles, the number of rows of lateral roots is indicative of the number of protoxylem poles in the parent root, so, for example, three rows of laterals indicates a triarch primary.

CHAPTER 5

The stem

Introduction

The primary stem, together with the tissues making up the first stages of secondary thickening, is considered here. The primary stem, like the root, has an epidermis and cortex, but often a distinct boundary layer such as the endodermis is not visible. An exodermis is rare, but a hypodermis is frequently present. Sometimes there is a layer that looks like an endodermis, but this normally lacks the Casparian strips, and is normally called an endodermoid sheath or starch sheath. The vascular system follows, and this starts off as individual vascular bundles in both monocots and dicots. Unlike the root, the protoxylem poles are directed towards the centre and the protophloem poles to the outer side in most cases. The centre is often parenchymatous ground tissue, but may be hollow. Again, unlike roots, the vast majority of species have stems with nodes, where leaves emerge and axillary buds may be present; there are variations in the monocots. Buds and lateral shoots arise in the outer tissues (exogenous), unlike lateral roots which are endogenous. In the following account, terms are explained, and some of the rich variety of stem anatomy is discussed.

Most dicotyledonous plants and gymnosperms, including annuals and even ephemerals, exhibit a degree of secondary thickening in their stems. This can start very early on, and may appear a few centimetres below the shoot apex. The secondary xylem and the secondary phloem are described in more detail in Chapter 3. It is easy to overlook that different parts of a plant are of different ages, and show varying degrees of secondary development. Students are thus cautioned, to ensure that comparative studies are carried out on material of comparable ages, when making comparisons between species.

An epidermis delimits primary stems, which is often very similar to that of the leaf of the same species. This is followed internally by cortical tissues, the outer layers of which, together with the guard cells in the epidermis, may contain chloroplasts. Chloroplasts in the epidermal cells are extremely

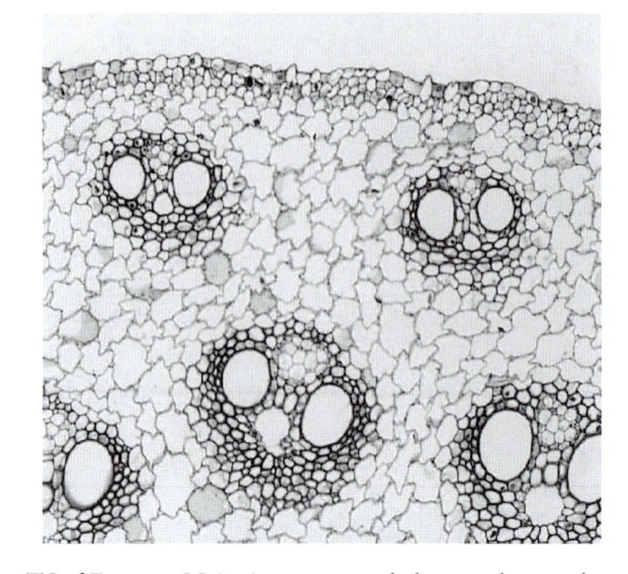

Fig. 5.1 Stem TS of *Zea mays*. Maize is a monocotyledonous plant, and resembles other grasses in the arrangement of tissues in the stem leaf and root. The stems of monocotyledons generally have a single ring of vascular bundles immediately beneath the epidermis, and internal to this a system of vascular bundles that are scattered throughout the pith. ×225.

rare in the seed plants except for the guard cells. This is a character which separates them from the ferns. Some cells acting as a physiological boundary between the cortex and stele are often present, forming a cylinder (endodermoid sheath, starch sheath). They may be morphologically distinct as a true endodermis, but are generally thin-walled, axially elongated parenchyma cells and sometimes they cannot be discerned as a separate layer.

Strengthening tissues can be present in the cortex or around the periphery of the stele (usually associated with the phloem), or in both positions. These tissues are usually in the form of axially arranged, rod-like groups of cells (fibres, and sometimes axially elongated sclereids), with gaps between them. Only in stems with very limited growth in thickness do they form a complete cylinder, and this only when primary and secondary growth have ceased. A good example of the latter is to be found in *Pelargonium* species, where the inner limit of the cortex is clearly demarcated by a ring of sclerenchymatous perivascular (= surrounding the vascular region) fibres.

The vascular bundles can take up a variety of arrangements. In dicotyledons and gymnosperms, they usually occupy one ring, just to the inner side of the cortex. In monocotyledons they may form one ring, or may appear to be scattered in several to many rings, or lie without apparent order in the central ground tissue (Fig. 5.1). The possession of several rings of vascular bundles is not the prerogative of monocotyledons. Several families of dicotyledons have this type of arrangement, notably those with climbing members, and also in the Piperaceae.

When vascular bundles are not scattered, the centre of the stem is usually parenchymatous, the cells are often thin-walled but rarely may be lignified in mature stems in some species. This central, pith-like region may contain some sclereids or parenchyma cells with thickened (lignified) walls. In some plants the central parenchyma breaks down to form a canal. Diaphragms of specialized stellate cells may traverse such canals, and in some plants axially arranged diaphragms are also present.

Most dicotyledonous stems have nodes, where leaves are attached and with leaf gaps in the axial vascular system. Leaf gaps become more apparent when some secondary thickening occurs. Each leaf usually has a bud in its axil. If the axillary bud develops, a branch gap is formed also. The internodes do not normally bear buds, unless these arise adventitiously. Monocotyledons have a range of types of shoot organization. They may have nodes where leaves are attached, as in grasses and sedges, or no formal node may be discernible in the internal structure, although the leaves appear to be attached to the stem in a similar way to those of the nodal plants when viewed from the exterior, as in palms, and axillary buds are usually present. Because there is no cambium developed within or between the individual vascular bundles in monocotyledons, no leaf or branch gaps form. Because anatomy and gross morphology are so closely related, it is important to study the morphology of a plant as a whole organism before cutting it up to look at the cells and tissues. Indeed, for intensive and comprehensive studies, development should also be followed. Only if you examine morphology and development is it possible to be sure that you can locate similar parts of various species for comparative anatomical studies.

Stems – cross-sectional appearance

The cross-section of a primary stem may have a more or less angular to circular outline. However, it can take on one of a wide range of forms, some of which assist in the identification of a family, as in Labiateae, where the section is square, or may help to distinguish genera; for example many *Carex* species have stems with a triangular cross-section. Often the outline is modified near to nodes or in regions of leaf insertion. Sometimes a wing or ridge of tissue in line with the sides of a petiole may continue down the internode as in, for example, *Lathyrus*. In general, the outline of the section taken in the middle of an internode would be described for comparative purposes.

As indicated in the introduction, stems have all or most of the following tissues, working from the outside inwards: epidermis, hypodermis, cortex

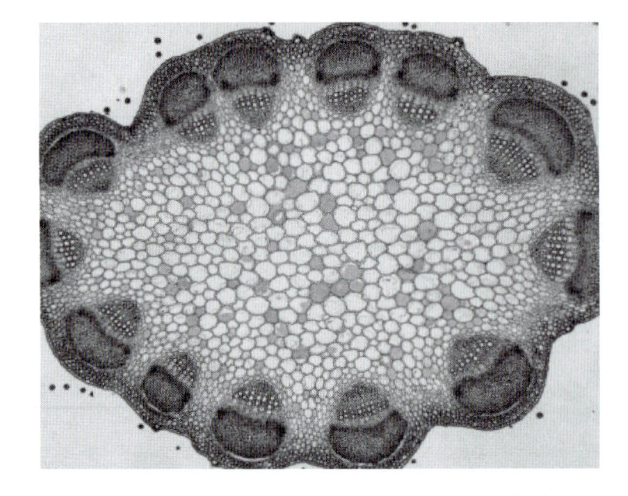

Fig. 5.2 *Trifolium* (red clover) stem TS: a mature stem at the end of primary growth, that is, the vascular bundles contain very limited amounts of secondary xylem and secondary phloem. The cortex is very narrow and is composed of chlorenchyma. The cortex is separated from the vascular bundles and the underlying pith by a starch sheath. The pith is parenchymatous. ×125.

(with both collenchyma and chlorenchyma, or either alone), an endodermoid layer (or a well-defined starch sheath), vascular bundles in one or more rings, or apparently scattered, and a central ground tissue or pith (Fig. 5.2).

Sometimes a pericycle can be distinguished, but this is normally regarded as part of the phloem. A true endodermis with Casparian strips is rarely present.

Epidermis

The epidermis can be one or more layered. The cells may be similar in form to those of the leaf of the same species, but more often they show more elongation in the direction parallel with the stem axis, and their anticlinal walls as seen in surface view are often not markedly sinuous. The outer wall is usually thicker than the anticlinal or inner walls. The proportion of cell length to cell width, or height to width as seen in TS, can be used with caution as a diagnostic feature, but actual measurements are not normally sufficiently constant to be used in identification or classification. A range of the normal measurements of a cell, with an average figure, should be quoted in diagnostic descriptions where a sufficiently large sample has been examined in order to obtain sound data.

Cuticular sculpturing, fine surface features which include very small papillae (micropapillae) and fine ridges (striae) in a variety of arrangements, may be similar to that in the leaves of the same species, but frequently differs in details.

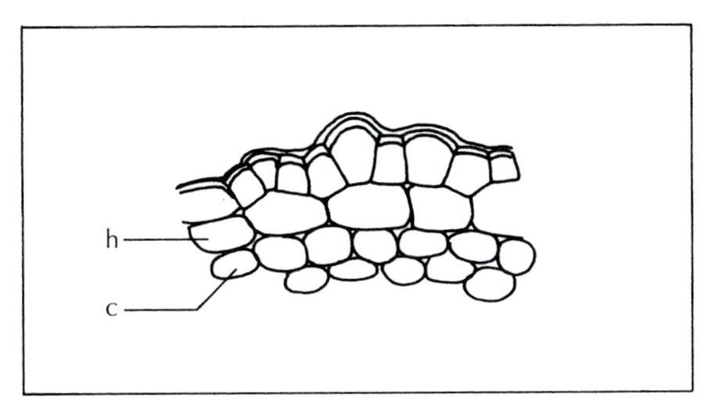

Fig. 5.3 Hypodermis in stem of *Salvadora persica*. c, cortex; h, hypodermis. ×290.

Hypodermis

The layer of cells under the epidermis is called the hypodermis. This may show distinct differences from the next cortical layer, but in many taxa is defined only by its position. When it is not different in appearance it is commonly described by systematic plant anatomists as being absent. A distinct hypodermis is found in *Salvadora persica*, for example Fig. 5.3. However, because of the sporadic occurrence of the distinct hypodermis in the taxa of vascular plants, its presence or absence is of little taxonomic and small diagnostic value (except at the species level).

Stomata

In taxa with well-developed leaves the stem stomata tend to be much more sparse, but of the same type as in the leaf of the same species. When the stem is a principal photosynthetic organ, either supplementing or replacing leaves, stomata tend to occur in a higher frequency. Often the guard cells are aligned parallel to the long axis of the stem and the stomata appear in rows.

Trichomes

Hairs, papillae and scales exhibit the same wide range as on the leaf, and examples are shown in Figs 6.14–6.17. The type of hair can be of diagnostic value at species level, sometimes also at genus level, but rarely at family level.

Silica bodies

Silica bodies are often present in the stem epidermis or in other parts of the stem in species which have them in the leaves. There are also leafless species of some families which have silica bodies in the stems for example *Lepyrodia scariosa*, Restionaceae. In the epidermis, the cells most likely to have silica bodies are those above fibre strands or girders. The range of form is shown in Fig. 6.24.

Cortex

The cortex can be very narrow, and composed of few cell layers, or wide and multilayered. The cortical zone is traditionally regarded as extending from epidermis or hypodermis to an inner boundary inside which vascular bundles are present. The inner boundary is often very indistinct, and sometimes vascular bundles from leaf traces may be present among cells which clearly belong to the cortex itself. Again we have an example of a tissue which is difficult to define clearly.

Chloroplasts may be present in the collenchyma cells of the outer cortex, or in more or less well-defined layers of parenchymatous cells, as well as in palisade cells or cells of various shapes. Stems of leafless plants, for example some *Juncus* species, often have a very formal and regular chlorenchyma arrangement. There are a few herbaceous plants which lack chloroplasts in cortical tissues, and these plants normally figure among those with abnormal modes of nutrition, for example *Orobanche*.

In species with a wide cortex, the cells of inner layers are normally larger than those of outer layers and have few, if any, chloroplasts. In aquatic plants, large, formal air spaces may be present in the cortex which itself merges with central parenchymatous tissues.

Fibres and sclereids are a prominent feature in the cortex of many species. Often the grouping of fibres into strands with well-defined cross-sectional outlines and in characteristic positions in the cortex will help in the identification of a plant. Fibres can show individual peculiarities, which enable one to identify even isolated strands. This is particularly true for fibres which are of economic importance, many of which are cortical in origin, for example *Linum*, flax, and *Boehmeria*.

Crystals and tannin often occur in cells of the cortex and the central ground tissue. Cluster crystals or druses are probably the most common type, but solitary crystals of various shapes and sizes, similar to the range exhibited in leaves, are of widespread occurrence. Raphides are not of widespread occurrence in dicotyledons.

Endodermis

In some plants the inner boundary of the cortex is well differentiated into an endodermis and shows Casparian strips, for example Hydrocharitaceae. In *Helianthus*, the cells are well-defined and are rich in stored starch but lack Casparian strips, thus they constitute a 'starch sheath'. In many other plants where the cells in this zone are morphologically distinctive but lack Casparian strips or stored starch, they are probably best called an 'endodermoid layer'. Some people prefer to use the term endodermis for this cell layer even when Casparian strips cannot be seen.

Vascular and strengthening tissue

Many monocotyledons, but few dicotyledons, have a well-developed cylinder of sclerenchyma to the inner side of the endodermoid layer. In it are embedded some of the vascular bundles, often all the small bundles, and sometimes some of the larger bundles as well. Most dicotyledons with secondary growth in thickness lack such a cylinder. Their 'open' vascular bundles, each with a cambium, are arranged in a ring. Each bundle may have a cap of fibres to its outer side, but the flanks of the bundles are not enclosed, so enabling unhindered development of the interfascicular cambium to produce a continuous cambial cylinder in secondary growth (see Chapter 3).

In many of those monocotyledons lacking a sclerenchymatous cylinder, individual bundles have a sheath of sclerenchyma. This is frequently only a few layers thick on the flanks, and several layers thick at the xylem and phloem poles. An outer parenchymatous sheath features in a number of these plants. Because there is no cambium as such, the enclosing of vascular tissues in these plants does not affect normal growth and development. In the monocotyledons which have a vascular plexus at the nodes, the sclerenchyma sheaths become discontinuous. If there is an intercalary meristem, the bundle sheaths are poorly developed in the meristematic zone.

In monocotyledonous stems the vascular bundles are either collateral, with one xylem and one phloem pole, or amphivasal, with the xylem encircling the phloem. Amphivasal bundles are frequent in rhizomes (which are modified stems) and less frequent in inflorescence axes (flower stems). Figure 5.4 shows a range of bundle types. Dicotyledons usually have open bundles, but climbers, for example Cucurbitaceae (cucurbits), may not develop interfascicular cambia, and the bundles remain as discrete strands. Since such bundles are present in a compressible parenchymatous matrix, twisting and distorting of such stems during the process of climbing does little damage to the vascular bundles themselves. In cucurbits and a number of other climbers the phloem is particularly well developed. It is present in two strands, one on either side of the xylem on a radial axis as seen in transverse section. Bundles of this kind are called bi-collateral. Because they occur in relatively few families their presence in a sample to be named becomes most helpful in reducing the field for further analysis.

Vascular bundles, which have phloem surrounding the xylem, are termed amphicribal.

There are plants with anomalous growth in which individual bundles have cambium and extend radially in secondary growth, without uniting

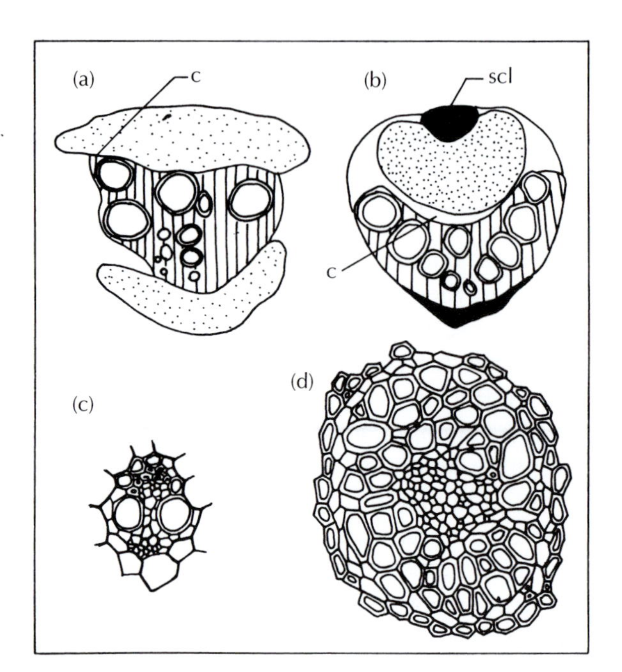

Fig. 5.4 Vascular bundle types from stems. (a) *Cucurbita pepo*, diagram of bicollateral bundle. ×15. (b) *Piper nigrum*, diagram of collateral bundle; cambium remains fascicular. ×15. (c) *Chondropetalum marlothii*, detailed drawing of collateral bundle, lacking cambium. ×110. (d) *Juncus acutus* detailed drawing of amphivasal bundle. ×220. c, cambium; scl, sclerenchyma.

laterally, as for example in the Piperaceae (Fig. 5.4). Several other abnormal forms of bundle arrangement are found among the dicotyledons.

It is a dangerous practice to attempt to define various types of vascular bundles on the observations made from a few transverse sections of monocotyledon stems. This is because over its length, a single bundle can exhibit changes in its cross-sectional appearance. A newly entering leaf trace can look different from a main axial bundle but the two are parts of the same strand, the first being the increment before bridging and connecting with the strands from other parts of the plant.

Transport phloem within the axial system

The plant vascular system plays a pivotal role in the delivery of nutrients to distantly located organs. Recent discoveries have provided new insight into a novel role for plasmodesmata and the phloem in terms of the transport and delivery of information macromolecules including proteins and ribonucleoprotein complexes. Ruiz-Medrano et al. (2001) suggest that the

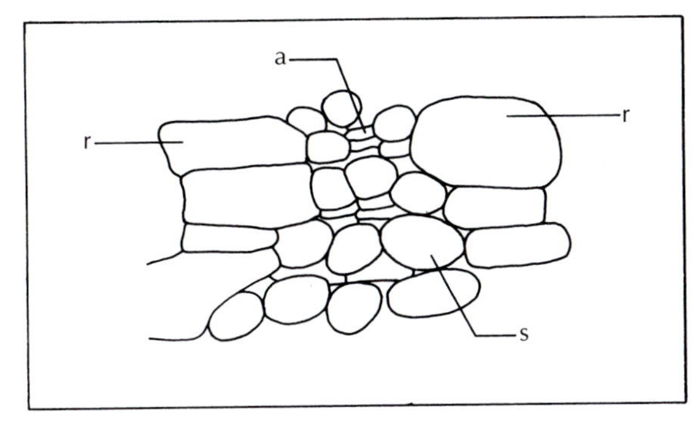

Fig. 5.5 Albuminous cells in gymnosperm phloem, *Acmopyle pancheri*. a, albuminous cell; r, ray; s, sieve cell. TS, ×290.

phloem may function as a conduit for inter-organ communication. Within stems, a great deal of assimilate is transported through the axial system, to local or distant sinks. Figure 6.28 illustrates the difference in size between loading transport and unloading phloem in *Nymphoides*. Within the transport phloem, sieve tubes are between 10 and 25 μm in diameter, and their associated companion cells between 15 and 38 μm in diameter. Clearly, sieve tubes in transport phloem are larger than corresponding sieve tubes in loading phloem in this species.

The course of vascular systems in monocotyledonous stems has been studied for many years and is under active investigation at present. Modern microscopy techniques, including fluorescence and confocal microscopy, stack frame imaging of whole and sectioned material have enabled researchers to understand for the first time the true complexity of many stems, including palms and Pandanaceae. With newer and more powerful techniques becoming available, a new area of comparative anatomy is emerging in the study of whole vascular systems. The results of this study might well show basic types which underlie the major phylogenetic divisions in the plant kingdom.

Within the vascular bundles, the phloem and xylem of primary systems show only axial systems of cells. The rays are a feature of secondary development.

Phloem in gymnosperms has well-developed sieve cells, with sieve areas and their associated albuminous cells. In angiosperms the albuminous cells are replaced by companion cells (Fig. 5.5). It is thought by those who study phloem that an evolutionary sequence can be observed, from systems in which companion cells are poorly defined and the sieve tube elements communicate by rather scattered sieve areas on oblique walls to the most advanced in which sieve plates are very well defined and constitute the transverse end wall between elements in a sieve tube, and in which the companion cells are very well developed. Since the advanced sieve tube member has no nucleus, the organization of the element is carried out by the nucleate companion cell adjacent to it. Damage to the companion cell in this system may bring about failure of the element which it directs. Phloem is not the easiest of tissues to study with the light microscope, and consequently it is only in recent years that the beautiful comparative studies such as those carried out in the laboratories of Katherine Esau and later by Ray Evert have contributed greatly to our understanding of the anatomy and structure of phloem.

Sieve plates can be seen easily in a number of plants, especially those that have large-diameter sieve tubes; a good example is the Cucurbitaceae.

As in the leaf, the first formed phloem in stems is termed protophloem. It is often functional for a short period only, as it usually differentiates in regions of rapid cell expansion and elongation and often becomes compressed, but not before the later-developing metaphloem matures.

Transport tissue – structural components

Primary phloem and xylem may contain sclereids and fibres. Primary xylem is composed of protoxylem, in which the tracheary elements usually have helical (spiral) or annular wall thickenings. In metaxylem the wall thickenings can be more extensive, and breached by pits (with membranes) arranged in scalariform, alternate, reticulate or less regular ways. The protoxylem has to be capable of considerable extension, without breaking, during the first phases of primary growth in length of the stem. However, the elements often do rupture, leaving a protoxylem canal, termed a lacuna. The more rigid metaxylem matures after this extension phase and as a consequence is less liable to damage. Its structure does not have to allow for axial extension.

It is often difficult to decide if the protoxylem and those metaxylem fractions with annular or spiral thickenings are tracheids or vessel elements, because perforation plates can be very obscure and may even appear to be present in damaged, macerated preparations when they are, in fact, absent. Even with cells which have scalariform, alternate or reticulate pitting it may be hard to be certain if they are perforated or not, since perforations may be very small. This is of some importance, however. It is now widely believed that the narrow elongated imperforate tracheid is ancestral to the wider, shorter perforate vessel element. Therefore, plants which are vessel-less are thought to have primitive wood. The hinge of many phylogenetic systems swings on this delicate frame. Several methods are employed to try and determine if a cell is perforate or not. The tissues are usually macerated to separate the individual cells. These can then be examined with phase contrast illumination, when intact pit membranes show up well. Another method involves flooding the macerate with Indian ink, on a microscope slide. A coverslip is placed over the cells and gently tapped. The ink is then replaced by 50% glycerine, drawn under the coverslip using filter paper from the opposite side. The Indian ink contains solid particles, and if the cells are perforate, particles should be visible inside the lumina of the vessel elements.

More recently, macerated cells have been examined using the SEM, where membranes are readily observed. As with primary phloem, no radial system of cells is present in primary xylem, but fibres or sclereids and, occasionally, parenchyma cells can be present.

Central ground tissue

The central ground tissue or pith is composed of cells which are usually parenchymatous, with simple, more or less circular pits in their walls. The walls may be thin, and composed largely of cellulose, or thickened with lignin. This matrix of cells can in various plants contain sclereids, tannin cells or crystalliferous cells, or combinations of the three.

Certain groups of plants have special cells or tubes containing latex. *Landolphia* has latex cells; most Euphorbiaceae have latex tubes. In instances where members of the Cactaceae and Euphorbiaceae have evolved to appear similar externally, it is very easy to distinguish the two anatomically. These plants of dry areas have only to be cut: latex will ooze from all the euphorbiaceous representatives and very few of the Cactaceae and the liquid is watery not latex. Incidentally, this latex in euphorbs can be poisonous and irritating to the skin, and even lethal. Many members of the Compositae (Asteraceae) contain latex and *Taraxacum* was even grown experimentally in search of a *Hevea* substitute for rubber latex during the Second World War. *Hevea brasiliensis* itself is probably the best-known latex producer. Among the monocotyledons, many members of the Alismatales contain latex canals with a secreting epithelial layer.

Latex tubes are among the longest cells and are usually coenocytes, that is long tube-like structures, with several to many nuclei; they often go on growing during the whole life of a plant.

Because the incidence of latex-forming cells or tissues is restricted in the angiosperms and, further, because there are various types of cell, tube or canal, and since the laticifers can be articulated or unicellular, their presence in a plant can be a great help in identification. Also, the distribution of these cells or tissues in cortex, phloem, xylem or ground tissue can be diagnostic.

The centre of the pith may be composed of parenchymatous cells, or it may be hollow, either as a single tube, or variously divided by transverse or longitudinal septa. The cells in the septa or diaphragms can be of a simple, more or less isodiametric type, or they may be stellate or armed or branched

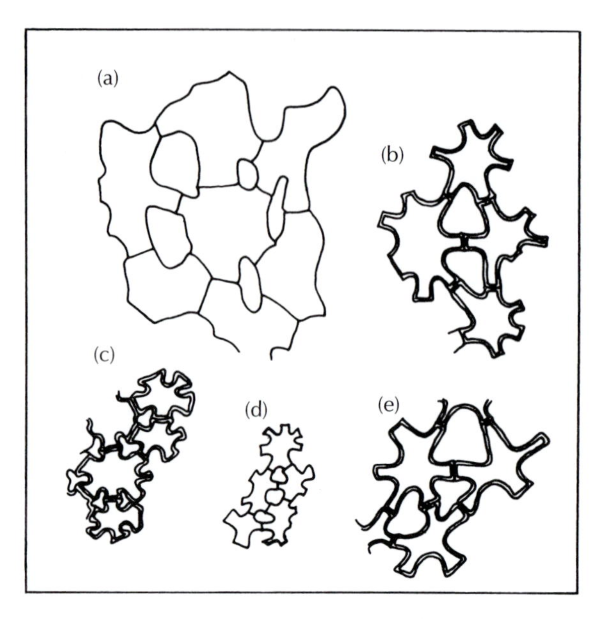

Fig. 5.6 Diaphragm cells in leaves of Cyperaceae. (a) *Becquerelia cymosa*. (b) *Mapania wallichii*. (c) *Chorisandra enodis*. (d) *Mapaniopsis effusa*. (e) *Scirpodendron chaeri*. All ×218.

cells of various descriptions. A range of such cells (from leaves) is shown in Fig. 5.6. Diaphragmed or chambered pith may be diagnostic of some genera such as *Juglans* (walnut) and *Sambucus* (elderberry) or species such as *Phytolacca americana* (pokeweed).

Concluding remarks

Primary stems show a wide range of structure in their histology, often characteristic of the species, genus or family. They are widely used in physiological studies, regularly occur in archaeological remains and can sometimes be found in circumstances where forensic examination is necessary, so a proper understanding of their anatomy is important.

Dicotyledonous stems do not remain in the primary state for long, and some secondary growth in thickness is usually present in most species. This is also true of the primary stems of gymnosperms. These are usually distinguishable from angiosperm stems because they lack any vessels. The xylem is composed mainly of tracheids with conspicuous bordered pits.

Monocotyledonous stems also show a wide range of structure. Some have a conspicuous outer ring of sclerenchymatous fibres associated with the outer ring of vascular bundles, others do not, for example many *Juncus* species. Cereals tend to have a soft parenchymatous core within which vascular bundles form a characteristic spiral arrangement (as seen through serial sectioning), with large and smaller bundles intermingled. In most cases, the vascular bundles in the stem are very similar to those in the leaf, with a similar arrangement of large metaxylem and protoxylem elements or lacunae evident. In stems, the vascular bundles are delimited by an inconspicuous parenchymatous sheath which may become lignified at maturity.

CHAPTER 6

The leaf

Introduction

Leaves grow on stems below their growing points, and are developed from leaf buttresses in buds. All leaves are lateral organs which develop from a foliar (or leaf) buttress, which, in simple terms, is a meristematic projection above the general surface of the protoderm (a primary meristem or meristematic tissue that gives rise to the epidermis; see Chapter 2).

Leaf buttresses are initiated near the shoot apex and in lateral buds, in regular sequence (phyllotaxis), and lead to the formation of mature leaves. Although we commonly think of leaves as being thin, flat and green, their form is closely related to the habitat in which the particular species grows. The outline and general shape can also be characteristic of a genus, and sometimes of a family.

In general terms, most developing leaves contain two groups of initials or meristems – the marginal and submarginal initials. Marginal initials generally give rise to the adaxial and abaxial epidermis of the leaf, whilst the submarginal initials generally give rise to all the internal leaf tissues, including the procambium, from which all vascular tissues are subsequently differentiated. In dicotyledonous plants, the transition from a condition where the leaf is a photoassimilate sink to one in which it is a net exporter of photoassimilate (source status) begins shortly after the leaf has begun to unfold, at which point, the major morphogenetic events that determine leaf shape are, to all intents and purposes, over. Maturation of the phloem and xylem in the midrib and the higher-order veins, which occurs in an acropetal direction, is largely complete before the transition begins. During leaf unfolding, the functional maturation of the minor veins generally begins in a basipetal direction. There is thus a degree of maturation of the leaf from the base to the tip of the lamina during the sink to source transition. The minor venation network forms the distribution network of the leaf which provides first an importing and then an exporting network as the leaves continue to expand.

Leaves show a surprisingly wide range of form when it is considered that in the majority of plants they perform three basic physiological functions. These involve the manufacture of food materials through the process of photosynthesis, the transport of assimilated material and the evaporation of water, a process that drives the transpiration stream and, concomitantly, aids cooling of the leaf in hot conditions when available transpirational water is not limiting. Each of these functions is either initiated or takes place directly in the mesophyll of leaves.

Attempts to visualize the steps by which the more unusual leaf forms arise must originate from a study of the development of the leaf itself. In other words, development must not be inferred but rather observed. Figure 6.1 shows the possible evolutionary pathways of several of these leaf forms, without suggesting that foldings and partial fusions, leading to total fusions, really occur during the growth of the present-day owners of a particular type.

To emphasize the danger of thinking of a sequence of folding and fusion processes, look at the vascular bundle arrangement of *Thurnia*, a plant from South America (Fig. 6.2). The relative position of the small vascular bundle system to the large bundle system, and the inverted orientation of these bundles, would require a great deal of tortuous folding to achieve, if in fact they did not arise at the same time, from a meristem!

That leaves have vastly differing internal structure is demonstrated by the mesophyll cells (photosynthetic and other parenchymatous tissues of the leaf blade contained between the epidermal layers), the photosynthetic cells of which are arranged in different patterns and locations and which may be ascribed directly to the functional processes of the photosynthetic cycle occurring within the leaf (see below). Leaves may be isolateral, isobilateral, dorsiventral, pseudodorsiventral or even needle-like in cross-section (see Glossary). Whatever the leaf shape, chloroplasts are concentrated within the cytoplasmic matrix of the chlorenchyma cells and, for the most part, the majority of the chloroplasts are to be found in the upper palisade mesophyll cells, or their equivalent. Mitochondrial populations in these obviously photosynthetic cells may be high as well.

Leaves can be classified in various ways, for example by their shape and size, their texture and colour, the degree of hairiness, to name but a few. These variable features are frequently reflected in different internal tissue arrangements. Some modifications are typical of plants that grow under particular conditions, but others may owe more to the genome than to the habitat which the plant occupies.

The shape of the leaf lamina and the nature of its margin are outside the scope of this book. It should be remembered that fragments of leaves, whilst not showing the shape of a leaf blade, may often have the characteristic dentations of the leaf margin and this can be a big help in identification.

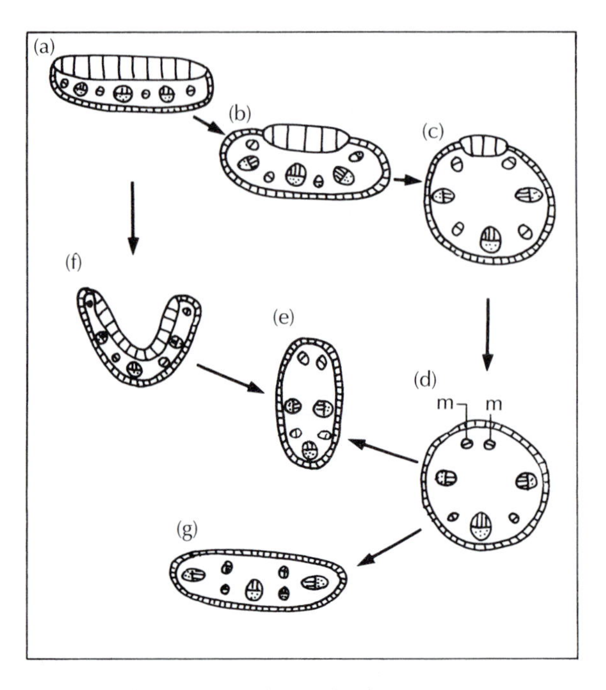

Fig. 6.1 Some possible evolutionary pathways leading to variations in vascular bundle arrangements in leaves. See text for fuller commentary. (a) One row of bundles; adaxial and abaxial surfaces distinct. (b) Adaxial surface much reduced. (c) Smaller adaxial surface, leaf becoming cylindrical. (d) Loss of adaxial surface, leaf cylindrical, bundles in one ring, but 'marginal' bundles still distinct (m). (e) Lateral compression, leaf of this type could arise from (d) or from (f), where the adaxial surface is progressively lost. (g) Could arise from secondary dorsiventral compression of the form in (d).

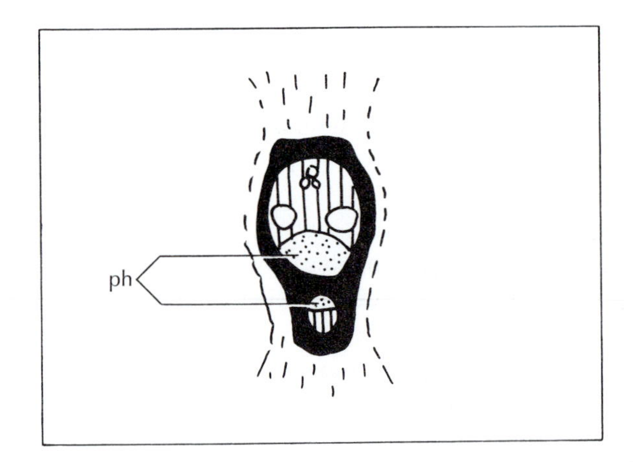

Fig. 6.2 Diagram of vascular bundle pair in *Thurnia sphaerocephala*, leaf. The small bundle is inverted so that the phloem poles (ph) of the pair are opposite one another. TS, ×57.

Although ecological adaptations are discussed in Chapter 8, they also warrant brief mention here. Leaves differ from most stems and roots in that they are almost all primary organs composed of primary tissues. However, some secondary growth can occur, as, for example, in the vascular supply of some gymnosperm leaves, in the petioles and midveins of some dicotyledonous foliage leaves, and in the leaf bases of some monocotyledons which display a form of secondary growth in their stems. Secondary growth in the dicotyledonous leaves differs from that in monocotyledons, just as it does in their respective stems. However, large changes do not occur in the shape or thickness of dicotyledonous leaves after primary growth has ceased. Primary growth occurs at the basal (intercalary) meristem of many monocotyledonous leaves, often for a long time after maturation of the distal portions. Grass in lawns would not recover from cutting and continue to grow if this were not so!

Some leaves are ephemeral, and are quickly shed, leaving nothing but a scar or perhaps a basal, membranaceous sheath as in *Elegia* (Restionaceae) or *Equisetum*. The stem, which remains green, then takes over the functions of the leaf in such xeromorphs. Other plants shed their leaves at times of physiological drought, for example when the ground is frozen, as is demonstrated by north and south temperate mesophytic trees and shrubs. The leaves on herbaceous perennials and annuals last for one growth season only. In biennials, it is common for rosette leaves to live for two years. Some trees and shrubs of temperate regions (or at temperate altitudes in the Tropics) have reduced leaves, for example, as in *Pinus* and *Cedrus*.

Many families have members in which leaves thrive for more than one season and are thus termed 'evergreen'. Plants with long-lived leaves are not confined to particular climatic or altitudinal zones. The evergreens include such plants as the conifers mentioned above, as well as broad-leaved plants like *Camellia*, *Borassus*, *Phoenix*, *Rhododendron*, *Ilex*, some *Quercus* species, *Coffea* and *Ficus*. In *Araucaria*, a gymnosperm, the leaves can live for decades and the green tips can be seen in the bark of the trunk.

It is interesting to note that some desert plants develop their leaves only after rain, for example *Schouwia* of the Brassicaceae, and then lose them during periods of sustained drought. Many bulbous plants, for example *Narcissus*, *Tulipa*, *Albuca* and plants with corms, for example *Gladiolus*, *Watsonia*, *Crocus*, have leaves that grow during or after the wet season and die down, leaving the underground storage organs safe from desiccation during the dry or cold period. Aquatic plants may have required hibernation periods, where leaves die at the end of a growing season, for example *Potamogeton*, *Stratiotes* and *Nymphaea* species.

By now it should be apparent that there is no such thing as a 'typical' monocotyledous or dicotyledonous leaf. Except in extreme cases of reduction, as, for example, in some aquatic plants or xeromorphs, most plants have leaves which are made up of a combination of various essential components – the mechanical and supply systems, the tissue in which photosynthesis is carried out, and the outer skin or epidermis.

Leaf structure

Many leaves are flattened dorsiventrally. The leaf of *Ilex aquifolium*, in Fig. 6.3 in transverse section and surface view, serves to illustrate general dicotyledonous foliage leaf anatomy. In this example, the epidermis forms the boundary between the atmosphere and the underlying mesophyll and vascular and non-vascular tissues. Its cells are specialized for this function. Note that the adaxial epidermal cells (Fig. 6.3c) have thickened outer walls. The epidermal cells are covered by a thin cuticular layer. In this dorsiventrally compressed leaf, the upper and lower surfaces are different, as can be seen from Fig. 6.3(a–c). Stomata occur among the cells of the lower (abaxial) surface only; the leaf is thus hypostomatic (see Fig. 6.3d). The mesophyll consists of chlorenchymatous, palisade-like cells on the adaxial side with few intercellular spaces, and of more loosely arranged spongy cells with larger intercellular spaces on the abaxial side (Fig. 6.3c). Part of the vascular system is shown, including the large midrib bundle (Fig. 6.3a) and in more detail in Fig. 6.3(b). A smaller secondary vein is illustrated in Fig. 6.3(c). In all leaf blade vascular bundles, phloem occurs to the abaxial and xylem to the adaxial side of the leaf. The larger and many of the smaller vascular bundles frequently have a cap of sclerenchyma cells associated with the phloem pole only.

In general terms, all leaves have similar features – an epidermis with stomata, mesophyll and vascular tissue. However, the arrangement of these components is, to a large extent, dictated by the physical environment such as water availability, light intensity, ecological niche and herbivores. Through selection pressure, it is the interplay of these environmental parameters which serves to drive the evolution of leaf structure. The epidermis may, for example, be one or more layers thick, there may be either a thick or a thin cuticular covering, there may be a hypodermis associated with the epidermis. Stomatal distribution may be on both surfaces of the leaf or on one surface of the leaf only, and they may be raised above the general leaf surface, flush with the leaf surface, or in some cases, sunken into crypts. The mesophyll may be specialized or unspecialized. If specialized, a palisade or spongy layer may exist and in some leaves, palisade tissue may exist on both sides of the leaf (isobilateral) as is the case in many succulents. The mesophyll may be compact, with numerous small intercellular spaces as in xerophytes, or may contain a large intercellular space volume, as in some mesophytes and most hydrophytes.

The general mechanical requirements of a typical leaf are thus: (i) adequate gas exchange and (ii) functional transport pathways.

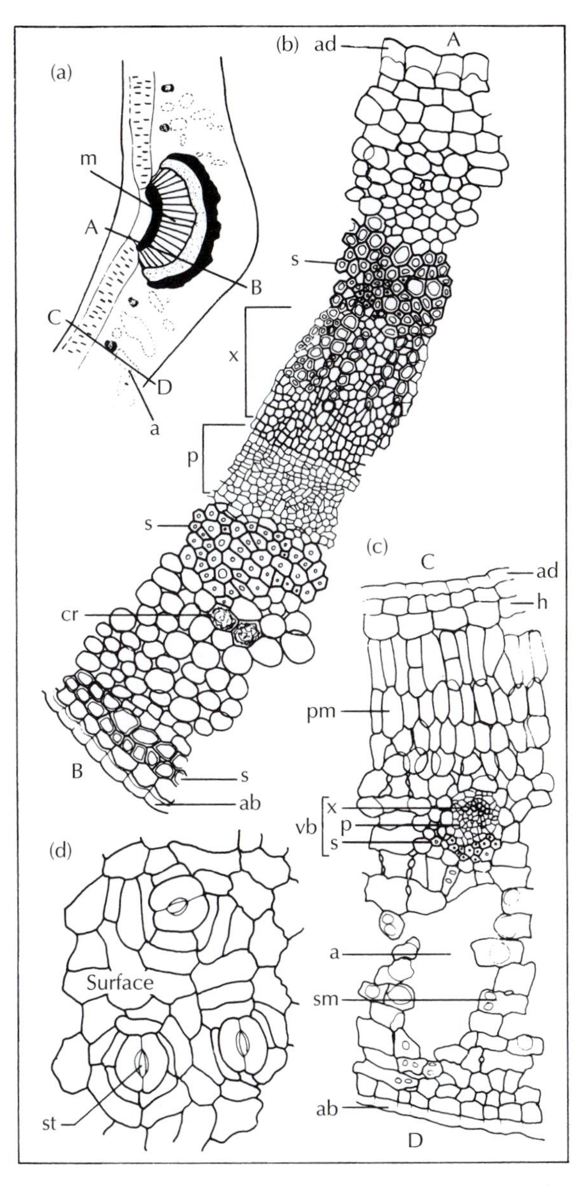

Fig. 6.3 *Ilex aquifolium* leaf TS and surface. (a) Low-power (×22) diagram of midrib region, A–B and C–D indicate where detailed drawings (b) and (c) were taken. (b) Detail of midrib. TS, ×130. (c) Detail of lamina. TS, ×130. (d) Abaxial surface. ×200. a, air space; ab, abaxial epidermis with thick outer wall; ad, adaxial epidermis with thick outer wall; cr, crystal; h, hypodermis; m, midrib bundle; p, phloem; pm, palisade mesophyll; s, sclerenchyma; sm, spongy mesophyll; st, stoma; vb, vascular bundle; x, xylem.

Leaf surfaces must be mechanically adapted to meet environmental stresses, but translucent, to allow photosynthetically active radiation to pass through them to reach the pigment chlorophyll in cells beneath.

Cuticle and cuticular sculpturing

In all but the wettest environments, leaf surfaces must also be capable of helping to reduce water loss. This is helped in many species by the presence of a transparent outer layer, the cuticle, which retards water loss. This tends to be thinnest in species not normally subject to water stress, and thickest in those that are. Angiosperms with submerged leaves may have an exceedingly thin cuticle, or it may be absent. The cuticle may also give added mechanical strength. It helps resist abrasion by blown sand particles, or in the case of some conifers, blown ice crystals. The main component of cuticle is cutin, which may permeate the walls of epidermal cells, or just the outer walls. It is most developed in species confined to extremely arid habitats. Low availability of water to the plant can be induced by saline soils, so plants growing on these often show adaptations similar to those from dry habitats.

The cuticle and the outer part of the wall of the epidermis that it covers and grades into is patterned or sculptured in many plants. If the sculpturing is of low relief it will not show strongly in sections and may be faint in surface view. The strong sculpturing in *Aloe*, for example, may often be obscured by the granular appearance of the interface between cuticle and epidermis (Fig. 6.4).

Although many patterns can be seen with the light microscope, either on intact cells or with detached cuticles or surface replicas, the scanning electron microscope is important in surface studies.

In aloes and haworthias, the range of cuticular outer cell wall patterns is such that individual species or groups of species can often be identified by their particular pattern. Striations are quite common, as are micropapillae. Some patterns are shown in Fig. 6.5.

Sometimes the cuticle and its markings are masked by a covering of wax. Most people are familiar with the waxy 'bloom' on apples and plums, and have noticed that some leaves have a dull sheen on them, for example, as in the cabbage (*Brassica*) or in *Agave*. Few may realize that many other plants also have a waxy crystalline covering, because this may be very thin and easily removed. Other chemicals such as flavenoids are sometimes involved. Surface wax may be smooth, or may show varying degrees of roughness. It may function in helping reduce water loss, but it has reflecting and other properties.

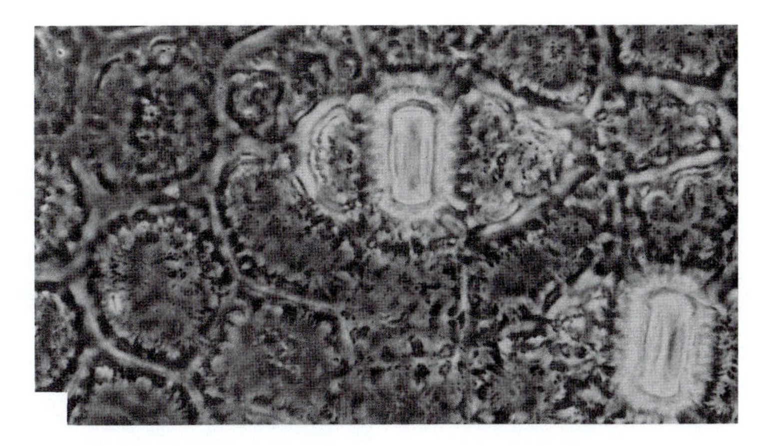

Fig. 6.4 Anoptral contrast, *A. branddraaiensis*, the granular interface between the cuticle and cell wall makes interpretation of the cuticular pattern difficult. ×400.

Many plants with xeromorphic characteristics have a waxy covering, which retards cuticular water loss (see Chapter 8).

Wax takes on many crystalline forms, and may also be present as a melted-down layer. Wax in a few species may go through a daily cycle in which wax crystals of one form melt, and recrystallize into another form, but this is rare. Many xeromorphic monocotyledons have large numbers of their stomata plugged by wax. The scanning electron micrograph in Fig. 6.6 shows some wax flakes in *Aloe lateritia* var. *kitaliensis*. Wax embellishment is often associated with sunken stomata.

The appearance of leaf surface sculpturing can be complex. However, by using a straightforward procedure, the sculpturing can be broken down into four elements, for description.

Primary sculpturing defines the overall arrangement of cells, generally visible at low magnification.

Secondary sculpturing defines: (i) the orientation and shapes of the cells and describes the number of anticlinal walls, whether they are straight, or if not, the degree of sinuosity; how distinct they are, as ridges or channels, for example (e.g. cells six-sided, as long as to twice as long as wide, with straight sunken anticlinal walls; Fig. 6.5b) (ii) details of the outer (periclinal) wall (e.g. flat, concave, convex with a pronounced central papilla; in Fig. 6.5c they are low-domed); (iii) the position, type and frequency of stomata.

Tertiary sculpturing is the finer detailed sculpturing found on the outer periclinal wall, superimposed on the primary sculpturing. It may be absent, and the surface is then described as smooth. It includes

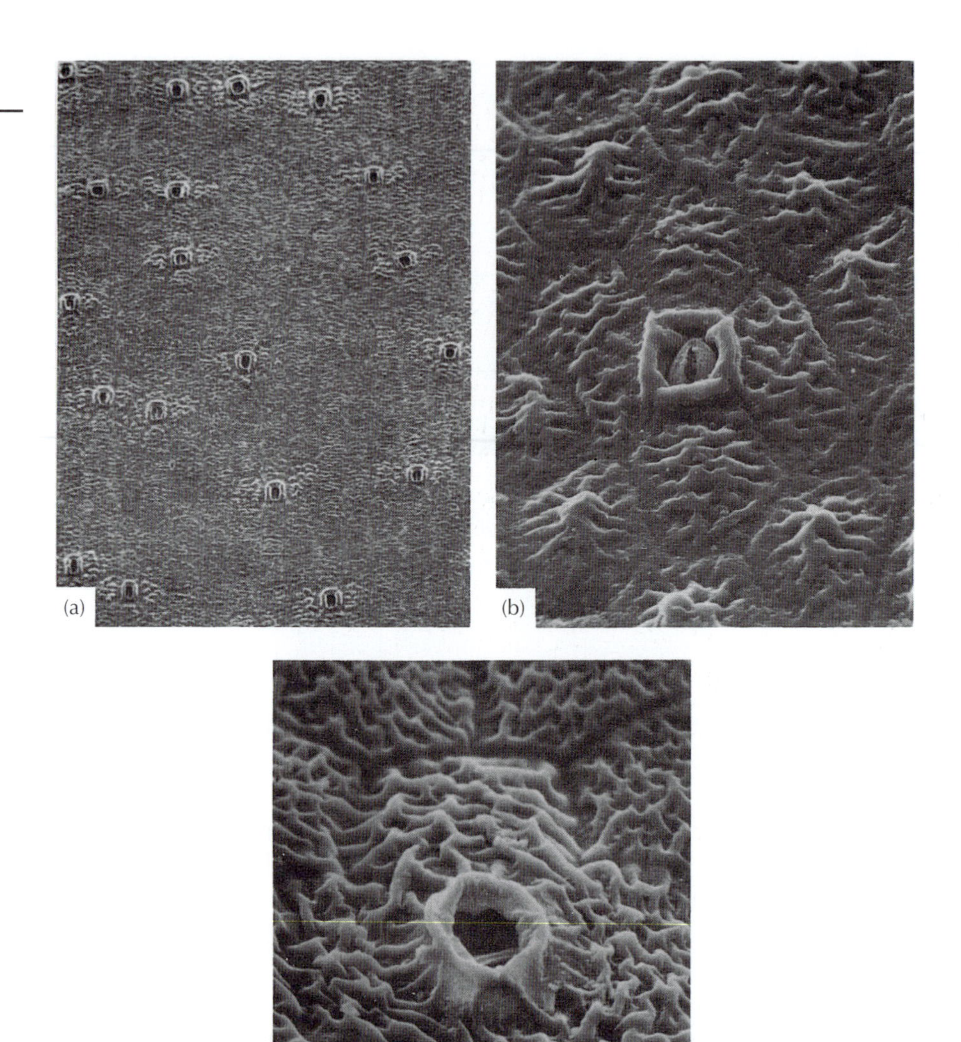

Fig. 6.5 Cuticular patterns are more easily seen using the scanning electron microscope. (a) Low power view (×50) of *Aloe rauhi* × *A. dawei* showing distribution of stomata. (b) *Gasteria lutzii* × *Aloe tenuior* var. *rubra*. (c) *Haworthia cymbiformis*. Note that the rim to the stomatal pore is four-lobed in the hybrid plants, an *Aloe* characteristic. *Haworthia* belongs to the group of very succulent species within the genus, and has lobes which are fused into a cylindrical collar. (b, c) ×600.

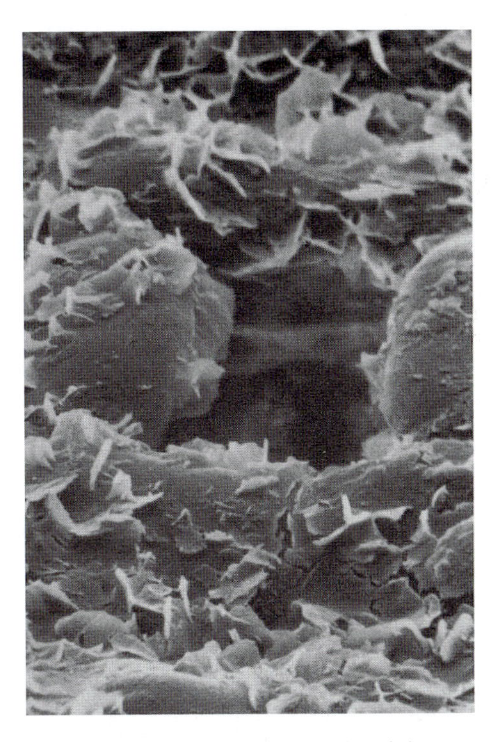

Fig. 6.6 *Aloe lateritia* var. *kitaliensis* wax flakes on the four lobes surrounding a stoma. The guard cells are deeply sunken and can just be seen. ×2200.

micropapillae, and defines their size and distribution (e.g. fine, covering the whole cell surface), striae, their thickness, distribution and orientation (e.g. coarse, longitudinally oriented striae, along the long axis of the cell, or striae forming a reticulum as in Fig. 6.5c).

Quaternary sculpturing includes epicuticular secretions, for example wax, farinose material (some *Primula* species). This may be present as smooth layers, upright flakes with random (e.g. Fig. 6.6) or defined orientation; coarse or fine amorphous particles, filaments or tubes, for example. See the references in the further reading.

Epidermal cells vary a great deal from species to species, particularly as seen in surface view. Many monocotyledons, and in particular those with strap-shaped or axially elongated leaves, have elongated cells that are arranged in well-defined longitudinal files. These cells may be 4–6 or more sided; their anticlinal walls may be straight, curved or sinuous. Sometimes the outlines of these walls are more sinuous near the cell outer wall than near the inner wall. Figure 6.7 shows a range of named examples of cell forms.

The epidermal cells of leaves of grasses fall into two distinct classes, described in the literature on grass anatomy as 'long' and 'short' cells. These two size classes should not be confused with variations in cellular dimensions that are to be seen over veins (costal cells) and between veins

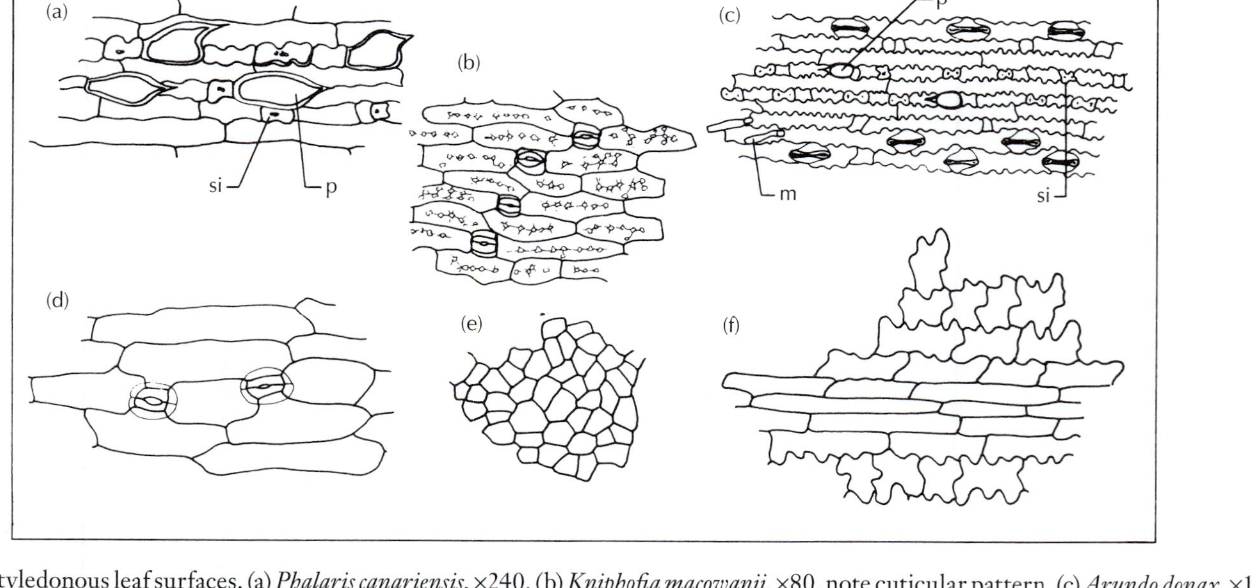

Fig. 6.7 Monocotyledonous leaf surfaces. (a) *Phalaris canariensis*, ×240. (b) *Kniphofia macowanii*, ×80, note cuticular pattern. (c) *Arundo donax*, ×120, note microhairs. (d) *Clintonia uniflora*, ×70. (e) *Smilax hispida*, ×150. (f) *Gloriosa superba*, ×54, note elongated costal cells over vein and cells with sinuous walls between veins (intercostal cells). m, microhair; p, prickle hair; si, silica body.

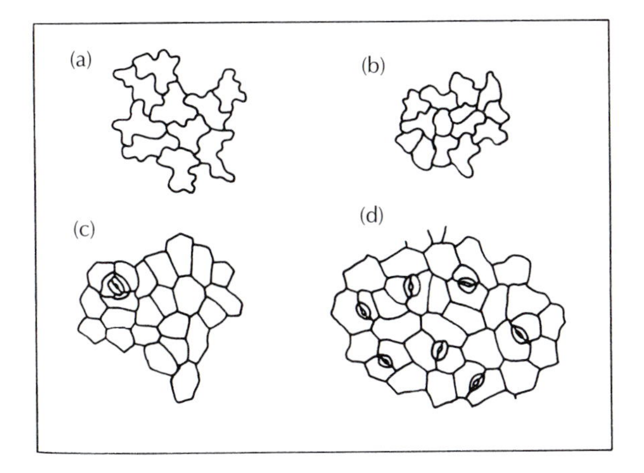

Fig. 6.8 Dicotyledonous leaf surfaces (abaxial): (a) *Acacia alata*; (b) *Aerva lanata*; (c) *Plumbago zeylanicum*; (d) *Cassia angustifolia*. All × 120.

(intercostal cells). The true 'short' cells are frequently suberized, or they may contain silica bodies. Even small fragments of leaf from a member of the Poaceae can often be identified to the family level, based upon the epidermal cell characters.

The majority of dicotyledons and many monocotyledons without axially elongated leaves (strap-shaped), for example *Smilax*, *Gloriosa*, tend to have epidermal cells of irregular shape and size. They have straight, curved or sinuous anticlinal walls. Because dicotyledon leaves lack a basal meristem, but grow in area by regions of cell division, their epidermal cells are rarely arranged in clear rows. Figure 6.8 shows a range of cell types from named plants.

Anticlinal walls of the epidermal cells of both monocotyledons and dicotyledons can be very thin and hardly visible from the surface, or they may range through degrees of thickness to very thick, so that the lumen of the cells appears from the surface to be very reduced (Fig. 6.9). In dicotyledons, as in monocotyledons, the costal cells frequently differ from those of intercostal regions; they tend to be elongated in the direction of the veins.

Sometimes the cells of the upper and lower surfaces of leaves may be similar in size and structure, but more often they are not alike. In monocotyledons with true dorsiventral leaves, adaxial and abaxial epidermal cells may differ markedly in size, and the adaxial epidermis, may contain bulliform ('motor') cells that are considerably larger than the normal epidermal cells. The dissimilarity may be in cell size and wall thickness, or merely the absence of stomata from one surface.

Cells at the margins and the tip of the leaf are often narrower than the rest, and have thicker walls. Some marginal cells may develop into unicellular or multicellular prickles.

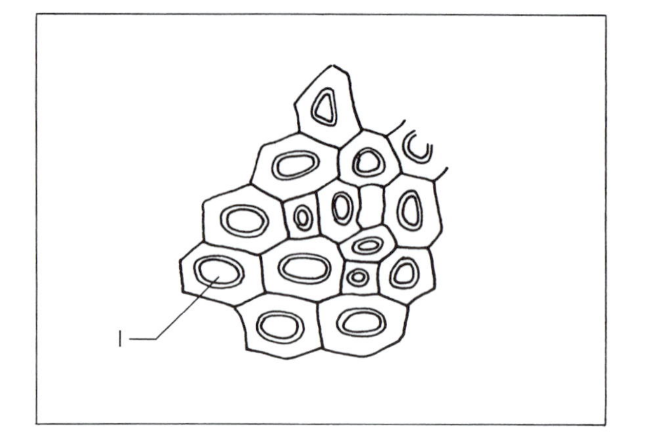

Fig. 6.9 *Gasteria retata*, leaf surface, showing very thick anticlinal cell walls and small lumina (1). ×145.

Measurements of epidermal cells have been made to try and distinguish between closely related species. If enough careful measurements are made and a statistical analysis carried out, significant differences may be detected. Unfortunately, this seemingly valuable method is limited in usefulness by the natural variation in size within different specimens of the same species, or even among cells from different leaves on the same plant. Sun and shade leaves, for example, can differ in this respect.

Even if absolute size differences may seem to be unreliable in many instances in distinguishing between species, the proportion of length to width of epidermal cells can often give useful data for comparison. This length: width ratio can be fairly constant in a species, even if the cell size varies phenotypically. The importance of selecting leaves for comparison from comparable positions on the various plants under study cannot be overstressed. Normally, one would select mature, vigorous leaves. The eye can play tricks, and it is easy to be misled about length: breadth ratios unless they are actually measured – look at the diagrams in Fig. 6.10. Here, diagrams showing various height: width ratios are illustrated, together with diagrams of the epidermis of named plants in transverse section.

Many leaves capable of rolling up in dry, unfavourable conditions, and reopening again under conditions when there is no water stress, have special, thin-walled water-containing cells that enable them to make these movements. These are the bulliform or motor cells. Examples may be found in many grasses, for example, marram grass, *Ammophila arenaria*, and many members of the bamboosoid grasses. The shape, size and disposition of such cells can be used as an aid to classification and identification. Cells with similar properties are present at the pulvinus and at the attachment regions of the leaflets to the rachis in many plants whose leaves fold at night.

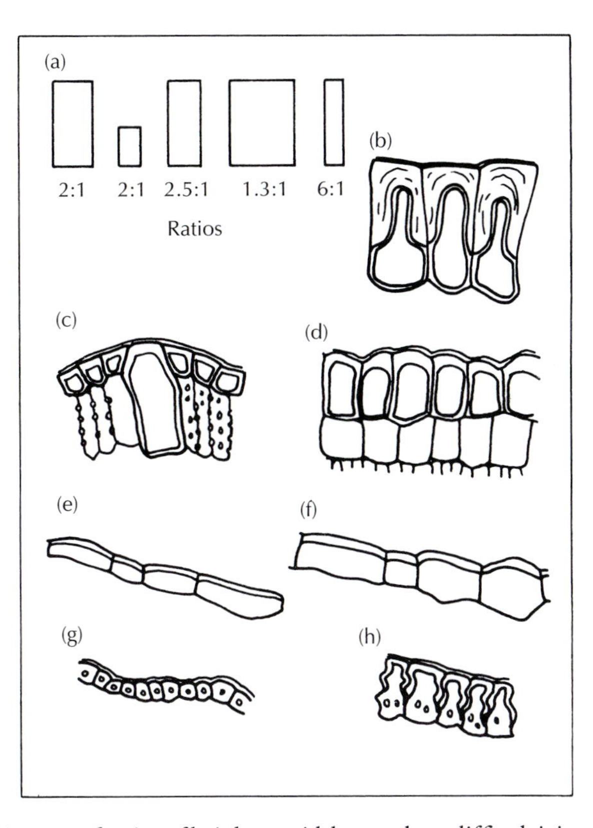

Fig. 6.10 (a) Diagram of ratios of height to width; note how difficult it is to judge these by eye. (b)–(h) Epidermis of selected plants in TS. (b) *Gasteria retata*, note the thick outer wall and the outer part of the anticlinal walls. (c) *Dielsia cygnorum*, note that some cells are larger than others. (d) *Elegia parviflora*, note the double epidermis. (e) *Cistis salviifolius*. (f) *Gloriosa superba*. (g) *Pinus ponderosa*, note the very thick walls. (h) *Thamnochortus scabridus*, note the wavy anticlinal walls; pits are also visible. ×145.

Stomata

The main control of water movement is provided by stomata. These consist of a pair of guard cells (often kidney-shaped) with a pore between them. The size of the pore is regulated by changes in shape of the guard cells, and is under active control, unless the plant is so dehydrated that it wilts. As hydraulic pressure is altered, the cells deform in a regulated way, aided by specialized, uneven wall thickening. When plants wilt, the stomata may open, and this can lead to damage.

Stomata may be present on both surfaces (amphistomatic), or only on the upper (hypertomatic) or only on the lower (hypostomatic) surface. They may occur at the same general level as surrounding epidermal cells, or they may be sunken below the general surface of the leaf as in cycads. In broad-leaved plants, stomata tend to have a scattered distribution, whilst in narrow leaved species, stomata are generally arranged in rows which are parallel to the longitudinal axis of the leaf blade. In some xerophytic plants (e.g. *Nerium oleander*) stomata are sunken beneath the abaxial leaf surface within stomatal crypts. In some angiosperms with aerial leaves, the distribution may vary from species to species, depending to some extent on the degree of xeromorphy or mesomorphy. They are characteristically absent from submerged aquatic leaves, but are present on the upper surface of floating leaves, for example *Nymphaea* and *Victoria*.

As mentioned, stomata may be superficial, that is, with the guard cells level with the surface of the leaf, or sunken, with a small outer chamber above the guard cells. Although many xerophytes have sunken stomata, and the majority of mesophytes superficial stomata, this is not invariably the rule. There may be particular adaptive advantages in each arrangement under certain circumstances; it may not be clear why some apparently 'unadapted' species survive while others around them are modified to a greater or lesser degree, but the timing of leaf emergence and their fall, or physiological adaptations, for example, may also play a part.

Of great interest to the taxonomist, or to the person wishing to identify a small leaf fragment, is the arrangement of subsidiary cells where these are present. Some of the various common types are illustrated in Fig. 6.11. Those stomata that lack subsidiary cells are called anomocytic, where the cells surrounding each stoma are not recognizably different or distinct from the remaining cells in the mature epidermis. Such stomata occur in the Ranunculaceae, for example. Stomata with two subsidiary cells, one at either pole, are called diacytic; *Justicia* and *Dianthus* species have such stomata. Stomata with two subsidiary cells, one on either flank (i.e. laterally), are termed paracytic; these occur in for example, *Juncus*, *Sorghum*, *Carex* and *Convolvulus* species. The paracytic type also includes species with a number of subsidiary cells in a parallel arrangement on either flank. Tetracytic stomata, with four subsidiary cells, can be seen readily in *Tradescantia*; here one cell occurs on either flank and one at either pole. If there are three cells of unequal size surrounding the guard cell pair the stoma is called anisocytic, as in *Plumbago* and members of the Brassicaceae. Cyclocytic stomata have a ring of subsidiary cells of approximately equal size and in which the individual cells are not very wide, whereas in the actinocytic type the subsidiary cells radiate strongly. Naturally enough, there can be intermediate forms that cannot easily be classified. Aberrant forms are also frequent, for example, two paracytic stomata may share one of the subsidiary cells.

Although occasional species exist which have several types of stomata on a leaf, most have one type only. This means that by noting the type of stoma present, the identity of a plant can be narrowed down. Of course, many families share the more common paracytic and tetracytic types, so the combination of all characters available must be seen to fit with reference material before identification can be made. There are other stomatal types,

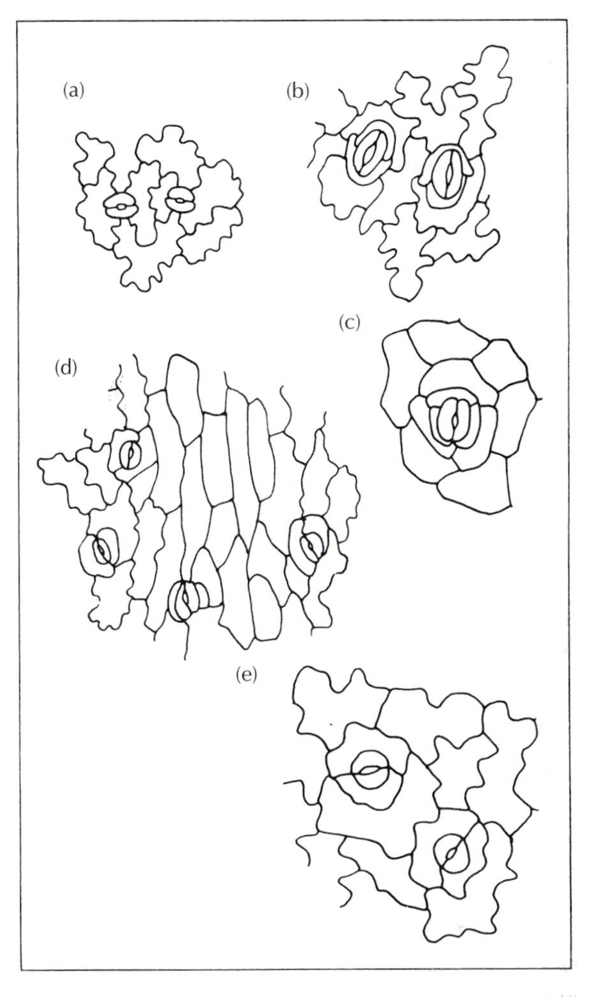

Fig. 6.11 Adaxial leaf surfaces, showing various stomatal types. (a) *Chrysanthemum leucanthemum*, anomocytic. ×109. (b) *Justicia cydonifolia*, diacytic. ×218. (c) *Plumbago zeylanicum* anisocytic. ×218. (d) *Convolvulus arvensis*, paracytic. Note elongated cells over veins. ×109. (e) *Acacia alata*, paracytic. ×218.

and indeed the ferns provide some interesting forms, the polocytic with the guard cell pair towards one end of a single subsidiary cell and the mesocytic type, where the guard cell pair is in the centre of a subsidiary cell are two such examples.

It is all too easy to think that some forms of arrangement of subsidiary cells must be primitive, and some more advanced. By speculating, phylogenetic sequences can be postulated, and interrelationships suggested. One great danger in doing this arises because a mature stomatal type may be formed by more than one developmental sequence in different groups of plants. Perhaps we need two systems of naming stomatal types, the first taking into account the mature form and used only for identification, and

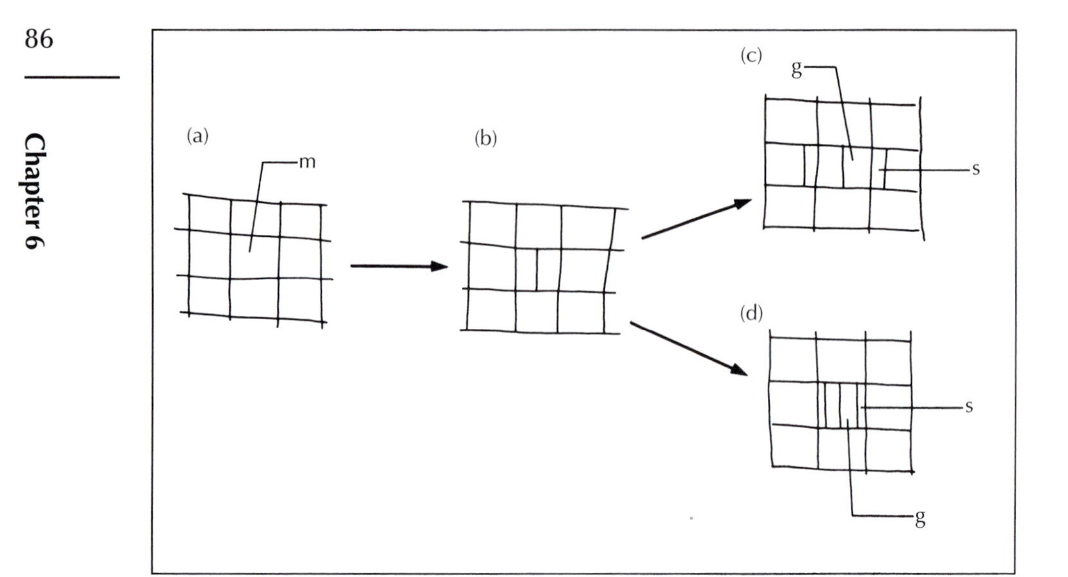

Fig. 6.12 Two routes for formation of the paracytic type of stoma. In (a) → (b) → (c) guard cells are derived from the cells flanking the guard cell mother cell. In (a) → (b) → (d) the guard mother cell divides to produce two cells, each of which divides once more. g, guard cell; m, guard mother cell; s, subsidiary cell.

the second derived from a study of the ontogeny of the stomata and used by the phylogeneticist or taxonomist. Figure 6.12 shows two possible ways by which paracytic stomata may arise. In the first route the guard cell mother cell (meristemoid) divides first to produce two cells, then each of the flanking cells divides to form one subsidiary cell. The second pathway involves the division only of the guard cell mother cell. Two flanking cells may divide to form one subsidiary cell each, either before or after division of this cell.

Sometimes mature stomata may appear at first sight to have no subsidiary cells. A study of the early stages of development could show that cells surrounding the guard cell mother cell divide in a particular way that differs from that which normally occurs among the other epidermal cells. Many aloes appear to have four subsidiary cells, whereas up to eight cells may surround the stomata. These subsidiary cells have oblique anticlinal walls. Most other cells in areas not adjacent to stomata have transverse walls. The oblique walls are the product of additional divisions in cells next to the guard cell mother cell.

Sometimes stomata are specialized to exude droplets of liquid water. They may simply be 'giant' stomata, larger than the others on the leaf, as in some members of Anacardiaceae. They might be specialized, and elevated at the end of a small mound situated at the termination of a small veinlet. Structures through which droplets of water may exude but which have non-

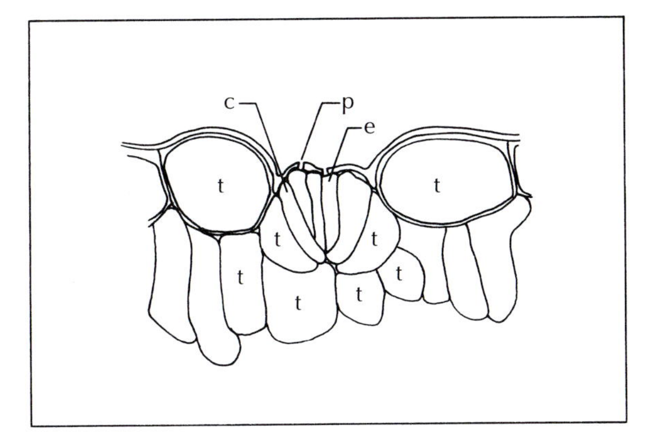

Fig. 6.13 *Limonum vulgare*, TS of salt gland from leaf. c, cup cell; e, excretory cell; p, pore; t, tannin-filled cells. ×330.

functional guard cells are called hydathodes. Salt glands are a type of hydathode modified for the exudation of salt water. They are often surrounded by an encrustation of salt. Examples of hydathodes may be found in saxifrages and salt glands in *Limonium* (Fig. 6.13).

Crystals and silica bodies can occur in the epidermis but for convenience will be described in the 'mesophyll' section.

Trichomes

Hairs and papillae (and scales) are collectively called trichomes. Taxonomists use their occurrence and cellular structure extensively as an aid to identification, because there is such a wide range of form. When a plant possesses hairs or papillae, they are usually of a type or types characteristic of that species. It should be noted that various samples of a plant of a given species may range from being glabrous (hairless) to very hirsute (hairy). This means that the numbers and density of hairs can be a poor character to use taxonomically, except, perhaps, in defining subspecies or varieties, if there are other, linked characters to support those divisions. Also, although hairs are so diverse in form, there are very few types that can be used even in the diagnosis of a family, that is, are of taxonomic significance.

The greatest value of hairs is in identification, that is, they have high diagnostic value. They are constant in a species when present, or show a constant range of form. Consequently, small fragments of leaf with hairs can often be matched with known material. If you look at the descriptions of powdered drugs of leaf origin in the European Pharmacopoea or your national pharmacopoea you will see the hairs carefully defined. Examination of hair types can help in quality control, for example of dried herbs, like mint, where cheaper substitutes may have been added (Fig. 6.15).

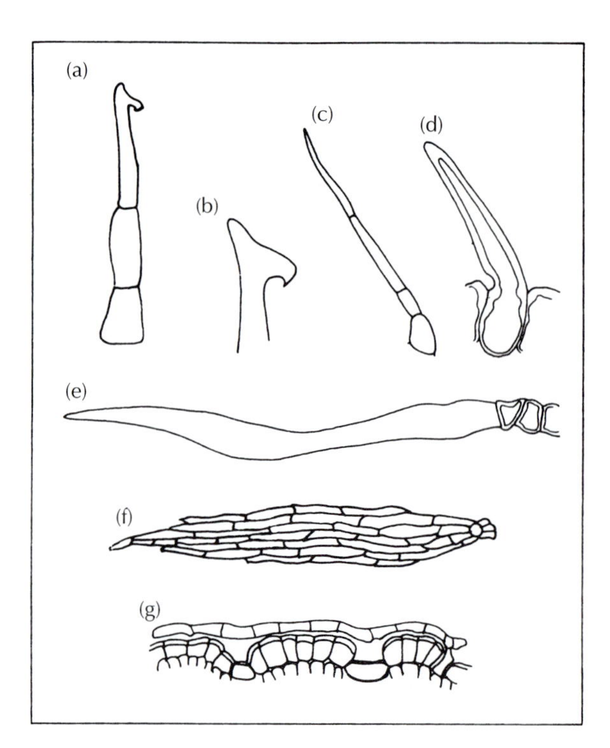

Fig. 6.14 Hairs in Centrolepidaceae (a)–(c) and Restionaceae (d)–(g). (a, b) *Aphelia cyperoides*. ×75 and ×150 respectively. (c) *Centrolepis exserta*. ×75. (d) *Thamnochortus argenteus*. ×218. (e) *Loxocarya pubescens*. ×218. (f, g) *Leptcarpus tenax*. Surface view, ×113; longitudinal section, ×120.

In some families, individual species can be defined on the form of their hairs alone. Among these are Restionaceae and Centrolepidaceae which show good examples of simple, unbranched hairs (Fig. 6.14). The relative sizes of the basal cell and the cells of the free portion vary from species to species. The curious 'boathook' end of the cell in *Aphelia cyperoides* is diagnostic. *Gaimardia* has complex, branching filamentous hairs.

In the Restionaceae, *Leptocarpus* from Australia, New Zealand, Malaysia and South America, flattened, shield-shaped stem hairs occur. These hairs are multicellular, diamond-shaped plates, held closely to the surface of the stem on short, sunken stalks as shown in Fig. 6.14. Until recently it was thought that *Leptocarpus* also occurred in South Africa, but the hair type and other internal histological differences show that the South African plants really belong to a distinct genus, which at that stage in the investigation, had not been named. A close relative to *Leptocarpus* in Australia is *Meeboldina*, which have diamond-form hairs, and two large, thin-walled translucent central cells, together with a border of thick-walled cells with recurved micropapillae which effectively zip adjacent hairs together so that they will strip off in a sheet if you try to pull one away.

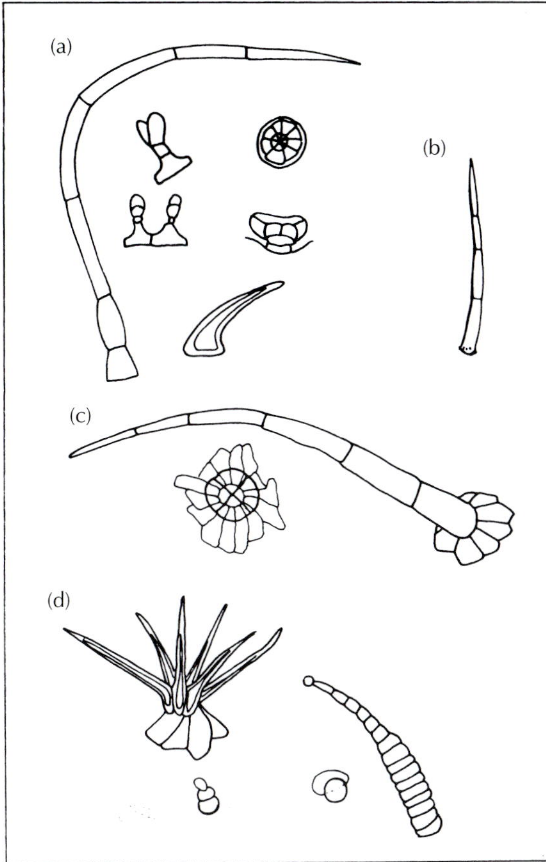

Fig. 6.15 (a) *Mentha spicata*, range of hair types. (b) *Corylus* hair (*Corylus* also has multicellular outgrowths). (c) *Origanum vulgare* (marjoram) hair and sunken gland. (d) *Cistus salviifolius*, range of hair types, one dendritic, the other glandular. All ×200.

Some plants have hairs on both upper and lower surfaces but in many cases they are confined to the lower surface. Examples of leaves with hairs on both surfaces are to be found among the silver-leaved composites. The air in the hairs masks the chlorophyll in the leaf, giving a highly refractive, silvery or white appearance.

Hairs are divided into two major categories, the glandular and non-glandular (or covering) hairs. Glandular hairs (Fig. 6.16) include the stinging hairs of plants like the nettle, *Urtica*. Less familiar are the irritant hairs from the pods of *Mucuna*, the 'cowitch' (Fabaceae), from the West Indies (Fig. 6.16a).

Simple glandular hairs are present on the leaves of plants that can trap and digest small insects and other small animals. Some of these are sticky, and some specialized to secrete digestive enzymes. *Pinguicula* and *Drosera* are examples.

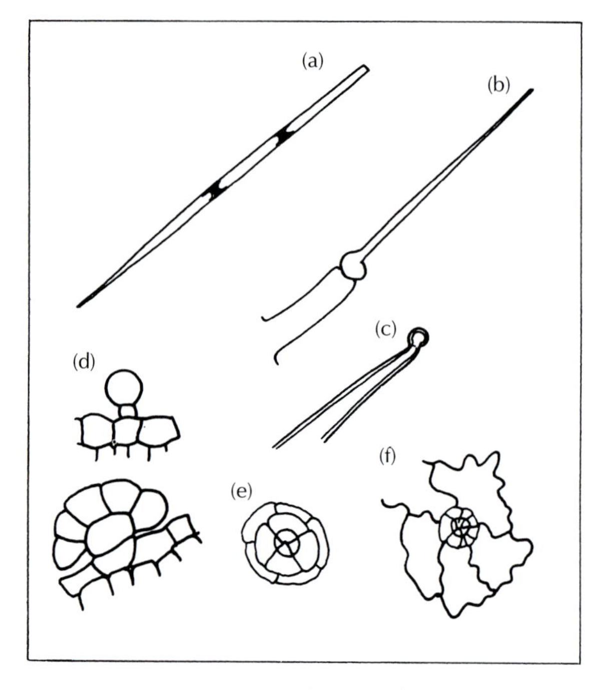

Fig. 6.16 Glandular hairs. (a) *Mucuna*, brittle, sharp hair containing irritant oil droplets. ×145. (b, c) *Urtica dioica*: (b) low-power hair on multicellular base, ×20; (c) fragile, sharp tip which can be broken off easily, ×290. (d) *Salvia officinalis*, multicellular and bicellular hairs. ×290. (e) *Justicia*. ×290. (f) *Convolvulus*. ×145.

Some of the fragrant (or less pleasant) essential oils occur in glandular hairs or leaves. The non-glandular hairs are much more varied and diverse than the glandular. A range of types are illustrated in Fig. 6.17, and the plants on which they occur are identified in the caption. One or more of these plants or its relatives should grow near your home. The larger hairs will be visible with a hand lens.

As mentioned earlier, usually the hair type is only one of many characters that may be used in identification. However, some families are easily recognized by their hairs, for example the T-shaped hairs of Malpighiaceae (Fig. 6.17h). Rhododendrons have been classified on the basis of leaf hairs, as an aid to the identification of species. Here, not only form, but also hair colour is used in the keys.

Microhairs are very short, two-celled hairs that are present on the leaves of some grasses, mostly from the tropics. Prickle hairs, which are usually prominent on margins and veins of grasses, are normally unicellular. They have thick walls that can be silicified. This is why it is easy to cut your hands on some grasses, and why cattle are selective in their grazing.

A number of hairs are of commercial value. These are not leaf hairs, but usually come from the fruit or seed, for example cotton (*Gossypium*) and kapok (*Bombax*).

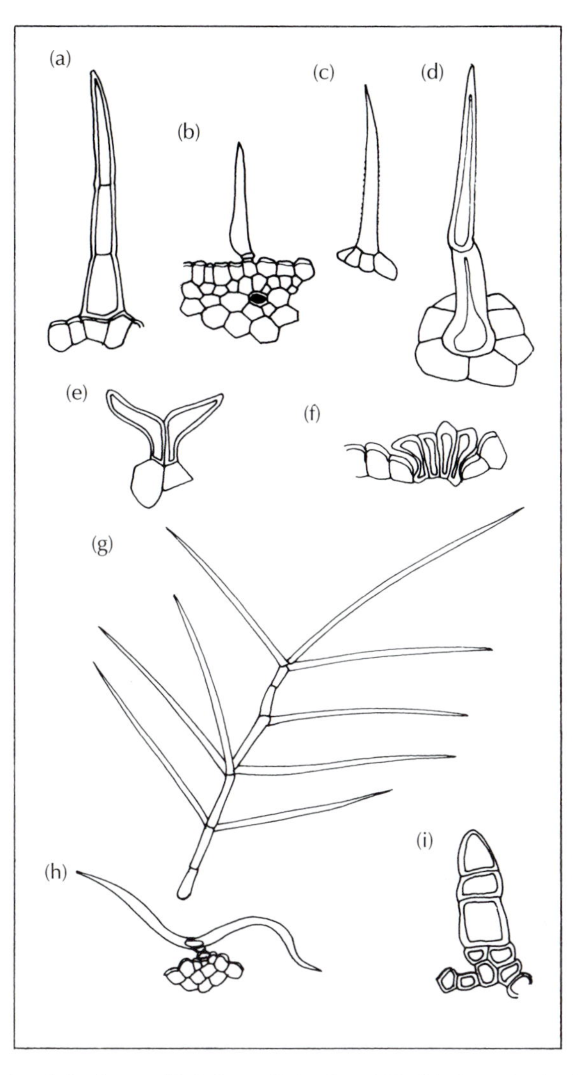

Fig. 6.17 Non-glandular hairs. (a) *Salvia officinalis*. ×220. (b) *Convolvulus floridus*. ×108. (c) *Coldenia procumbeus*. ×220. (d) *Justicia*. ×220. (e, f) *Trigonobalanus verticillata*. ×220. (g) *Verbascum bombiciforme*. ×54. (h, i) *Artemesia vulgaris*, ×220 and ×300, respectively.

The function of hairs is generally thought to be related to the water relations of a leaf. A densely hairy surface would tend to restrict the rate of flow of drying air. In xerophytes, hairs frequently have a suberin band in the wall towards their base. This prevents water leakage from the leaf through the cell wall of the hairs (apoplastic movement). Hairs, of course, increase considerably the potential surface area for evaporation. The converse requirement is true of hairs on epiphytes like *Tillandsia* (Bromeliaceae). In these plants, which are not rooted in the ground, the hairs are able to absorb water from rain or mist. They lack suberin bands. It is easy to test for 'waterproof'

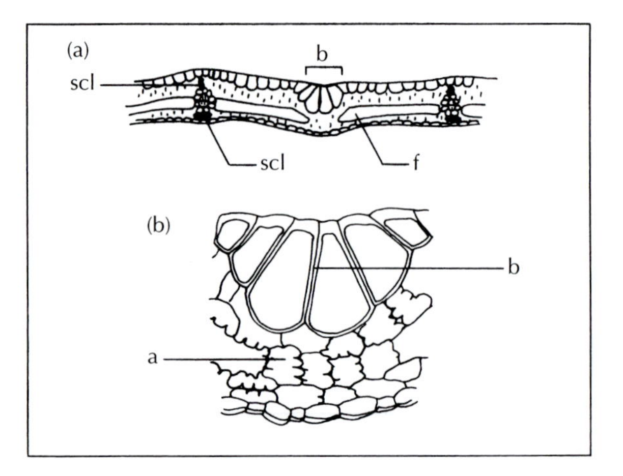

Fig. 6.18 *Pariana bicolor*, showing bulliform and fusoid cells. (a) Low-power (× 54) diagram of leaf TS, to show location of (b), detail drawing, ×218. a, arm cells of mesophyll; b, bulliform cells; f, fusoid cell (typical of certain bamboos); scl, sclerenchyma girders.

hairs. A piece of hairy leaf surface is cut carefully from a leaf and floated on a Petri dish containing a solution of calcofluor white for an hour, then removed and viewed in UV light (while wearing protective goggles). If there are suberin bands, the hairs will not fluoresce. If bands are absent, the hairs will fluoresce brightly. However, other hairs seem to have primarily an anti-herbivore function as in the grasses described above.

Scales have a wide base, are usually one to a few cell layers thick, and lack any vascular tissue. Their form, size and position can be used diagnostically. They are frequent on fern fronds.

Bulliform cells

Many leaves that are capable of rolling up in dry, unfavourable conditions, and reopening again under conditions when there is no water stress, have special, thin-walled water-containing cells that enable them to make these movements. These are the bulliform or motor cells. Examples may be found in grasses, for example marram grass, *Ammophila arenaria*, and many members of the bamboosoid grasses. The shape, size and disposition of such cells can be used as an aid to classification and identification. (Fig. 6.18).

Cells with similar properties are present at the pulvinus and at the attachment regions of the leaflets to the rachis in many plants whose leaves fold at night.

Hypodermal layers

Xerophytic floras typically may have a large number of species in which a hypodermal layer is well-developed. As the name suggests, these specialized cells are present to the inner side of and next to the epidermal cells. The cells within hypodermal layers typically contain few chloroplasts and are often thick-walled. Hypodermal cells are derived from cortical cells, not the epidermis. Some species with xerophytic anatomy such as *Nerium oleander* have a multiple layered hypodermis.

The mesophyll

The mesophyll usually consists of the thin-walled parenchymatous cells containing chloroplasts, called chlorenchyma, and other thin-walled cells concerned with water, food or ergastic or so-called 'waste product' (e.g. crystals, tannins) storage. Leaves of dicotyledonous plants differ greatly from those of monocotyledonous plants and from those of gymnosperms and ferns. Of course, there is some degree of intergradation, but generally, it is possible to separate these leaves, using some basic diagnostic criteria. Dicotyledons generally have a mesophyll which is composed of two differing photosynthetic cell types – palisade and spongy mesophyll cells; parenchyma cells may be present between these. Some monocotyledons are also like this, but there is a wide range of cell forms in the chlorenchyma, and frequently palisade cells are not present. The classical division of mesophyll into palisade-like cells and spongy cells may be misleading in its oversimplification. There are many intergrading cell shapes between the extremes. Figure 6.19 shows paradermal views of arm cells, part of the spongy tissue in *Clintonia*. Because some leaves lack a distinction of layers and others have very well marked layers, the mesophyll can be used as an aid to identification. It cannot often be used as a guide to the taxonomic position of a plant, but within a group of related plants there may be close similarities of arrangement. Environmental variations will not alter arrangements that are rigidly controlled by the genome. For example, palisade cells can be present next to the upper or to the lower surface, or to both. There are, however, striking changes that can occur to the layers themselves. In some cases, the numbers of layers of palisade cells have been counted and this figure used as a diagnostic character. Because in some plants the leaves growing in bright light may be thicker and have more layers of palisade cells than those leaves that have developed in the shade, this is not a sound diagnostic character and is clearly an effect of the environment.

Pharmacognosists (who, among other things, study plants and animals for natural products that might be applied in medicine) use a measurement called the 'palisade ratio'. This is particularly useful in defining small leaf fragments in powdered leaf products. This measure indicates the number of palisade cells that can be seen beneath an epidermal cell in surface view. An average figure is produced after many cells are counted. A statistically

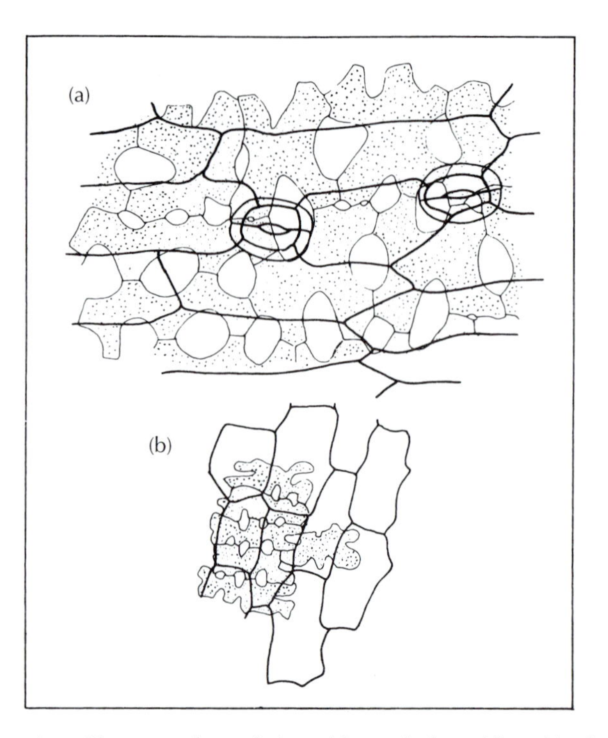

Fig. 6.19 *Clintonia uniflora*, paradermal views (through the epidermis) of arm cells, part of the spongy mesophyll (stippled). Note large air spaces between the cells. (a) Abaxial, ×115; (b) adaxial, ×80.

sound count will produce a fairly reliable typification and hence identification of the material.

The arrangement of mesophyll cells may indicate whether a plant has the normal, C_3 photosynthetic pathway (Fig. 6.20) or a C_4 (Fig. 6.29a) photosynthetic pathway. In Kranz, or C_4 plants, the mesophyll consists of radiating, elongated mesophyll cells surrounding a (usually) parenchymatous but often lignified bundle sheath, which, in turn, surrounds the vascular bundles. The radiating mesophyll is chloroplast-rich, and it is here that CO_2 is incorporated into malate or aspartate as the first step in the C_4 photosynthetic process. The parenchymatous bundle sheath cells on the other hand usually contain large, prominent, generally agranal chloroplasts. In C_4 plants, malate or aspartate produced in the mesophyll cells is thought to be transported via the numerous plasmodesmata which occur at the interface between the mesophyll and bundle sheath cells, where CO_2 is liberated and immediately fixed via the C_3 photosynthetic cycle, becoming incorporated into sugars, other carbohydrates and amino acids essential to sustain the rapid growth common to C_4 plants.

Among the Poaceae, there are a fairly large group of plants that are neither C_3, nor C_4, but display intermediate leaf anatomy to that of the 'typical'

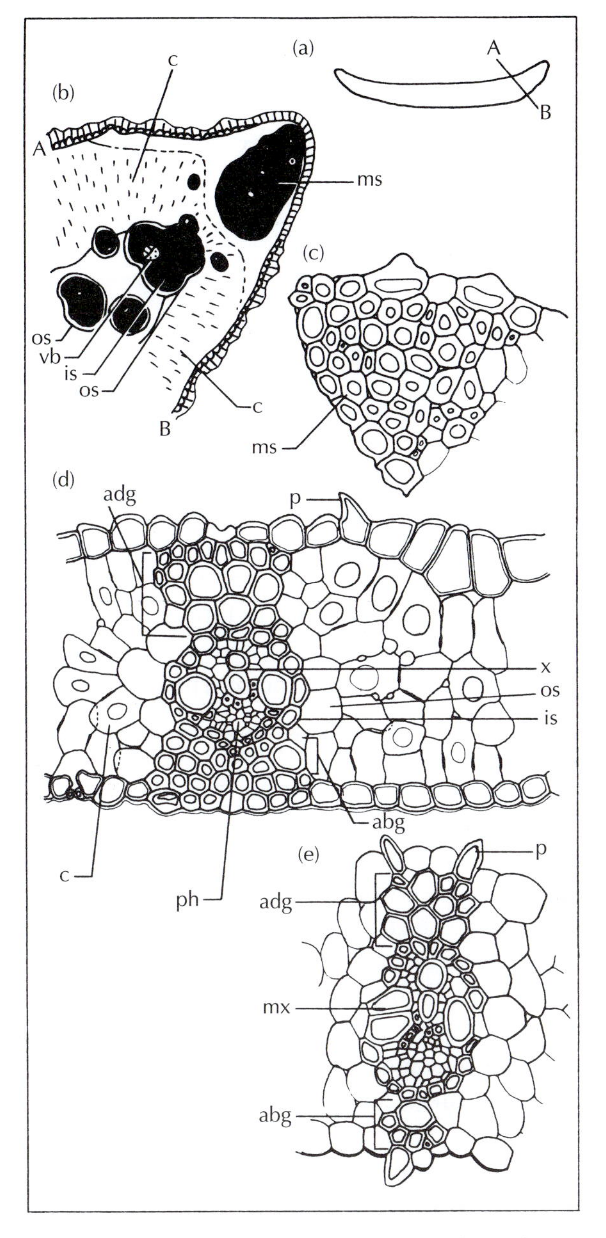

Fig. 6.20 Strengthening tissue in the leaf, as seen in TS. Sclerenchyma in *Agave franzonsinii* (a, b), *Aegilops crassa* (c, d), *Phalaris canariensis* (e). (a) Outline of leaf TS to show location of diagram (b, ×40). (c) Leaf margin, ×109. (d, e) Vascular bundles and their associated bundle sheaths and girders. (d) ×109; (c) ×230. adg, adaxial sclerenchyma girder; abg, abaxial sclerenchyma girder; c, chlorenchyma; is, inner bundle sheath; ms, marginal sclerenchyma; mx, metaxylem; os, outer bundle sheath; p, prickle hair; ph, phloem; vb, vascular bundle; x, xylem.

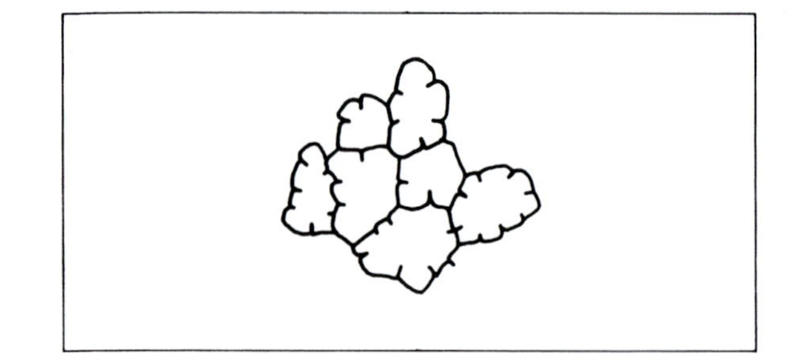

Fig. 6.21 *Pinus ponderosa*, plicate mesophyll cells from TS. ×145.

C_3 and C_4 species. Here, it is the arrangement, structure and position of the mesophyll and bundle sheath chloroplasts that yields clues as to the degree of 'intermediacy'. Whilst these C_3–C_4 intermediates are biochemically neither C_3, nor C_4, they seem to be able to follow a pathway that is dependent on several factors, including light intensity, air temperature, relative humidity, soil water availability and the nutritional status of the soil for example.

Some dicotyledonous foliage leaves contain a specialized, longitudinally orientated mesophyll, called the paraveinal mesophyll, which separates the upper palisade from the lower spongy mesophyll. As noted above, in many monocotyledonous plants, the mesophyll is not differentiated into spongy and palisade layers. The vascular bundles are surrounded by an initially parenchymatous bundle sheath, which may undergo lignification as the cells mature. There may be a specialized, concentric arrangement of the photosynthetic mesophyll surrounding the bundle sheath cells as in C_4 plants.

In many gymnosperms and some angiosperms the mesophyll cells are plicate, with inwardly directed wall foldings (Fig. 6.21). The infoldings increase cell wall surface area and probably therefore make up, to some extent, for the smaller number of chlorenchyma cells that are often found in such leaves.

Sclereids

Sclereids can occur as isolated cells in the mesophyll, or in well-defined positions relative to other tissues such as within vascular bundles. Sclereids perform a mechanical supportive role, more especially in leaves, which lack well-developed girders or strands. They range in size and form, as described in the Glossary. Some of the types found, and the plants in which they occur, are shown in Fig. 6.22.

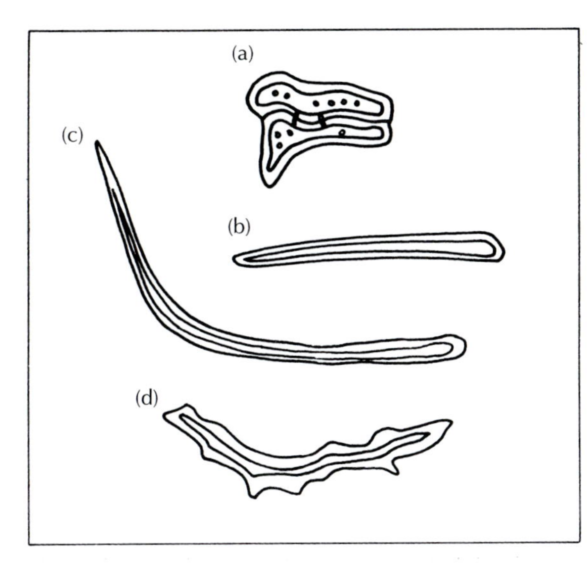

Fig. 6.22 Selected sclereids from leaves. (a) *Olivacea radiata*. (b, c) *Olea europaea*. (d) *Camellia japonica*. All ×290.

Air spaces

These may be present in the mesophyll, between veins. These are much larger and usually more formal than the air cavities between cells of the spongy mesophyll, and often form by the lysigenous (dissolving) or schizogenous (splitting) breakdown of thin-walled parenchymatous cells between veins. In monocotyledons, especially the grasses, the intercellular spaces are greatly reduced, particularly in more xerophytic species. Reduction of intercellular airspace volume is greatest in C_4 xerophytic grasses.

Water storage cells

Water storage cells are large, colourless and thin-walled, and usually lacking in conspicuous cell contents. Sometimes areas of the wall may be thickened in such cells. Water storage cells occur in many families, notably those which have representatives growing in arid conditions. Further details are given on page Chapter 8.

Ergastic substances

Specialized cells in the mesophyll may be used in making identifications. Firstly, there are those cells containing 'ergastic' substances. These are products related to the physiological activity of the plant and may constitute stored food materials, such as starch, oil, protein and fat. They also include substances that cannot be related yet to a particular function. If the

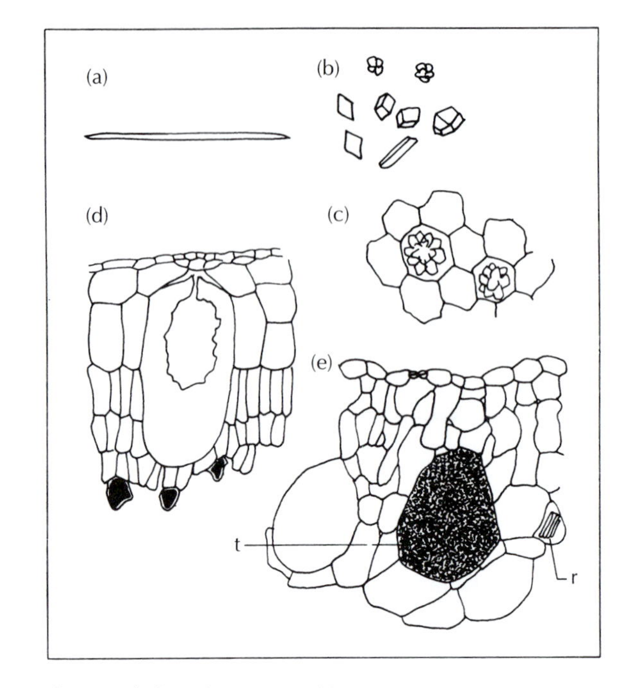

Fig. 6.23 Crystals, cystolith and tannin and latex cells. (a) Styloid crystal, typical of many Liliaceae. (b) *Acacia alta*, crystals from leaf. (c) Cluster crystals in *Passiflora foetida* leaf. (d) *Ficus elasitica*, leaf TS showing cystolith; dark cells contain latex. (e) *Oscularia deltoides*, leaf TS with large tanniniferous idioblasts (t) and raphides, (r). All × 125.

function of such a substance is not clear, it is often simply called a waste product. This is a rather lazy way out of the problem, particularly since many of these substances are currently being identified as physiologically active by chemists. Chemists often do not know which cells of the plant contain them and it could be that some of the so-called 'waste products' are really important to the plant. Clearly, there is an obvious need for closer co-operation between morphologists and those extracting these potentially important and interesting plant products.

Crystals

Probably the best known of the ergastic substances, crystals, are very commonly thought of as waste products, again without sound evidence. Crystals are usually composed of calcium oxalate and more rarely of calcium carbonate. Because they are of widespread occurrence, they are of limited value to the applied anatomist. However, some families have never been recorded as having crystals, for example Juncaceae, the rush family. Others very frequently have a particular type, for example families within the Asparagales frequently have styloids (Fig. 6.23). Using this information, it should be possible to separate leaf fragments from families such as the Con-

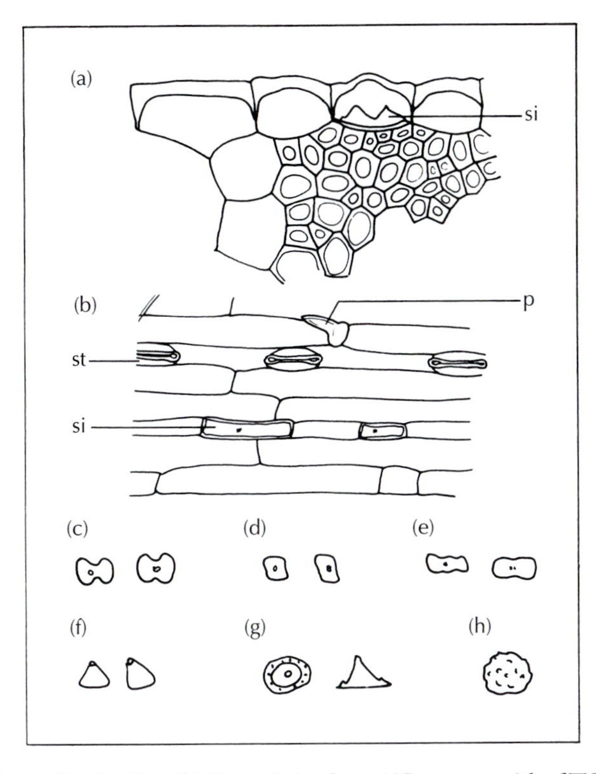

Fig. 6.24 Various silica bodies. (a) *Cymophylus fraseri* (Cyperaceae) leaf TS. Note location of silica bodies (si) in epidermal cells above sclerenchyma girder. ×218. (b) *Aegilops crassa* (Poaceae) abaxial leaf surface. p, prickle hair; si, silica body; st, stoma. ×109 (c–h) Isolated silica bodies. (c) *Zea mays* (Poaceae). ×200. (d) *Bambusa vulgaris* (Poaceae). ×200. (e) *Agrostis stolonifera* (Poaceae). ×200. (f) *Evandra montana* (Cyp.). ×200. (g) *Cyperus diffusus* (Cyp.). ×200, first body in surface view, second in side view. (h) Typical of many palms and Restionaceae. ×300.

vallariaceae and Juncaceae. However, other monocotyledonous families, such as the Iridaceae, have crystals similar to those of the families in the Asparagales and such diagnostic characters must be used carefully and always in conjunction with others!

In the dicotyledons a particular 'saddle-shaped' or twin crystal is common in Leguminosae (Fig. 6.23b) and its presence along with other features can help in distinguishing members of that family from others. Various prismatic and cluster crystals also shown in Fig. 6.25 have a very wide and scattered distribution through many families. Their presence in a fragment to be identified is only of real diagnostic value if they match exactly the type found in properly named reference material that has been narrowed down, using other characters, as probably being the species concerned.

Crystals can be associated with particular tissues, for example in the parenchymatous bundle sheath surrounding the veins, or they may occur in special idioblasts within the mesophyll. Sometimes there are no large crystals, but merely fine 'crystal sand' in the lumen of certain cells. Cystoliths are a special example of idioblasts; they occur in relatively few plants, for example *Ficus elastica*, and are illustrated in Fig. 6.23.

Silica bodies

These are similar in appearance to crystals. They normally occur in special cells (stegmata) next to fibres or other lignified tissues, or in the epidermal cells, particularly those near to fibrous cells associated with bundle sheaths. However, since silica bodies are amorphous and not crystalline in structure, they can be distinguished from crystals by simple tests. In reality they are small opals! Silica bodies do not show birefringence (i.e. they do not shine brightly, as crystals do) when viewed between crossed polars in the polarizing microscope. In addition, they often turn pink when treated with a saturated solution of carbolic acid, and we know of no crystals that do that. If you should use this histochemical test, be careful to keep the carbolic acid off your skin and wear protective glasses!

Silica bodies often occur in epidermal cells, usually one, but occasionally more to a cell, in a limited range of families. Because they are easy to see – it is worth examining a simple epidermal strip or scrape from one of the grasses, the Cyperaceae, particularly *Carex* species or a palm leaf surface, for example in *Borassus* species. In the bamboos, as in *Bambusa vulgaris*, they are almost cuboid, as shown in Fig. 6.24. Those of *Zea* and *Agrostis* (dumb-bell-shaped to oblong) are also illustrated, together with some others from grasses or sedges that may be easily available to you.

There are silica bodies of many shapes and sizes in the grasses and palms and extensive taxonomic use is made of them. Their form can be of help in the identification of fragments of cereal or grass that may have constituted part of the diet of an animal whose feeding habits are under investigation. They survive digestion and can be found in quite remarkable situations. For example, it was recent practice to use horse dung in the clay when bell founding and it was thought that medieval bell founders also used dung to reinforce the clay of their bell moulds. Fragments of bell moulds from ruins of a thirteenth-century chapel at Cheddar were examined for such evidence and there were leaf or chaff surface fragments together with silica bodies, probably of oats, (Fig. 6.25), which had survived being eaten, fired in the clay by the molten bell metal and then several hundred years of burial!

The silica bodies of sedges are cone-shaped or conical, with flat bases. They often have small satellite cones around them as shown in Fig. 6.24. No grasses have this type of silica body.

Closely related families can sometimes be distinguished through the presence or absence of silica bodies. For example, among the Juncales, the

Fig. 6.25 SEM photograph of a fragment of chaff from one of the Gramineae, found in fragments of a bell mould from ruins at Cheddar. Note the outline of silica bodies. ×1000.

rush family, Juncaceae and the Centrolepidaceae, which is a very small family of semi-aquatic plants from the southern hemisphere, lack silica bodies. On the other hand Restionaceae, which are rush-like plants mainly from Australia and South Africa, typically have silica bodies shaped like small, spiky balls. In the Restionaceae, the silica bodies rarely occur in epidermal cells, but more frequently in stegmata, specialized cells with thickened inner and anticlinal walls and thin outer walls. The thickening is often lignified and sometimes also suberized. However, most species of Restionaceae lack leaves, and as the silica bodies occur in cells in the stem this is probably not the place to be discussing them. However, the stem contains chlorenchyma and carries out many of the physiological functions of leaves in that family.

The function of silica bodies is not understood. It is thought that plants cannot prevent the uptake of silicon with other elements, and that silicon in excess is deposited in an inert form; hence the proximity of silica bodies to veins. However, this does not explain why many plants that must surely also take up silicon in excess do not form silica bodies. They do cause wear in teeth of grazing animals. However, in the process of co-evolution, such animals have developed teeth that continue to grow during their lifetime, thus counteracting the deterrent.

Ecologists have used silica bodies persisting in peat layers to determine the nature and species composition of earlier vegetation at a range of sites.

Tannins

Tannins generally have a scattered distribution through various plant families. Polyphenolic substances are usually characterized by their reaction with ferric chloride solution, when they turn blue-black. Their chemical diversity is a phytochemical problem.

The presence of tannins in special cells or cell layers can, nevertheless, be used as a diagnostic character even if their chemical identity is not known. However, a word of caution is necessary. Tannin may appear at certain seasons in some plants, such as the Poaceae, so lack of tannins at a particular time of year is not a reliable feature, and the plants cannot be assumed to lack them totally. Some tanniferous idioblasts are illustrated in Fig. 6.23. Some *Lithops* species owe their mottled brown appearance to tannin cells. One very familiar family rich in tannin is, of course, the Theaceae to which the tea plant belongs.

The function of tannins is also little understood. They may act as an ultraviolet light shield, perhaps like the xanthophyll components in many other plants. Normally tannins occur in epidermal cells. High-intensity sunlight can damage chloroplasts, so such a 'screen' may impart physiological and ecological advantages. Additionally, the astringent taste (a warning of the harm they do in binding with the stomach wall?) may protect leaves from being eaten.

Aleurone grains

Aleurone grains may be present as may starch grains.

Strengthening systems in the leaf

Mature leaves may contain additional marginal strands of sclerenchyma, and some fibre strands or girders may be associated with the vascular bundles (Fig. 6.20; *Aegilops crassa*, *Phalaris canariensis* and *Agave franzosinii*). Collenchyma is frequently present in the raised ribs above and below the midrib bundle, and is also occasionally found in similar positions in relation to the large and intermediate vascular bundles in monocotyledons as illustrated in Fig. 6.20(d), where ad- and abaxial hypodermal sclerenchymatous girders subtend this large bundle. Figure 6.20(e) shows an intermediate vascular bundle, in which the adaxial hypodermal sclerenchymatous girders extend, and are associated with prickle hairs on the upper and lower leaf surfaces. It is important to note that the hypodermal strands or girders found in leaves of the Poaceae and Cyperaceae, for example, may become sclerified in mature leaves.

The vascular system

The specialized cells which conduct water and salts upwards from the roots, and the cells involved in the transport or translocation of the substances synthesized in the leaf mesophyll and other tissues, are grouped together in well-defined strands called vascular bundles. In the leaf, these are seen as the midrib and vein system. Vascular bundles are continuous, either directly or if developed, through the petiole, with the primary system of vascular tissue in the stem. Alternatively, if secondary growth has occurred, leaf bundles may be continuous with the secondary xylem and phloem. The vascular system in leaves may be likened to the system of tributaries which feed into a major river. The xylem system functions in reverse, with larger veins essentially supplying water and dissolved solutes to the smaller veins. The phloem, on the other hand, is the pathway through which assimilated materials are translocated. This pathway functions from the smaller towards the larger veins. Assimilates often take particular pathways through a series of small, intermediate and larger veins, before ultimately discharging into the midvein system, which connects to the phloem in the stem. Assimilated material usually follows a pathway from the source (where synthesized) to sink (where utilized). According to the source of demand, phloem can transport materials in either direction. Translocation in plants involves movement of water and dissolved inorganic nutrients through the xylem from roots to the aerial parts of the plant, and the transport of photoassimilated material from sites of synthesis (source) to sites of utilization (sink) via the phloem.

The xylem is responsible for apoplastic transport in vascular plants, which is not limited totally to water transport, but in addition, the transport of various macro- and micronutrients, amino acids and other important inorganic substances, from the roots to the stem and, ultimately, the leaf via the apoplastic continuum.

The phloem is responsible for the transport of the major proportion of soluble carbohydrate as well as other essential products. The phloem forms the major long-distance symplastic transport pathway in all vascular plants. Translocation usually takes place from a site of synthesis of assimilated material (called a source) to a site or sites of utilization (called sinks). The assimilated material is translocated in a water-based medium, which emphasizes the essential inter-relationship between the xylem and phloem, more particularly so in the leaf where most of the phloem loading takes place in mature plants.

Transpiration is the driving force that facilitates the movement of solutes through the xylem. Transpiration requires that water entry is facilitated at the roots. Efficient transport must occur in a conducting system, in order that water may be moved to other regions of the plant where it is utilized in numerous biochemical and growth-related reactions. The xylem is also the principal pathway through which water is moved from point of entry, to a point of exit, which in higher plants is via stomata, by the process of transpiration. Transpiration itself facilitates leaf cooling by evapotranspirational heat loss to the atmosphere. Thus the physiological necessity of regulating transpirational water loss, via the cuticle, the epidermis and the stomata, have led to variations in the surface sculpturing of the cuticle, changes in the size of epidermal cells, and alteration of stomatal cells, related to the habitat preference of the particular species.

The vascular tissue within the leaf blade is arranged in a pattern that appears to be under strong genetic control. The phenotypic expression of the genotype varies very little in overall pattern, although the number of bundles may vary in the leaves of plants of any one species found growing under a range of conditions. However, the main features that characterize a particular type of venation are constant enough for use in identification of fragments. It is rare for a family or genus to have a 'unique' pattern, but some families can be distinguished by the constancy of a particular type. For example, Melastomataceae have a vein running parallel with the margin in most species.

Vascular patterns have been classified by several different authors. The system proposed by Hickey has proved to be the most popular. Hickey (1973) published his classification system, which details a classification of the architectural features of dicot leaves. It was updated with Wolf, and presented again in Metcalf and Chalk's Anatomy of the Dicotyledons, volume 1, 2nd edition. (1979). Hickey and Wolf's system deals with the placement and form of those elements which contribute directly to the outward expression of leaf structure, including shape, marginal configuration, venation and gland position, and was developed from an extensive survey of both living and fossil leaves. This system partially incorporates modifications of two earlier classifications: that of Turrill for leaf shape and that of Von Ettingshausen for venation patterns. After categorization of such features as shape of the whole leaf and of the apex and base, leaves are separated into a number of classes depending on the course of their principal venation. According to Hickey, identification of order of venation, which is fundamental to the application of classification, is determined by size of a vein at its point of origin and to a lesser extent by its behaviour in relation to that of other orders. This classification includes a description of the areoles, i.e. the smallest areas of leaf tissue surrounded by veins which form a contiguous field over most of the leaf. Perhaps the most useful aspect of this classification system lies in the fact that most dicot taxa possess consistent patterns of leaf architecture, allowing a rigorous method of describing the features of leaves, which is of immediate usefulness in both modern and fossil plant taxonomic studies. Some of the key aspects of the system are shown in Fig. 6.26.

Vascular systems can be studied best in cleared material, which have subsequently been stained with safranin. A range of main vein systems is readily observed. The angle at which the veins depart from the midrib can be a useful and relatively constant feature for a species. The nature of the vein endings of the final or smallest order of branching is also used taxonomically. Figure 6.27 shows the 'open' type of vein ending in *Plumbago zeylanicum*.

The commonly held idea that all monocotyledons have parallel venation is quickly dispelled by examination of such leaves as *Bryonia* or *Smilax*. In addition, monocotyledonous leaves may have large numbers of cross veins, joining the more prominent longitudinally arranged (parallel) vein network. These cross veins may be longer than the sum of all the parallel veins. There are also dicotyledons that do not have the so-called net-like type of venation.

In most dicotyledonous leaves (except those which are greatly reduced and needle-like, e.g. *Hakea*), the phloem pole in a vascular bundle faces the lower (abaxial) leaf surface and the xylem pole the adaxial surface. This is generally true for most species, which have a collateral vascular bundle arrangement. However, there are at least 27 families which have bicollateral vascular bundles, that is, they have phloem at either side of the xylem. Examples here are to be found in the pumpkin family (Cucurbitaceae), the Solanaceae (potato, tobacco, tomato), the milkweeds (Asclepiadaceae) and others, such as the Chenopodiaceae and Apocynaceae. In mature leaves of the potato, the smaller minor veins (fifth and sixth order) may contain few living adaxial sieve tube members, whilst all the corresponding abaxial sieve tube members in the same veins are living.

Lamina vein orders

Most leaves have two components in their vascular networks – a major and minor vein system. What makes them different? Simply, the major veins in dicotyledonous foliage leaves occupy much of the cross-sectional area of the leaf, and are often associated with hypodermal collenchymatous or sclerenchymatous strands. Viewed in cross-section, they may even sometimes show signs of a cambial zone. However, this cambial zone displays only limited secondary growth, which is more evident nearer the base of the leaf and (if present) down in the petiole, where the vasculature of the main vascular supply to the leaf assumes a more cauline appearance (that is, it is more

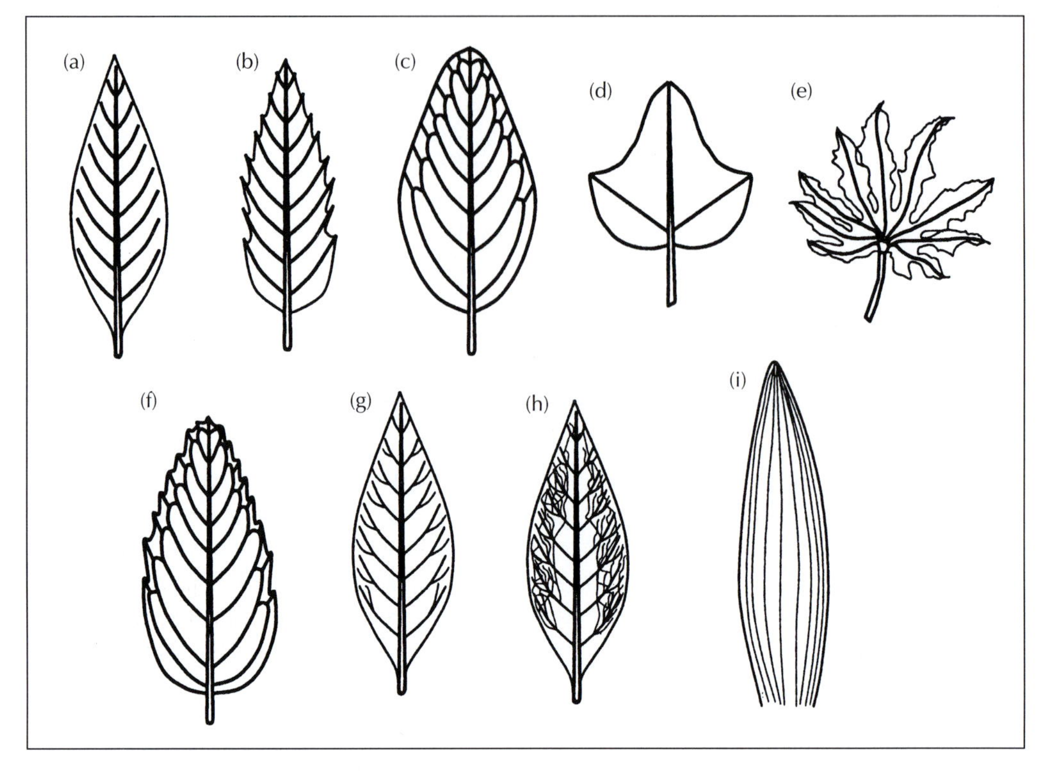

Fig 6.26 Aspects of Hickey's classification system. Many species have first order lateral veins that emerge from the midrib vein, which then arc outwards and upwards towards the lamina margin, without terminating at the leaf margin. This is defined as eucamptodromous. In (b) the first order lateral veins arc outwards and upwards from the midrib, and terminate at the leaf margin, where their endings often form teeth. This is defined as simple craspedodromous. In (c) the first order lateral veins overarch beneath the leaf margin, forming a primary interconnected network. This is defined as brochidodromous. In (d) the leaf contains three similar-sized veins, which effectively form three compartments in the leaf. This is an example of an actinodromous leaf. In (f) the first order lateral veins overarch, and extensions continue to the leaf margin, which is serrated in the semicraspedodromous leaf. The first order lateral veins branch several times near the margin of the lamina in the cladodromous leaf illustrated in (g). In reticulodromous leaves, the first order lateral veins branch many times towards the margin of the lamina. In parallelodromous leaves the three orders of lamina vein make up a longitudinal system of parallel veins. In palinactodromous leaves (e) the leaf is dissected at its base into a number of arms. In (d) and (e) each of the basally attached veins is the same size. (Redrawn from Hickey, 1973.)

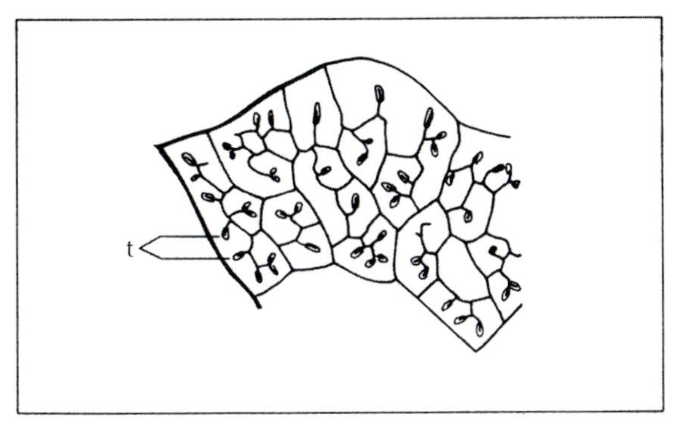

Fig. 6.27 *Plumbago zeylanicum*, paradermal view of veins to show open type of venation. Note enlarged tracheids (t) at veinlet ends. ×20.

stem-like). By contrast, minor veins lack associated mechanical supporting tissue. Unlike the major network veins, the minor veins are usually embedded within the interface between the palisade and spongy mesophyll layers. As mentioned, minor veins are embedded within a horizontally orientated mesophyll, termed the paraveinal mesophyll in some dicotyledons, which, judging by the relatively high plasmodesmatal frequencies recorded between adjacent cells, is the principal symplastic solute conduction pathway from the palisade and spongy layers, into the surrounding parenchymatous bundle-sheath cells, terminating at the sieve tubes within major and minor veins.

Major vein differentiation

The sequence of events that takes place within the dicotyledonous foliage leaf may be summarized as follows. Once the blade or lamina starts to expand due to anticlinal and periclinal cell division within the marginal and submarginal initials, the procambial strands begin to form – the first of these to become evident is the midrib or main vein. This vein is blocked out acropetally, and vascular tissues differentiate in regular sequence (protophloem followed by protoxylem, then metaphloem followed by metaxylem) towards the tip of the still-immature expanding leaf. The major veins of the lamina follow suite, initiating at the base of the leaf, they too, differentiate acropetally. Thus the first-formed of the major lamina veins matures first, and the last-formed (apical) major veins differentiate and mature last.

The minor veins

In dicotyledonous foliage leaves, the minor veins differentiate basipetally, from the apex and the leaf margin, back towards the major vein network. Thus, it is quite feasible for the tip of the developing leaf to mature with respect to transport, before the base of the leaf. The apex could therefore conceivably export photoassimilated material to the still immature basal part of the leaf, during the overall maturation and development process.

The phloem

Broadly, the phloem consists of a series of conducting elements associated with vascular parenchyma elements. In angiosperms, the conducting elements are referred to as sieve tube elements or sieve tube members, and these are almost always associated with specialized parenchyma cells called companion cells. Companion cells are ontogenetically related to the sieve tube members. Among the gymnosperms and ferns, the phloem is composed of less specialized conducting cells, called sieve cells, and these are associated with albuminous cells. The phloem is physiologically characterized by the zones within which it is found – the source (where assimilate is loaded) and the sink (where assimilate is unloaded). Source and sink are connected through the plant's axis, via the transport phloem, which has characteristics not shared by either source or sink phloem.

Loading phloem

Within the leaf, the phloem is responsible for two distinct physiological activities. Firstly, the actual process whereby the sieve tubes are loaded (phloem loading), and secondly, they are involved with the transport of the loaded assimilates from the source to sinks in other parts of the plant (phloem transport). Clearly, the physiological activities of the sieve element–companion cell, the sieve element–transfer cell, or the relationship of the sieve elements to intermediary cells, in angiosperms, or the sieve cells to albuminous and other contiguous parenchyma cells, may influence the structure of these important cells within the leaf profoundly. Cross-sections of leaves, particularly of the minor veins in dicotyledonous, and of many monocotyledonous foliage leaves, reveals that the sieve element–companion cell complex (where the companion cell is clearly recognizable) consists of larger diameter companion cells and relatively narrow diameter sieve elements. Figure 6.28 illustrates this point for *Nymphoides*, where the sieve tube diameter in the smallest minor veins is approximately 6 μm and the companion cell is much larger, at approximately 30–35 μm in width. The companions cells and, for that matter, intermediary cells, (in the phloem of leaves of *Coleus* species for example) usually have a dense cytoplasmic matrix, and contain large populations of mitochondria, and as illustrated in Fig. 6.28, are much larger than their associated sieve tube members, reflecting their role in the phloem loading process.

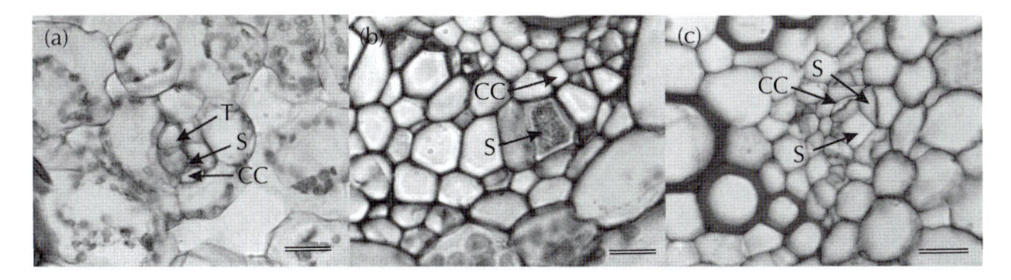

Fig. 6.28 Micrographs showing the size change relationships between the companion cell and the sieve tube cell in loading, transport and unloading phloem, in *Nymphoides*. (a) Minor vein in leaf lamina. (b) Phloem in central vascular bundle in the submerged petiole. (c) Phloem strand from a root. Scale bars: (a) 20 μm; (b, c) 10 μm. CC, companion cell; S, sieve tube member; T, tracheary element. (a) ×250; (b, c) ×500.

Plasmodesmatal frequencies vary between the cells along the entire phloem loading pathway, i.e. from the mesophyll cells to the functional vascular parenchyma including companion cells, intermediary cells and albuminous cells. In particular, plasmodesmatal frequencies (assuming that all plasmodesmata are functional) may strongly influence both the mode and the rate at which phloem loading occurs. The presence of plasmodesmata between cells is an indicator of potential symplastic loading, whilst the absence of plasmodesmata is an indication of a potential apoplastic phloem-loading pathway.

Sieve element loading may therefore be either symplastic or apoplastic. Whether loading is symplastic or apoplastic, there is ultrastructurally a great difference between the companion cell–sieve tube complex in the leaf, the stem and the root (Fig. 6.28). Oparka and Turgeon (1999) have suggested that the cell interactions between the sieve element and its companion cell must rank among the most complex and mysterious – they suggest that the companion cell functions as a 'traffic control centre' facilitating many and varied transport processes across the companion cell–sieve tube interface.

Structural and functional specialization of companion cells

The term 'companion cell' is used to describe that cell, or group of cells, which is derived from the same mother (procambial) cell as the sieve tube member. However, the identification of the 'companion cell' may be problematical in some species, more especially so in monocotyledons. By contrast, transfer cells are relatively easy to identify, as they always have wall ingrowths that enhance uptake from the apoplast due to the vastly increased cell wall and plasmalemma surface area. Intermediary cells occur in several families, and function in the phloem-loading pathway by conversion of the sugars into larger molecules, through a 'polymer trap' mechanism which effectively concentrates larger polymers in the intermediary cells, increasing their concentration, and facilitating phloem loading via the intermediary cells. Photoassimilates and other substances usually accumulate against concentration gradients in the phloem, a process known as loading. In mature leaves, the sieve element–companion cell complexes (SE–CCs) of minor veins, where loading occurs, are connected to surrounding cells by plasmodesmata. These pores appear to participate in loading in plants that translocate raffinose and stachyose, but in sucrose- and polyol-translocating species their function is less certain (Turgeon 2000). There is argument that large numbers of plasmodesmata between the SE–CCs and surrounding cells should cause dissolution of the concentration gradient unless the size exclusion limit of the pores is small enough to retain accumulated solute species. In leaves of willow, *Salix babylonica* L., a sucrose-translocating plant with a high degree of symplastic connectivity into the minor vein phloem, the sucrose concentration gradient is absent between mesophyll and phloem, leading to the conclusion that phloem loading does not occur. Once inside the SE–CCs, solute may be able to pass freely between sieve elements and companion cells, because they are also symplastically connected. However, Turgeon (2000) postulates that due to net flux into the sieve tubes in source leaves, a continual drain of metabolic intermediates out of companion cells should occur, and this transport step could be regulated in minor veins, to prevent continual loss of needed solute molecules to the translocation stream.

In a recent study, Hoffmann-Thoma et al. (2001) studied the minor-vein ultrastructure and sugar export in mature summer and winter leaves of the three broadleaf-evergreen species *Ajuga reptans* var *artropurpurescens* L., *Aucuba japonica* Thunb. and *Hedera helix* L. to assess temperature effects on phloem loading. Leaves of the perennial herb *Ajuga* exported substantial amounts of assimilate in form of raffinose-family oligosaccharides (RFOs). Its minor-vein companion cells represent typical intermediary cells, with numerous small vacuoles and abundant plasmodesmal connectivity to the bundle sheath. By contrast, the woody plants *Hedera* and *Aucuba* translocated sucrose as the dominant sugar, and only traces of RFOs were reported. Their minor-vein phloem possessed a layer of highly vacuolated cells within the vascular bundles, intervening between mesophyll and sieve elements, which were classified either as companion or parenchyma cells, depending on the location and ontogeny of these vacuolate cells. Both cell types showed symplasmic continuity to the adjacent mesophyll tissue although at a lower plasmodesmal frequency compared to the *Ajuga* intermediary cells. *p*-Chloromercuribenzenesulfonic acid did not reduce leaf sugar export in any of the plants, indicating a symplasmic mode of phloem loading.

The companion cell, or specialized intermediary cell, are thus physiologically significant in facilitating and regulating phloem loading, which appears in all species to be regulated in the minor (or small) veins of leaves. In the Magnolid *Liriodendron tulipifera*, plasmodesmatal frequencies leading into minor vein companion cells are higher than in species known to load via the apoplast. According to Goggin et al. (2001), the companion cells are not specialized as 'intermediary cells' as they are in species in which the best evidence for symplastic phloem loading has been documented. Furthermore, application of the inhibitor, chloromercuribenzenesulfonic acid was demonstrated to largely, but not entirely, inhibit exudation of radiolabelled photoassimilate. The findings of Goggin et al (2001) are therefore most consistent with the presence of an apoplastic component to phloem loading in this species, contrary to speculation that the more basal members of the angiosperms load by an entirely symplastic phloem loading mechanism. Sucrose transporters are postulated to load photosynthetically produced sucrose apoplasmically into the sieve elements. For example, sucrose transporters like SUC2 have been localized in the companion cells (see Tazz and Zeiger, 2002). Mutant *Arabidopsis* plants containing DNA insertions in the gene encoding SUC2 have been recently identified which in the homozygous state, result in stunted growth, retarded development and sterility (Gottwald et al. 2000). Source leaves of mutant plants accumulate starch, and radiolabelled sugar was not efficiently transported to sinks such as roots and inflorescences.

Unloading phloem

Ma and Peterson (2001) have shown that in *Allium* roots the highest plasmodesmal frequencies were detected at the metaphloem sieve element–companion cell interface and all other interfaces had much lower plasmodesmal frequencies. In the pericycle, the radial walls had high plasmodesmal frequencies, a feature that could permit lateral circulation of solutes, thus facilitating ion (inward) and photosynthate (outward) delivery, and additionally, if the plasmodesmata are functional, a considerable symplastic transport pathway exists between the exodermis and pericycle as well.

Developing seeds are net importers of organic and inorganic nutrients and nutrients enter seeds through the maternal vascular system at relatively high concentrations in the phloem. They exit importing sieve elements via interconnecting plasmodesmata. During subsequent symplasmic passage, they are sequestered into labile storage pools within vacuoles and as starch. High densities of plasmodesmata could support symplasmic delivery of accumulated nutrients to underlying storage cells where polymer formation (starch, protein) may take place (Patrick and Offler 2001).

Specifics of the monocotyledonous foliage leaf

A commonly mentioned anatomical feature of C_4 plants is the orderly arrangement of mesophyll cells with reference to the bundle sheath cells, forming concentric layers around the vascular bundle as seen in transection (Fig. 6.29a). Bundle sheath cells of C_4 plants have few if any intercellular air spaces between them, which is in direct contrast to the sometimes large intercellular space volume between mesophyll cells in C_3 plants. Observations of the concentric arrangement of mesophyll and bundle sheath cells of certain grasses and sedges prompted Halberland to compare the mesophyll layer to a *Kranz* (wreath-like) structure. The structure of monocotyledonous foliage leaves depends to a large extent on the type of photosynthesis (i.e. C_3; C_4) and on the environmental conditions that the plants grow in (i.e. xerophytic, mesophytic or hydrophytic). Most monocotyledonous foliage leaves are basically parallel-veined, but large numbers of cross-veins serve to interconnect the parallel vein system. Parallel veins in monocotyledons are classified as large, intermediate or small.

The monocotyledons are more diverse and although many of them have the type of vascular bundle orientation described above, others have very different arrangements. In grasses, three orders, or classes, of vascular bundle can be recognized within the leaf blade:

1 **First order, or large bundles.** These bundles are characterized by the presence of large metaxylem vessels on either side of the protoxylem, which is often represented by a lacuna. Obliterated protophloem is evident on the abaxial side of the metaphloem sieve tubes, but is usually obliterated and non-functional in mature leaves. Sclerenchyma girders or strands may extend from these bundles, to the adaxial and abaxial epidermis.

2 **Second order, or intermediate bundles.** These bundles lack large metaxylem vessels and protoxylem lacunae, but may contain both proto- and metaphloem sieve tubes. Intermediate bundles can be supported by hypodermal sclerenchyma strands or girders that occur either on the ad- and abaxial sides of the vein, or on one side only. In the grasses and sedges, intermediate vascular bundles typically occur between successive large bundles.

3 **Third order, or small bundles.** Besides lacking large metaxylem or protoxylem lacunae, these bundles lack protophloem and are not normally associated with either hypodermal strands or girders and are embedded within the mesophyll.

Leaf blade bundle anatomy

Among the grasses, two anatomical variations are noteworthy – that is, the Panicoid (Fig. 6.30a) and Pooid (Fig. 6.30b) groups. In the Panicoid grasses, the mesophyll is radially arranged and surrounds a parenchymatous bundle sheath. Panicoid grasses contain dimorphic chloroplasts, with granal chloroplasts within the radiating (Kranz) mesophyll and generally -agranal chloroplasts within the parenchymatous bundle sheath cells. Bundle sheath chloroplasts are much larger than the Kranz chloroplasts and lack Rubisco – the Calvin cycle is thus not supported within Kranz mesophyll cells.

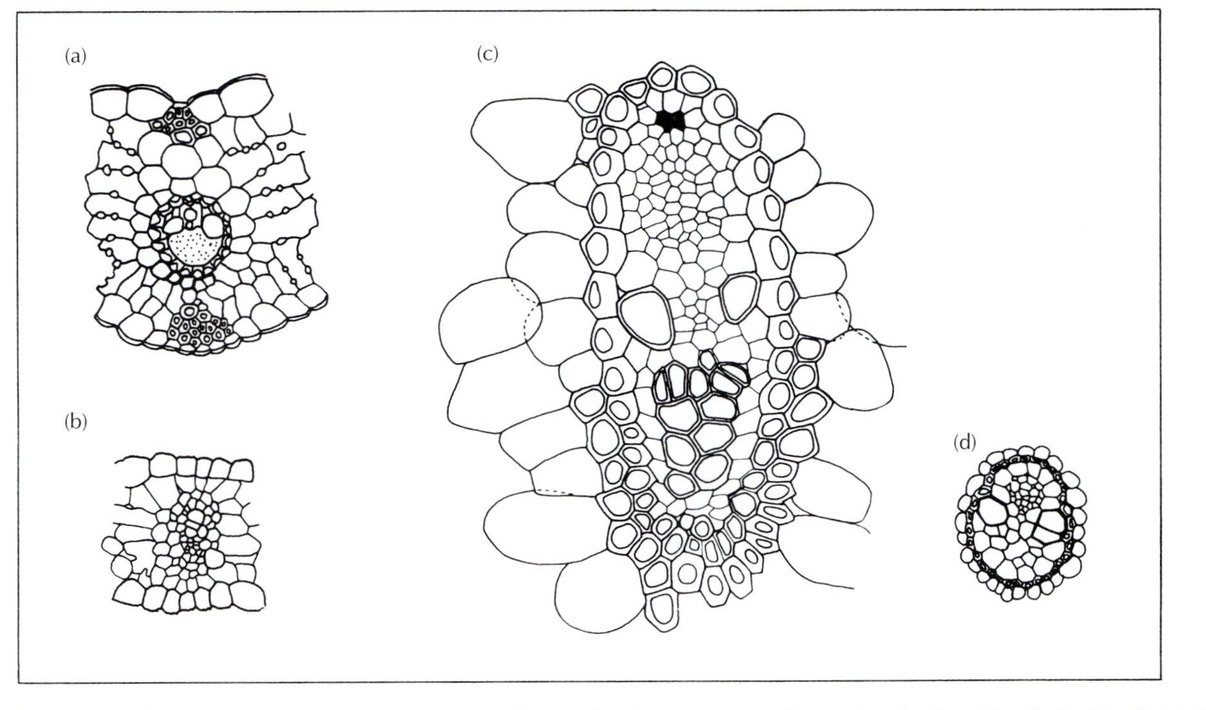

Fig. 6.29 Bundle sheaths. (a) *Briza maxima*, inner mestome (sclerenchyma) sheath, outer parenchyma sheath, abaxial and adaxial sclerenchyma strands and radiate chlorenchyma. ×120. (b) *Gloriosa superba*, parenchyma sheath only. ×120. (c) *Cymophyllus fraseri*, parenchyma, followed by mestome sheath and outer parenchyma sheath. ×128. (d) *Fimbristylis*, three sheaths, inner parenchyma, followed by mestome sheath and outer parenchyma sheath. ×218.

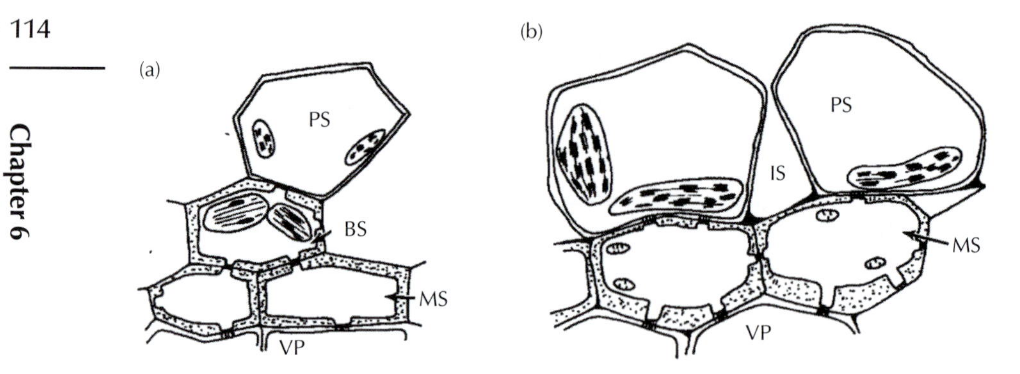

Fig. 6.30 (a, b) Line drawings based on electron micrographs of typical Panicoid and Pooid leaf blade bundle anatomy. BS, parenchymatous bundle sheath; IS, intercellular space; MS, mestome sheath; PS, parenchymatous (Kranz) sheath; VP, vascular parenchyma cell. ×1000.

Instead, these cells are associated with the initial incorporation of CO_2 as aspartate, which is transported to the bundle sheath cells, via numerous plasmodesmata, where malate or aspartate is decarboxylated, and the liberated CO_2 is immediately incorporated via Rubisco, into the Calvin cycle. Panicoid grasses are thus C_4 photosynthetic species. In many Panicoid species, an additional cell layer exists between the bundle sheath and the vascular tissues below. This layer, which consists of thick-walled lignified cells, is termed the mestome sheath. Ontogenetically, the mestome sheath is derived from the procambium. Mestome sheath cells may either completely surround the vascular tissue, or surround the phloem tissue only within the vascular bundles. The middle lamella between the bundle sheath cells and, in some species that between the mestome sheath cells contains a suberized layer, is termed the suberin lamella. The compound middle lamella has been shown to restrict the movement of solutes, forcing transport (i.e. photoassimilate inwards and water outwards) to take an entirely symplastic route, via plasmodesmata there. The suberin lamella may have important ecological consequences, preventing the excessive movement of water from the apoplast, under conditions of water stress. Pooid grasses, like the Panicoid species, may be associated with a mestome sheath (Fig. 6.30b). Unlike the Panicoid grasses however, the Pooid species do not exhibit chloroplast polymorphism, do not have compartmentalized Rubisco activity and all follow the C_3 photosynthetic pathway.

Whilst monocotyledonous foliage leaves are generally described as being parallel-veined, large numbers of transverse veins exist within the leaf blade. Figure 6.31 is an electron micrograph of a transverse vein in the leaf blade of *Saccharum officinarum*. In some cross veins, the radial walls of the parenchymatous element may contain a well-developed suberin lamella, which as stated, may have a regulatory role in solute loading and water loss from the xylem to the mesophyll.

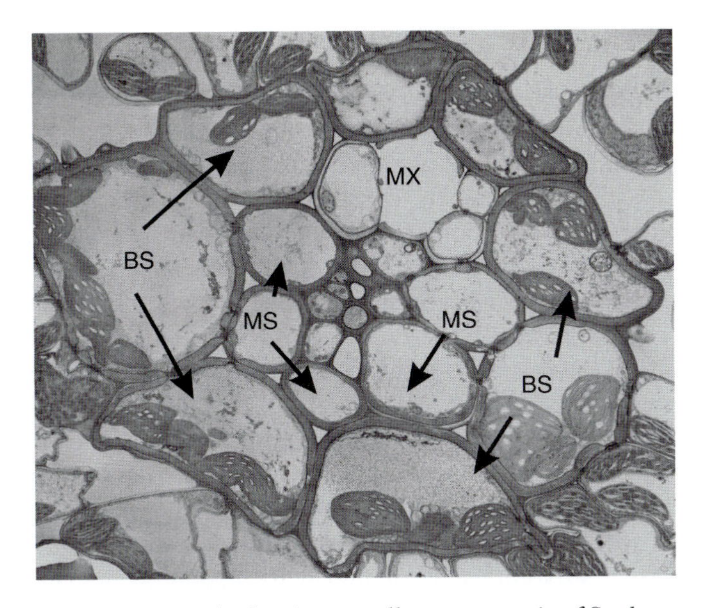

Fig. 6.31 Electron micrograph, showing a small transverse vein of *Saccharum officinarum* in transverse view. This vein is surrounded by two sheaths; an outer bundle sheath (BS) and an inner mestome sheath (MS). Bundle sheath mestome sheath interfaces are often associated with a suberin lamella. The vascular tissue consists of metaxylem vessel and associated parenchyma, whilst the phloem contains several sieve tubes and associated parenchyma and companion cells. ×1650.

Monocotyledonous phloem

As mentioned earlier, the phloem in mature monocotyledonous leaf blades including those of grasses and sedges differs from that of the dicotyledonous leaf blade bundles. Generally, in monocotyledons, the phloem within mature bundles is composed of functional metaphloem sieve elements, associated with vascular parenchyma cells, including companion cells. The phloem may contain specialized, late-formed metaphloem sieve tubes, which appear to lack the companion cell associations that exist with the early, thin-walled sieve tubes. The late-formed metaphloem sieve tubes have thickened, usually cellulosic walls, which, in some cases (e.g. barley and wheat) may undergo lignification. The thick-walled sieve tubes usually border on, or occur in close proximity to, the metaxylem vessels within leaf blade bundles. The thick-walled sieve tubes may be seen with a good light microscope. Physiologically, the thick-walled sieve tubes may be symplasmically isolated from other cells within the vascular bundle. In most monocotyledonous species that have been examined thus far, very few plasmodesmatal connections exist between the thick-walled sieve tubes and companion cells and associated vascular parenchyma cells. Instead, thick-walled sieve tubes may be symplasmically connected directly to vascular parenchyma cells.

The phloem and xylem are not the only tissues present in the veins. They form the central core, around which sheaths of specialized cells are formed, which separate the vascular tissues from the mesophyll. Two principal types of sheath exist, namely sclerenchymatous and parenchymatous sheaths. There may also be parenchyma associated with the phloem or xylem, and the phloem may contain fibres.

Sclerenchymatous sheaths are composed of fibres and/or sclereids. Sometimes the walls of these cells that face the phloem or xylem are more heavily thickened than the others, as is the case in some sedges. The parenchyma sheath is normally composed of much larger and wider cells, with thinner and (usually) relatively unlignified, walls. If both types of sheath are present, the sclerenchyma sheath is usually innermost. In grasses, this inner lignified sheath is referred to as a mestome sheath. In some genera and species, an additional sheath may be present – an inner parenchymatous sheath, which is in turn, surrounded by an intermediate, sclerenchymatous sheath and an outer parenchymatous sheath. When present, the inner parenchymatous sheath in leaf blade bundles of some sedges may contain large, agranal chloroplasts. This anatomical form is found in *Fimbristylis*, a member of the Cyperaceae, and indicates the presence of the Kranz syndrome, and that C_4 photosynthesis may be present. Named examples of the various types of sheath are illustrated in Fig. 6.29.

The Cyperaceae

Anatomically, there are distinct similarities between the C_4 Cyperaceae and C_4 Panicoid Poaceae. Like the Poaceae, the Cyperaceae are photosynthetically either C_3 or C_4 and the phloem within the leaf blade vascular bundles contains two types of sieve tube – early, thin-walled sieve tubes and late-formed thick-walled metaphloem sieve tubes, which are generally in close spatial association to the metaxylem, and lacking companion cells. The major differences in Cyperaceae compared to Poaceae lies in the distribution of the chloroplast-containing parenchyma surrounding the phloem (termed the border parenchyma in the literature) in the vascular bundles, and in the shape, and thickness of the walls of the cells which have been equated to the mestome sheath of grasses (Fig. 6.32). Four variants of Kranz anatomy occur in the Cyperaceae. Of these, three of these anatomical types (fimbristyloid, chlorocyperoid and eleocharoid) are unique among taxa with C_4 photosynthesis in that the photosynthetic carbon reduction tissue (PCR, functional equivalent of bundle sheath) is located within the vascular strand and is separated from the primary carbon assimilation tissue (PCA, positional equivalent of mesophyll) by the mestome sheath layer. In the rhynchosporoid group, the PCR tissue is located in the position of the

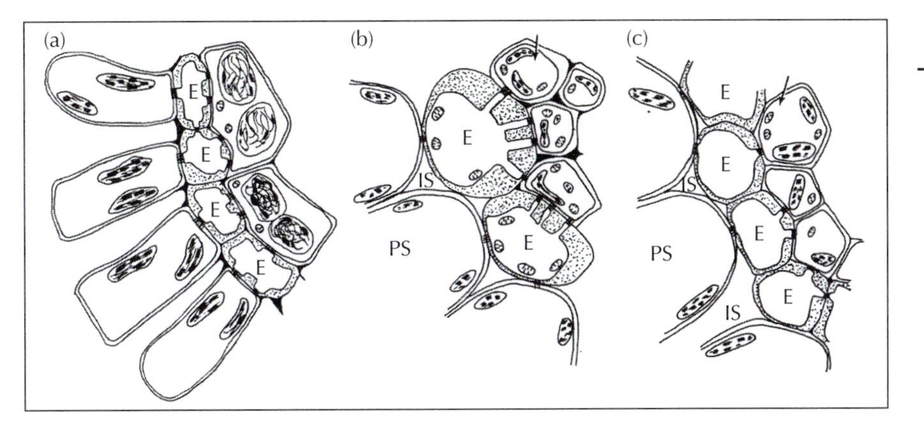

Fig. 6.32 Line drawings (a–c) showing the basic anatomical features of leaf blade bundle structure in the Cyperaceae. The variation of cell thickness is most notable in the cell walls of the endodermis. Note the distribution of chloroplasts in the border parenchyma and the presence of large chloroplasts (agranal in some species) in the border parenchyma. Examples are, left: *C. fastigiatus*; *C. esculentus*; *Mariscus congestus*; centre: *C. sexangularis*; *C. pulcher*; *C. accutiformis*; right: *C. albostriatus*; *C. textilis*; *C. papyrus*. E, endodermis; IS, intercellular space; PS, parenchyma sheath. ×850.

mestome sheath (Soros & Dengler 2001). Suberin lamellae may be present in either outer tangential and/or inner radial and tangential walls of the sheath layer. In C_4 species, chloroplasts within the border parenchyma, are large and obviously agranal. The border parenchyma zone commonly encircles both xylem and phloem, or only the phloem.

Chloroplast dimorphism and the lack of grana in the primary carbon reduction (PCR) cells are indicative of the C_4 syndrome. Experimental evidence exists for the positive localization of Rubisco in these large, agranal chloroplasts.

Bundle sheath extensions

Anatomically sheaths may be complete, or present at the bundle poles as caps only, or present on the flanks of the bundles only. Collenchymatous or sclerenchymatous girders may interrupt bundle sheaths. In the grasses bundle sheaths are often associated with a suberized, osmiophyllic compound middle lamella. In C_4 grasses, the location of the suberin lamellae can assist in separating the three C_4 photosynthetic subtypes. NADP-ME and PCK species contain a suberin lamella in the bundle sheath cells, particularly in the outer tangential and radial walls of the bundle sheath cells, or sometimes only in the inner tangential walls of bundle sheath cells (*Zea mays*). NAD-ME species do not have suberin lamellae associated with the walls of the sheath cells. In C_3 species, the suberin lamella (if present) appears to be confined to the mestome sheath (e.g. *Phalaris canariensis*,

(Fig. 6.20), or in *Bromus unioloides* for example. In some instances, suberin lamellae may be associated with the bundle sheath as well as the inner, mestome sheath (as in *Saccharum officinarum*).

There may be adaxial or abaxial extensions (hypodermal collenchymatous or sclerenchymatous strands) to the sheaths, reaching towards, or in contact with either epidermis. The outline of these girders as seen in transverse section can be used to distinguish species in some groups. In some plants, subepidermal strands of fibres may be present in line with vascular bundles. In a number of genera a hypodermis is present, composed of one or more layers of cells and occurring to the inside of the epidermis. The cells usually differ in shape or degree of wall thickening from both epidermis and mesophyll. It is a useful diagnostic feature. Its presence is often associated with plants adapted to grow in dry parts of the world.

Endodermis

Whilst few leaves contain a true endodermis, such as that which surrounds the vascular bundles in Gymnosperms such as *Pinus*, there are some families within which an endodermis-like or 'endodermoid layer' can be recognized. As with a true endodermis, the cells making up an endodermoid layer may contain a suberin lamella, which is present at least in the outer tangential and radial walls of these cells. The presence of Casparian strips and or suberin lamellae have led to the cells being incorrectly being termed 'mestome sheath'. Among the Cyperaceae (*Fimbristylis*, *Pycreus*, *Eleocharis*) there are species that show good examples of well-developed endodermoid sheaths, which appear exarch to the inner 'Kranz' mesophyll, which is completely opposite to the situation in C_4 grasses which display double-sheathed Kranz anatomy, where the mestome sheath is centripetal to the outer bundle sheath!

Secretory structures

External secretory structures

External and internal secretory structures exist in many plants. External secretory structures are of epidermal origin, and are usually glandular. Glandular trichomes are composed of a stalk and a head. The stalk may be unicellular or multicellular. In some species, several rows of cells make up the stalk. The head may, like the stalk, be unicellular or multicellular. Various secondary plant products are secreted within the gland, which, although usually covered by a cuticle, allows the essence through the apparently pore-less glands. Some trichomes can be regarded as being extra-floral nectaries whilst others are hydathode-like structures.

Secretion within the plant may be carried out either by single cells, small groups of cells or by tissues. Oil-secreting cells may be distributed throughout particular tissues.

In many instances, groups of thin-walled cells form an assemblage generally referred to as secretory cells, which surround a duct that is schizogenously formed, as for example in the sunflower (*Helianthus*) stem. In plants, a combined process of lysigeny and schizogeny may form ducts.

The resin ducts formed in conifers, are thought to be of schizogenous origin, as here, like the sunflower, the resin ducts are surrounded by a ring of clearly demarcated secretory (epithelial) cells.

Internal secretory structures

Laticifers are considered to be an important internal secretory channel. The latex may contain a combination of secondary plant metabolites, including carbohydrates, organic acids and alkaloids. Laticifers may be classified into non-articulated (originating from single cells) which are capable of potentially unlimited growth, and articulated laticifers, which being compound in origin, consist of longitudinal files of cells, in which the cross walls are hydrolysed, thus forming an extensive anatomising network of multinucleate cells.

Concluding remarks

From this brief discussion, it should be clear that leaves vary greatly between genera, families and, even sometimes, between species. Some characteristics may have significance in the identification of plants at the generic or even specific levels. Clearly, whilst one can separate most monocotyledonous plants from dicotyledons and gymnosperms, using fairly simple, easy to observe characteristics, 'typical' leaves do not exist, rather there exists an intergradation of anatomical characteristics. For example, most applied studies based on leaves, are concerned with the relationship between fine structure (mostly of phloem in minor veins) and function. Relatively few plants are amenable to this sort of study, because accessibility to the functioning cells can be so limited in many species. This means that although in due course we might expect to understand the mechanisms of translocation in a few species, we should not extrapolate and predict the same mechanisms for all plants. For example phloem in *Laxmannia* of the Anthericaceae has very small phloem elements in the vascular bundles of the leaves which are particularly narrow; these elements are embedded in a matrix of fibres and would thus appear to have very indirect communication with the mesophyll. Sieve element and companion cell pairs show particularly well in *Liriope* and *Ophiopogon* of the Convallariaceae, where they are embedded in sclerenchyma. Their obvious difference to the norm is what might make these examples good subjects for study.

Perhaps it is these differences, and the lack of an obvious norm, which makes plant anatomy such an interesting and fascinating study to us!

Flowers, fruits and seeds

Introduction

Apart from their ornamental or horticultural value, flowers have been studied mainly as a source of very important taxonomic characters and in relation to phylogeny and evolution. Their prime function in reproduction has naturally been the object of vast amounts of morphological and physiological investigation.

The extreme importance of fruits and seeds as food has provided the inspiration for a great deal of research. However this volume deals largely with vegetative anatomy, and only matters of particular applied interest relating to flower and fruit are outlined. The further reading list at the end of the book will assist those who wish to learn more on the subjects mentioned here.

Vascularization

Many of the anatomical features used in comparative studies are to be found in the arrangement and number of vascular bundles and their types of branching in inflorescences, flowers and floral parts. These patterns can be hard to interpret. Are all or most of the bundle pathways genetically predetermined? Could a significant proportion of bundles be present as responses to physiological demand, that is, related to physiological needs which are to be met rather than some archaic pattern recapitulating ancestral conditions? Despite the difficulties of observation and interpretation, many valuable studies have provided data on vascularization that enable us to understand the interrelationships between many genera and families of flowering plants.

Those whose interests lie in the phylogeny of flowering plants, or in the origins of angiosperm flowers, make considerable use of the results of the studies on vascularization. It is widely held that the vascular systems in flowers are conservative, that is, they may remain relatively unchanged even when the general shape of the flower has altered by evolution. This may lead to the formation of odd-looking loops or curves in some of the vascular strands to accommodate changes in the relative positions of the floral parts. In some flowers, small branches of the vascular system end blindly. This could be taken to mean that in one or more of the ancestors of the plant similar strands served some organs or appendages that are lacking in the present-day representative. For example, a modern, unisexual female flower might have remnants of a vascular system that would have served stamens in a bisexual ancestor.

When organs are adnate, for example, a stamen fused with a petal, it often follows that the vascular supplies of the two become fused into one strand. Bundle fusions can make it more difficult to interpret the vascular systems of flowers in comparative studies.

The number of traces to each floral organ can vary. Often petals have only one trace, but petals in certain families regularly have three petal traces. The number of traces to each sepal is often the same as that to the foliage leaves of the same plant. Stamens may have one or three traces, but one is by far the most common number. Carpels may possess one, three, five or more traces. Dorsal and marginal or ventral traces are distinguished in descriptions, when three or more are present.

Sometimes a flower may have very unusual morphology that is difficult to interpret. This may be related to adaptations to particular pollinators. It could be that examination of its vasculature would help one to understand the true nature of the various parts. If other members of the same genus or family have more normal flowers, then comparative studies could prove most informative.

A number of theories concerning the origin of the angiosperm flower have their basis in comparative studies of floral and vegetative vascular patterns of both living and fossil plants. Despite the amount of work done by numerous people, there is no common consensus of opinion. Doubtless, new theories will be proposed. Some think that we have all the evidence we need, if only we will interpret it properly. Others consider that there are such large gaps in the fossil record that no one will ever be able to prove their theories! Modern molecular studies have led to the development of much less subjective phylogenies, and it has become profitable to study flower morphology in the light of the new information these contain.

Because the classifications of plants give a great deal of prominence to characters of flower and fruits, there is a wealth of data on these parts. Consequently, it is normal to try to identify plants with flowers or fruit attached by reference to floras and herbarium specimens. Anatomical studies may help if the floral parts are in poor condition. Floral vascular anatomy may also give clues to the plant family that a species belongs to when the flowers or fruits are found separate from the rest of the plant. A good example of

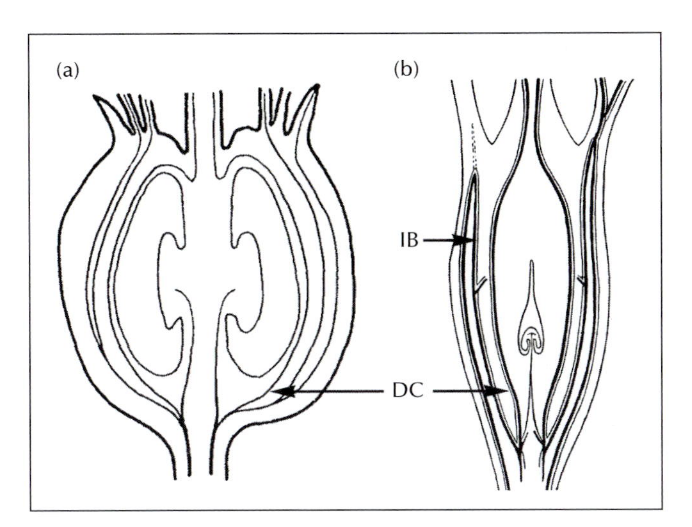

Fig 7.1 Floral vasculature of (a) *Gaylussacia frondosa* (Ericacaeae) with common bundles and (b) *Nestronia umbellulata* with inverted bundles that give rise to the carpel bundles. DC, dorsal carpel bundles; IB, inverted bundles.

this is the different vasculature patterns found in perigynous flowers and fruits from different plant families as shown in Fig. 7.1. There are two basic patterns. In one of them the vasculature to floral parts on the same radius arises from a common bundle in the ovary wall as in *Gaylussacia frondosa* (Fig. 7.1a) and all bundles as seen in cross-section have the xylem internal to the phloem. By contrast, in the other organization, the bundles that supply the carpel arise from a recurved bundle in the ovary wall as in *Nestronia umbellulata* (Fig. 7.1b), where the recurved bundles in the fruit wall are inverted, so that the phloem is interior to the xylem. These basic differences in vascular anatomy allow one to narrow the choices for unidentified fruits derived from inferior ovaries because the features are associated with particular plant families. Thus, the first condition is found, for example, in the Rosaceae and Ericaceae as well as in other families, and the second condition is found, for example, in the Cactaceae and Santalaceae as well as in other families. These features thus narrow the potential choices for identification of archaeological as well as palaeontological specimens.

SEM studies

In recent years studies of floral parts using scanning electron microscopy have added greatly to out understanding of the evolution of flowers. Small, fossilised flowers have been extensively studied. Developmental studies have helped separate analogous from homologous parts. Such characters as floral nectaries have been clarified by combined studies using SEM and thin sections for light microscopy. The further reading list has examples.

Palynology

Pollen studies have increased enormously with the advent of transmission and scanning electron microscopy. However, a great deal of foundation work was carried out with the light microscope; indeed a very sound basis was established. New tools have meant that details of fine surface patterning can be seen easily (Fig. 7.2). Surveys of families can now be carried out much more rapidly, and electron micrographs, particularly from the SEM, are easy to interpret.

Pollen grains are often readily identified to the genus level, and sometimes to the species level, if adequate reference material is available. In some families there is great variability in pollen grain morphology and surface features; in others there is uniformity.

Besides the taxonomic inferences which can be drawn from studies in comparative palynology, the subject has a number of other applied aspects. For example, honey purity and origin can be determined by a study of the pollen grains it contains – pure heather honey would not be expected to contain large quantities of *Eucalyptus* pollen. Adulteration can usually be detected microscopically. Pollen is often found on clothing and can provide useful evidence in forensic cases.

Pollen grains shed from plants remain in a recognizable form in peat deposits for very long periods of time. By careful analysis of the pollen grains in successive layers of peat, or at successive levels, it is often possible to build up a picture of the vegetation of previous ages.

Pollen–stigma interactions

Most plants have mechanisms by which they can 'recognize' pollen grains from their own and other species. Stigmas often possess a complex of chemicals that enable them to respond to chemicals contained in the outer layers of pollen grains.

The chemicals are usually proteins. Figure 7.3 shows a pollen grain germinating on a stigma. In species which are not self-fertile, the stigma rejects pollen from anthers produced by the same or other flowers on the plant. Many plants normally reject pollen from other species. However, some species are interfertile. In nature they might not normally be pollinated by other species. Their flowering periods might not coincide, or they may be too distant from one another. In some plants, the stamens are mature well before the stigma (protandry) and pollen is dispersed before the

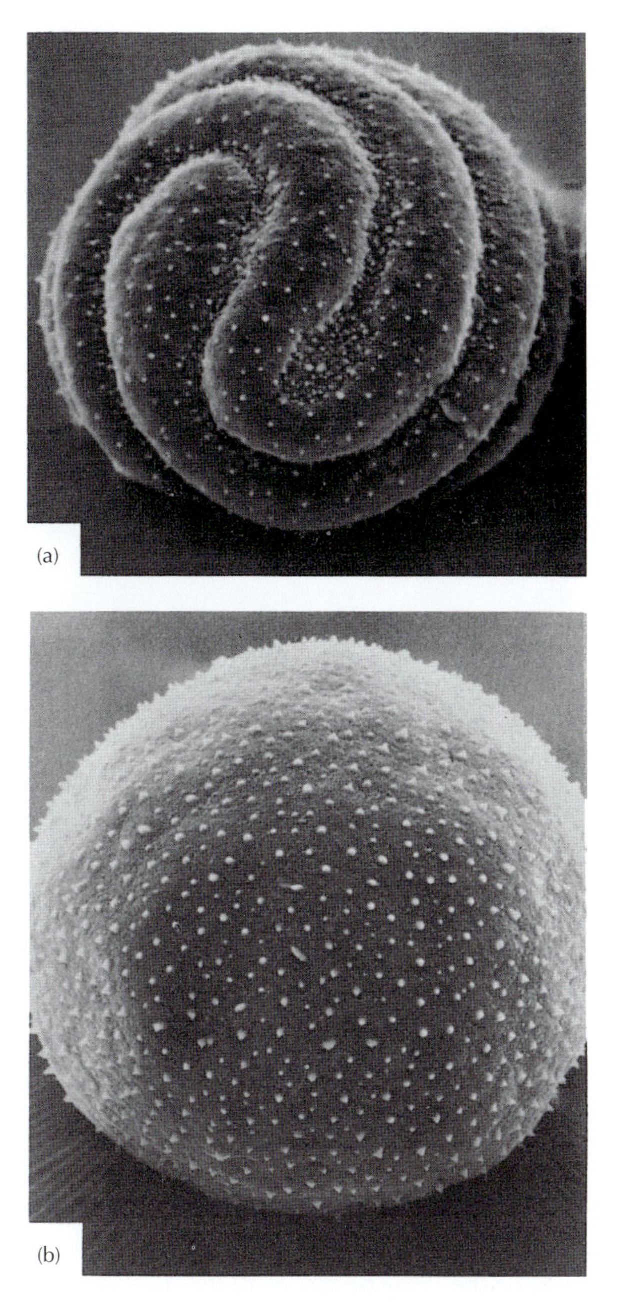

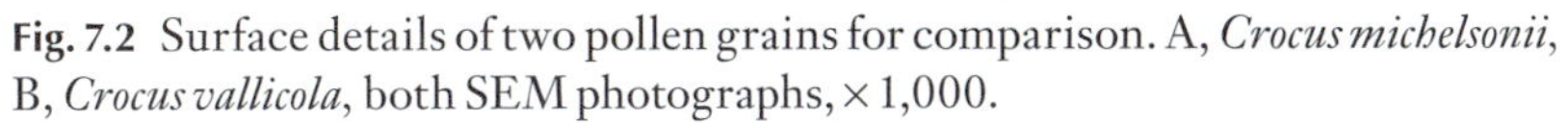

Fig. 7.2 Surface details of two pollen grains for comparison. A, *Crocus michelsonii*, B, *Crocus vallicola*, both SEM photographs, × 1,000.

Fig. 7.3 *Tradescantia pallida*, pollen grain germinating on stigma. p, pollen grain; pt, pollen tube; s, papilla on stigma. Freeze dried, viewed in SEM, × 1,000.

stigma is receptive. In others the stigma may mature and senesce before pollen from the same flower is released (protogyny).

Pollen can be stored alive at low temperatures, so we can try out crosses even if flowering periods are different. The horticulturalist attempting to maintain pure species is only too aware of the crossing which can go on between related plants brought together in one glasshouse.

Some of the rejection mechanisms result in obvious physical changes in the stigma. For example, reactions may occur which cause the stigma to callus over, so that the pollen tube cannot enter. Sometimes the size of the stigmatic papillae may be too great for small pollen grains to germinate on them successfully, or they may be too small for large grains, and also prevent effective pollination. Although pollen from closely related species may be accepted by a stigma, this is not always the case – incompatibilities may arise. In another mechanism, which prevents different species from crossing, the length of the style may be much longer than the potential length of the pollen tube, so that fertilization cannot take place.

Where it is desirable horticulturally or agriculturally to produce hybrid plants, anatomical and histochemical studies of the pollen–stigma interactions can help us to manipulate the process and override the blocking mechanism.

The system which promotes outcrossing may simply rely on different relative heights of anthers and stigma in the flower, as in pin-eyed and thrum-eyed primulas. In the first, the stamens are short and the stigma is elevated on a long style; in the second anthers are at the outer end of the corolla tube, and a short style keeps the stigma at a low level. Insects visiting a pin-eyed flower are more likely to deposit pollen on the stigma of a thrum-eyed than of a pin-eyed flower.

Pollen itself can sometimes be cultured and made to produce haploid plants.

Embryology

Embryo studies fall into two categories, firstly the comparative and developmental studies and secondly those aimed at embryo culture (or haploid, embryo sac culture).

Embryology and the sequence of cell divisions involved in embryo sac formation and following fertilization have become very specialized fields of study. There is a large body of comparative data available to the student, most of which is applied to evolutionary and taxonomic studies.

Embryo culture involves dissecting out embryos and growing them in culture media. This is sometimes done to ensure development and establishment of particular individual plants, but more often as a means of vegetative propagation.

Seed and fruit histology

The wide use of seeds and fruits in human food and animal feedstuffs has made knowledge of their anatomy of paramount importance. It is essential to be able to identify fragments of seeds and fruits in relation to possible adulteration and purity.

Although many species have been studied for seed and fruit structure, relatively few families have been carefully studied systematically and in detail, followed by documentation of the results.

Plants of economic importance have received most attention. The main cereals, oil seeds and edible leguminous seeds have been described anatomically, as have those of selected weeds and poisonous plants. Figure 7.4 shows some fruit walls and seed coats in transverse section. Good sources of information are the specialist books and reference books on the anatomy of food plants and only a very general overview is given here.

The fruit wall is termed the pericarp and it is divided into three regions, the outer or exocarp, the middle or mesocarp, and the inner or endocarp (Fig. 7.4). The surface of the fruit exhibits many of the features found in the epidermis of leaves and stems both in light microscopy and SEM. Some families have members with additional features that help in identification. An example of two features are the presence of slime cells and slime trichomes in some families such as the achenes of the Asteraceae (Figs 7.5, 7.6). The distribution of epidermal slime cells occurring either singly or in groups can be used to identify species as in *Anthemis* (Fig. 7.5). Slime trichomes (Fig. 7.6) are implicated in the adherence of the dry sclerified achenes to dispersal organisms and also in the uptake of water duration hydration of the fruits for germination of many Asteraceae.

Other features of the exocarp include the distribution and types of cells. As shown in Fig. 7.5, the ribs of the fruits of *Anthemis arvenis* are composed of tracheoidal idioblasts and those of *Anthemis perigina* are composed of sclerenchyma. Thus, even though the external morphologies are similar, the anatomies are quite different.

In the mint family (Lamiaceae), the nutlets have a unique anatomy in that the inner hypodermis appears as a heavily lignified sclerenchyma composed of elongated cells as seen in transverse section (Fig. 7.7). These sclereids appear as if they are lignified palisade cells.

Although the anatomy of the pericarp of fleshy fruits is histologically rather uniform with any given fleshy fruit, the anatomy of stone fruits, that is those with sclerified internal layers such as in peach or mango, shows histological differentiation, usually in the endocarp. The sclerified tissue of the endocarp can have different origins in different fruits as diagrammed in Fig. 7.8. As shown, some fruits have the endocarp derived from the inner epidermis (Fig. 7.8a) and in other cases from a multilayered epidermis derived from periclinical divisions in the epidermis (Fig. 7.8d). Still other fruits have the sclerified tissue derived from the hypodermis (Fig. 7.8b) or in other cases from a multiple hypodermis derived from periclinal divisions of the hypodermis (Fig. 7.8e). Two other patterns of development of the stony endocarp are those derived from the epidermis plus a multilayered hypodermis (Fig. 7.8c) and those with the stony endocarp derived from both a multiple epidermis and a multiple hypodermis (Fig. 7.8f). Although these patterns are known to occur, there is a paucity of data concerning their occurrence within the flowering plants and this is an area desperately in need of comparative studies. In families like the Apiaceae, Asteraceae and Lamiaceae for example, fruit anatomy provides many useful diagnostic and taxonomic characters. Undoubtedly, as more families are systematically studied, much of taxonomic and perhaps phylogenetic importance will emerge.

Enough is known for us to realize that good comparative studies of seeds can yield taxonomic characters of some significance. One excellent example

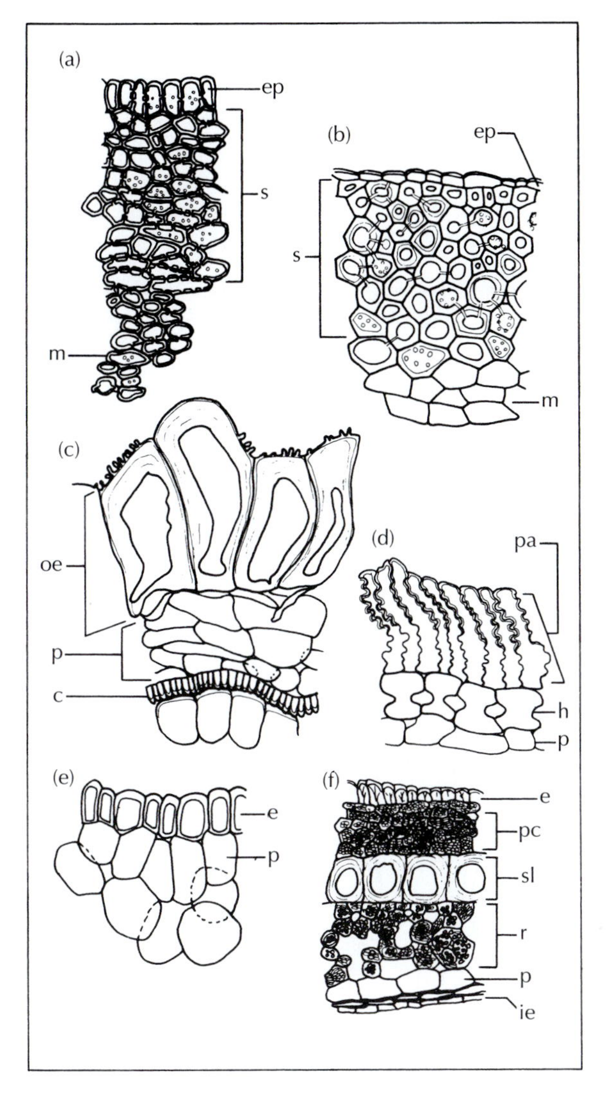

Fig. 7.4 Fruit wall and seed coat details in TS. (a) *Aesculus hippocastanum*, outer part of fruit wall. × 109. (b) *Fagus sylvatica*, outer part of fruit wall. × 109. (c) Outer part of seed coat of *Delphinium staphisagria*. × 109, note small outgrowths from epidermal cell walls. (d) *Cicer areitinum*, seed coat. × 218. (e) *Cola acuminata*, seed coat. × 218. (f) *Cucurbita pepo*. × 109. c, cells with U-shaped wall thickening; e, epidermis; ep, epicarp; h, hour glass cell; ie, inner epidermis; m, mesocarp; oe, outer epidermis; p, parenchyma; pa, palisade cells; pc, pitted cells; r, reticulate spongy parenchyma; s, sclerenchyma; sl, sclerenchyma layer.

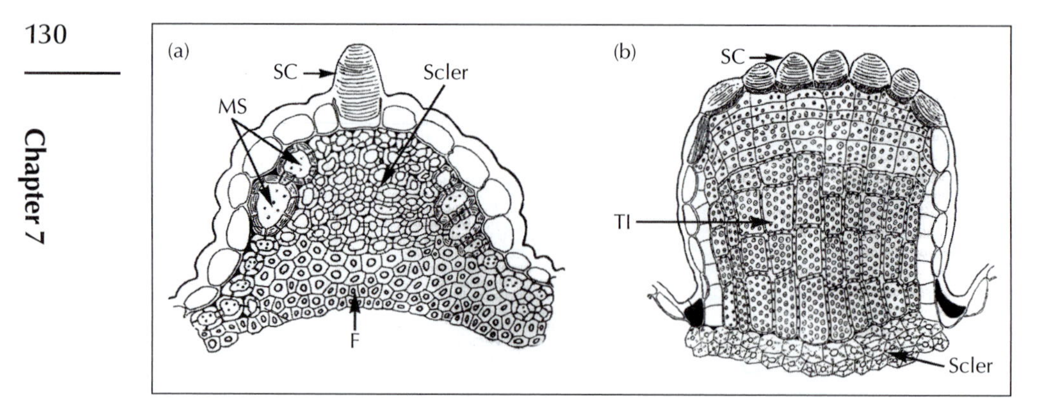

Fig. 7.5 Transverse sections of the exocarp of the fruits of *Anthemis* showing anatomical variation: (a) *A. perigina*; (b) *A. arvenis*. MS, macrosclerid; SC, slime cell; Scler, sclerenchyma; TI, tracheoidal idioblast. (After J. Briquet 1916.)

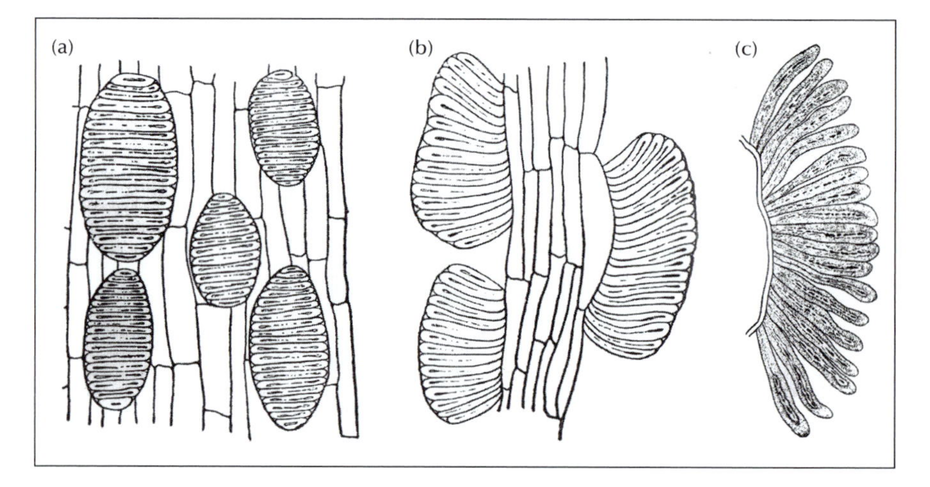

Fig 7.6 Epidermal slime trichomes on the exocarp of *Matricaria lamellata*. (a) Surface view of a slime trichome cluster. (b) Transverse section of a slime trichome cluster. (c) Longitudinal section of a slime trichome cluster. (After Alexandrov & Savcenko 1947.)

of this is the presence of ruminate endosperm (Fig. 7.9), which is diagnostic of some plant families such as the Annonaceae (the custard apple family) and the Myristicaceae (the nutmeg family) where the ruminate endosperm is easily seem in hand sections of whole nutmeg seeds.

Seed coats are usually composed of both the outer and inner integuments of the ovule. The mature seed coat is usually divided into three regions, an exotesta or outer seed coat layer(s), a mesotesta or middle seed coat layer(s), and an endotesta or inner seed coat layer(s). In thin seed coats, often, there are only an exotesta and an endotesta. Another example of the diversity of

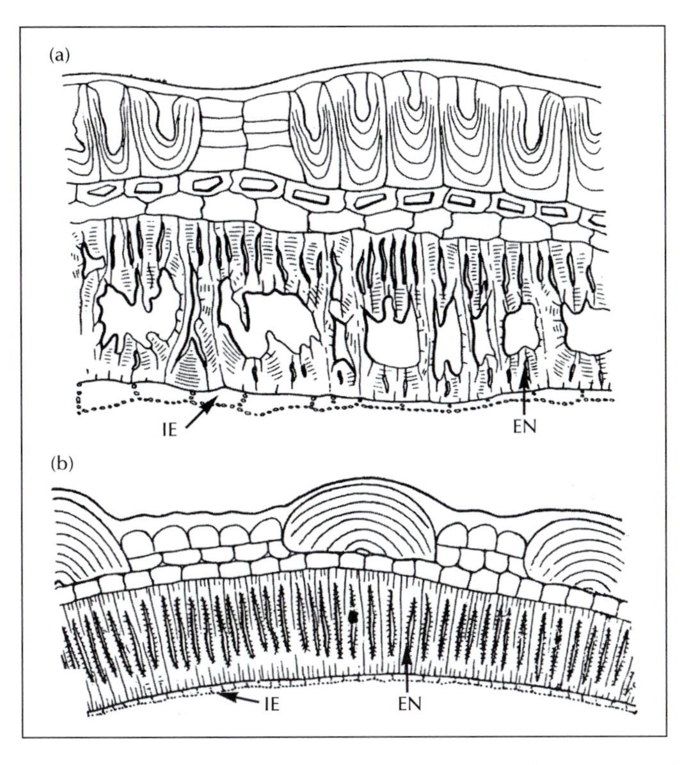

Fig. 7.7 Transverse sections of the pericarps of two species in the Mint Family. Note the tall pallisade like sclereids. (a) *Coleus barbatus*. (b) *Lavandula spica*. EN, endocarp; IE, inner epidermis. (After S. Wagner 1914.)

the anatomy of seeds can be found in the winged seeds (Fig. 7.10) of the brazil nut family (Lecythidaceae). Not only are the wings apparently different morphologically, they are also very different in their anatomy as is the body of the seeds. Thus, we see that the thickenings of exotestal cells of the body of the seed of *Cariniana legalis* are quite distinct as compared to those in *Couratari asterotrichia*, and that the exotesta is not present in the seed wing of the former but is quite distinct in the latter. Note also the difference in the anatomy of the mesotesta in the body of the seeds of these two species as well as in the wings.

As with pollen studies, interest in seed coat anatomy has been stimulated by the general availability of scanning electron microscopes. It is now often possible to detect minute differences in seed coat patterns which might enable us to define species characteristics. Seeds do tend to vary a lot in size and sometimes in shape within a species. Their surface patterns go through developmental stages and consequently only mature seeds should be studied for comparative purposes.

The adaptive or physiological significance of surface features is by no means clear, and valuable studies on the development of patterns to

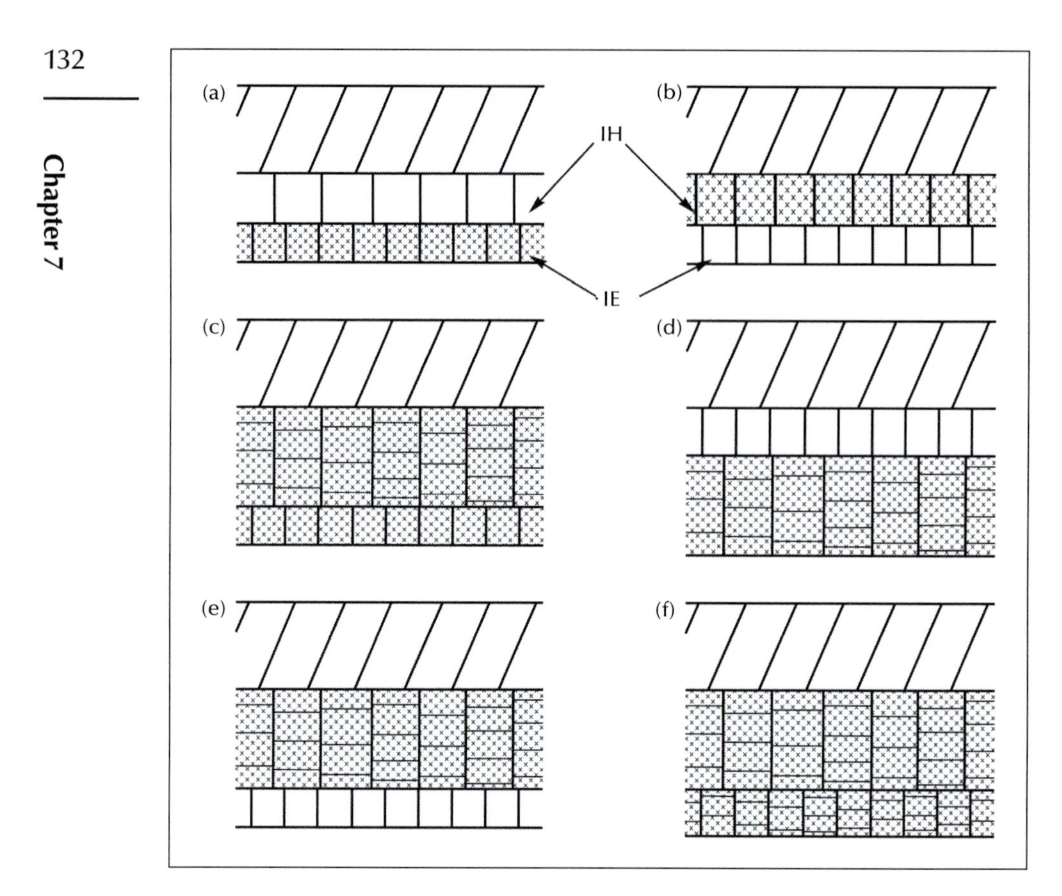

Fig. 7.8 Diagrammatic representations of the development of the stony endocarp in fruits from the inner epidermis (IE) and inner hypodermis (IH). (a) From the epidermis only. (b) From the hypodermis only. (c) From the epidermis and the multiple hypodermis. (d) From a multiple epidermis only. (e) From a multiple hypodermis only. (f) From both a multiple epidermis and a multiple hypodermis. Stippling indicates stony layer.

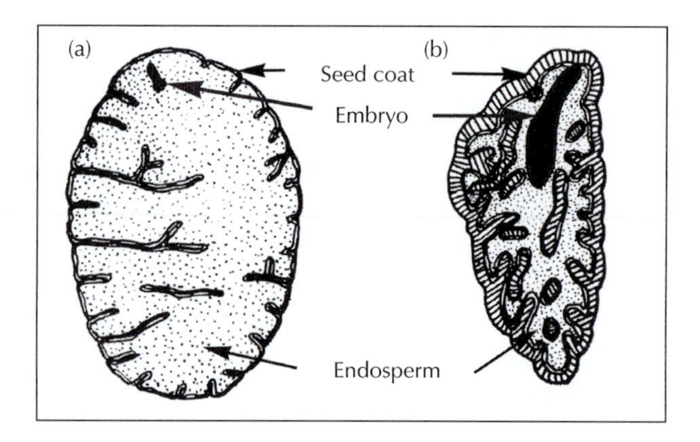

Fig. 7.9 Examples of ruminate endosperm in (a) *Asimina triloba* and (b) *Hedra helix*.

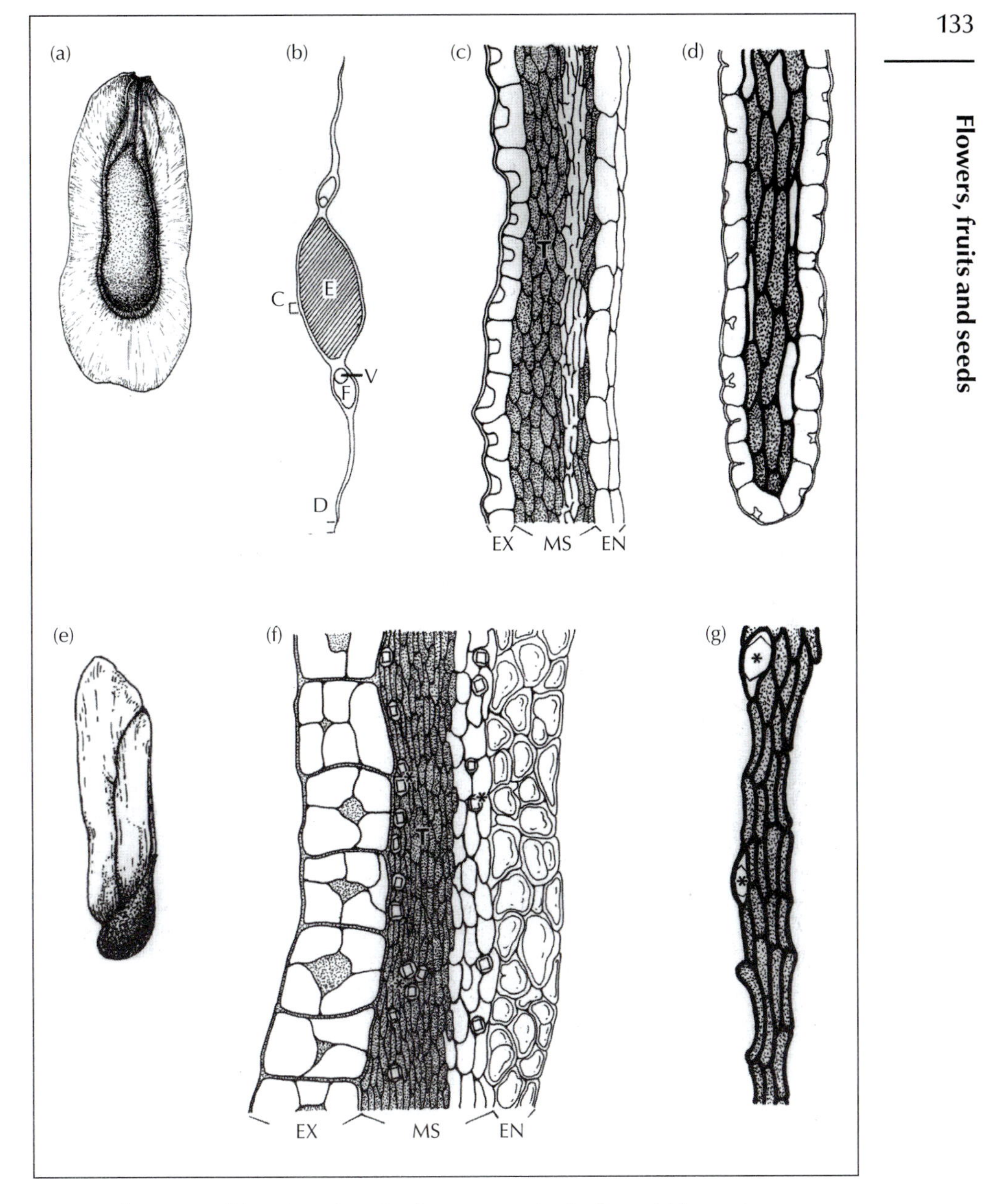

Fig. 7.10 Anatomy of the winged seeds of two genera in the Lecythidaceae in transverse section. (a–d) *Couratari asterotrichia*. (a) The whole seed. (b) Transaction of the whole seed. The areas labelled C and D in (b) are magnified in (c), the seed body, and (d), the seed wing. (e–g) *Cariniana legalis*. (e) The whole seed. (f) From the seed body. (g) From the seed wing. EX, exotesta; MS, mesotesta; EN, endotesta. (Drawings courtesy of Scott Mori.)

maturity are of current interest. Changes that take place during germination are also studied, as are changes under storage conditions that could lead to deterioration of the seed and loss of viability.

The conditions required for germination are so specific and specialized for some seeds that elaborate experiments have to be conducted in order to discover them. Parallel anatomical studies can help in the interpretation of the results.

CHAPTER 8

Adaptive features

Introduction

The relationship between plant structure and the environment in which the plant grows was a subject that fascinated the early plant anatomists, and continues to be of great interest today not only to those working in anatomy and histology, but also to physiologists, ecologists, plant breeders, biochemists and molecular biologists.

Before considering the apparent adaptations to habitat, it is worth remembering that there are basic mechanical considerations and special adaptations that influence the form of leaves and stems. Added to this, family characteristics are also frequently expressed in their members, unless they are positively disadvantageous and might have little to do with habitat preference.

Mechanical adaptations

In Chapter 1, a detailed account is given of the mechanical systems found in plants. Here, a brief reminder is given of the principles, with some information on the mechanical attributes of some adaptations. In leaves of most land plants, the xylem has relatively little mechanical strength, and in most species the vascular bundles are accompanied by bundle sheaths or rod-like arrangements of fibres, sclerified parenchyma or collenchyma. The mechanical tissue may be accommodated within the thickness of the leaf, so that both surfaces are smooth, or it may produce prominent ridges above or below or both. The arrangement of veins in leaves is very varied (see Chapter 6). In the monocotyledons and some dicotyledons, many species have strap-shaped leaves, with the main veins parallel to one another. Grasses are typical examples. The centrally placed vein may be the largest. The axial veins are connected at intervals by transverse veins, which are in general narrower than most of those of the axial system. This produces a net-like arrangement, with the softer, non-load-bearing green tissue suspended between the veins.

Some monocotyledons, such as many aroids, are more like the broad-leaved dicotyledons in leaf form. They have a petiole-like structure and an expanded flattened lamina. The veins in the lamina generally consist of a midrib which follows directly in line with the veins of the petiole, and a series of lateral veins. The first order lateral veins may depart from the midrib in a regular pinnate manner up its length, fairly evenly spaced, and extending towards the leaf margins on either side. They might all originate near the base of the lamina, and fan out towards the margins, as three, five or more branches. In some leaves, for example *Gunnera* and rhubarb, the ribbing is very pronounced, and parallels can be seen with fan vaulting in buildings. Cantilevers are common in nature. Some leaves are very large, and their strengthened, ribbed venation provides mechanical support with economy of materials, cantilevering the relatively delicate green tissue out into the light.

The arrangement of mechanical tissue in petioles is of considerable interest. Leaf arrangement on stems usually helps minimize self-shading, but the fine tuning is carried out by the petiole. In some plants, the petiole may assist the lamina in tracking the sun. The petiole also enables the lamina to twist and partly rotate in the wind, minimizing the damaging effects of wind on the thin, sail-like structure. The design is such that the elasticity in the system allows the lamina to return to its preferred orientation for light interception, when the wind falls. This remarkable self-righting flexibility comes about through the structure and properties of the mechanical tissue. The cross section can be U-shaped (like plastic guttering, which has similar recovery properties to twisting), with some variants, or it may consist of a cylinder, with thicker and thinner parts, or a cylinder with additional rods either internally, externally, or both. It is not so clear how those with closed cylinders function. All petioles have to be able to withstand vertical loading well. Clearly all types also withstand torsional forces, because they work.

At the petiole base and sometimes at the petiolule bases of leaflets, there may be an enlarged pulvinus which contains parenchymatous cells which when turgid hold the leaf up. When the internal pressure in these cells is reduced, the leaf and leaflets may hang down in a 'sleep' position. Wilting in such plants can have a similar effect, which causes the leaf and leaflets to present a reduced surface area to the sun.

In feathery, pinnate leaves, and other forms with several leaflets, wind can be spilled by movement of the individual components. Many palms have large leaves, which when immature are entire. During expansion, predetermined lines of weakness in the lamina split, and the mature leaf gives the appearance of being pinnate. In addition to strength provided by reinforced veins, palm leaves often contain fibre strands, and possess thick-walled epidermal cells. Coconut palm is a good example. This passive approach to survival is effective. The leaves may become more tattered, but they generally survive to function.

Thin leaves, with vascular bundles arranged in one row, often more or less equidistant from either surface, are often accompanied ad- and abaxially, by mechanical girders or strands. This makes them similar to **I** beams and girders in buildings and other structures. In manufactured **I** beams, the upper and lower flanges are more robust than the upright central part between them. Indeed, it is common to find holes to have been cut in the central part or for it to take on a lattice-like form, saving materials where they are not needed for mechanical strength. Not uncommonly in nature, the real strength is supplied in the 'flanges', and the tissues between them serve mechanically only to hold the flanges in their respective spatial arrangement. Commonly soft, turgid parenchyma cells may be found in such parts of the 'beam', but sometimes there may even be air spaces. Here we see an excellent parallel in the economy of use of materials in plants, and efficiency in engineering.

Tubes or cylinders have been described above for stems (see Chapter 5), and as a developmental stage as the stem ages or increases in diameter (see Chapter 2). It is common for the mechanical tissue to be concentrated towards the periphery, where it can provide considerable mechanical support, with an economy of material. Hollow trees are often a cause for concern, but providing the outer tissues are alive and flourishing, and sufficient wood remains at the points of branch insertion to hold them up, there may be little to be worried about. Tubes are commonly applied in engineering because of their particularly efficient use of material in relation to strength.

Climbers like *Vitis vinifera*, the grape vine, and *Clematis* species, often retain separate vascular strands, even though these become radially elongated as the stem thickens with age. Between these are radial bands of thin-walled cells. When the stem becomes compressed as it climbs, the thin-walled regions are deformed, but the main conducting cells in the vascular tissue do not compress and can continue to function effectively.

Adaptations to habitat

In the early history of studies on adaptations in plants to habitat constraints, correlations were made on a rather empirical basis. Plants found growing in dry conditions for instance have been studied and shown to exhibit anatomical modifications not normally associated with plants from more mesic localities. Without any attempt at experimentation, the authors of that period would ascribe specific properties to the structures they saw. For example,

Haberlandt's book, *Physiological Plant Anatomy* was written largely from observation with little or no experimentation, and must therefore be used with care. Many researchers subsequent to Haberlandt have adopted his ideas uncritically. Where people have taken the trouble to study the anatomy of a range of plants from one habitat, they have found some features which seem to vary so widely in expression, for example in thickness of the walls of epidermal cells, that their adaptive significance is put in doubt. There are, however, certain types of modification which crop up with such a degree of regularity, and in such taxonomically diverse plants, that they might really be related to survival in that particular habitat. Such features lend themselves to experimental studies in physiology and studies on genes that control both the development of the structure and its function.

Despite any adaptations found in the anatomy of plants which might be thought to be of 'ecological' benefit, it is normal for family or generic characters to be well expressed and often dominant.

Not all adaptations are evident at the anatomical and morphological levels. Some are physiological, and physiological races of plants have evolved which fit them for growth in extreme conditions. For example, some races of *Agrostis* species can grow in areas of high concentration of heavy metals (e.g. copper) where other plants fail. These adapted grasses have been shown to accumulate and immobilize heavy metals in their roots, preventing these metals from entering and damaging cells and organelles in other organs.

The duration of life of the plant might be a dominant feature, which helps a species to survive. Ephemeral species may grow in normally xeric conditions if they can germinate their seeds, grow, flower and fruit when water is available. During this short period of activity, the plant may have adequate water and would not need any other xeromorphic adaptations.

The issue is made more complex when it is realized that there are often many microecological niches even within a small area. Diversity in anatomy could relate to such differences, which are often hard to detect without prolonged study of the area concerned. Seasonal variability in environment may be overlooked by those making plant collections at particular times of the year. It boils down to the observation that if a species is found growing successfully under a given set of conditions, it is there as a result of selection, adaptation and ability to compete with other species for that niche.

Some of the main habitats and commonly associated plant modifications are outlined below. Despite the cautionary remarks that we have expressed above, it is often possible to find in plants anatomical features which do show a close correlation to the habitat type in which they normally occur and which are clearly a result of adaptation to special conditions and physiological needs.

Plants growing under very dry conditions normally show a reduction in evaporating surface area. When leaves are developed they may be small, or have various features which would appear to assist them in regulating or reducing potential water loss. Leafless plants, for example many Cactaceae and Euphorbiaceae, and others with non-functional leaves, for example many *Juncus* species and most Restionaceae (a family present mainly in the southern hemisphere, in low rainfall areas with mineral-poor soils in Africa and Australasia, and one species in South America), often have subspherical or more or less cylindrical stems modified to perform the photosynthetic and transpirational functions normally ascribed to leaves. A sphere has the smallest surface area possible for a given volume and cylinders also have a low ratio of surface area to volume.

Xerophytic plants may be divided into two categories, drought escapers and drought resisters. Only those that resist drought might be expected to have strongly modified anatomy, but even the escapers that survive as seed or bulbs or in reduced, leafless forms may have some adaptations for the semi-arid periods in which they are normally leafy.

The bulbous habit is often related to dry situations; flowers and leaves are present as aerial organs for a limited period each year, for example *Narcissus*, *Tulipa*, *Haemanthus*, *Scilla*. Swollen underground stems also occur, for example many Asclepiadaceae; rhizomes, for example *Iris* species; or corms, for example *Crocus*, *Watsonia*; woody xylopodia, in for example many Australian Anacardiaceae and Myrtaceae. As in ephemerals, these plants normally grow actively when water is available, and their leaves in consequence may show little adaptation to xeric conditions. If they are evergreens, like eucalyptus species, their leaves become tough and show xeric modifications.

In those plants which have persistent (perennial) leaves or stems, morphological and anatomical modifications are quite common. Stomata are often (but not always) sunken; they can have various antechambers and cutin-lined substomatal cavities which could play a part in water regulation. Certain *Aloe* and *Haworthia* species (Fig. 8.1) show some such modifications. The cuticle itself is often thicker in xerophytes than in mesophytes, but cuticle and epidermal cell wall thickness are not reliable guides to xeromorphy. Stomata may be very numerous and widely distributed, or they may be confined to grooves or channels in the leaf or stem. Some xeromorphic leaves are capable of inrolling (e.g. *Ammophila*, Fig. 8.2) and thus enclosing the stomata when dry conditions prevail. On the other hand, when adequate water is available, it has been shown that conifers with needle-like leaves can transpire as rapidly as mesophytes. Plants like many aloes, with thick cuticle, epicuticular waxes, and thick outer walls to the epidermal

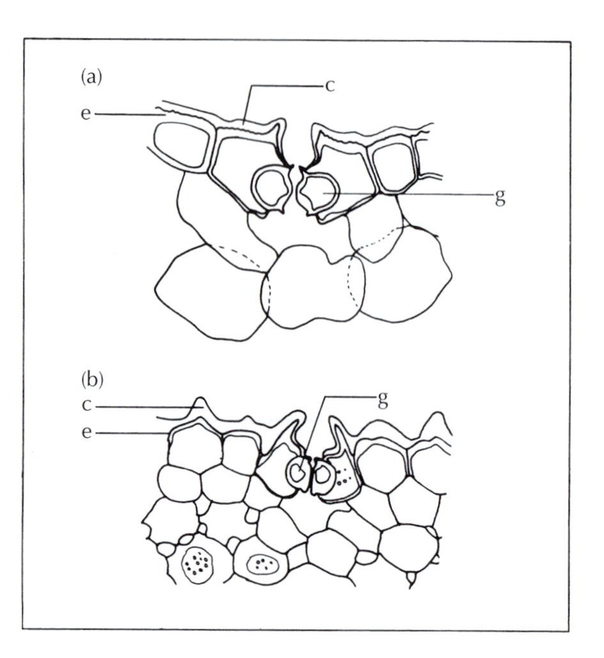

Fig. 8.1 (a) *Aloe somaliensis*, outer part of leaf TS. ×218. (b) *Haworthia greenii*, outer part of leaf TS. ×218. Note the sunken guard cells (g), the thick cuticle (c) and the thick outer wall to the epidermal cells (e). Both have succulent leaves, with little mechanical tissue.

cells, sunken and variously protected stomata seem to be quite able to regulate and minimize water losses during dry periods. The raised rim forming a suprastomatal cavity above each stoma may have a function enhancing evaporation when growth conditions are good. The structure could have a venturi effect, lowering pressure above the stoma and assisting transpiration under suitable conditions. In plants like these that have extensive protection from water loss, it may be that even when water is in adequate supply it is difficult to maintain enough evaporation from the leaves to drive the transpiration stream. In these circumstances, if the stomata at the bottom of the chimneys are open, external air flow above the chimneys could cause reduced pressure in them, enhancing the flow of water vapour from the plant.

Some plants, for example *Elegia* (Restionaceae), can grow in areas of adequate groundwater supply, but where strong drying winds could cause excessive water loss. These rush-like plants are not damaged physically by the strong winds since they are leafless and flexible. Many of the Restionaceae and some of the Juncaceae are remarkable in showing xeromorphic features in the stems but hydromorphic features in the roots. There is abundant mechanical tissue, usually sclerenchyma or other lignified cells, in the stems, but large air cavities in the cortex of the roots. It seems that the stems

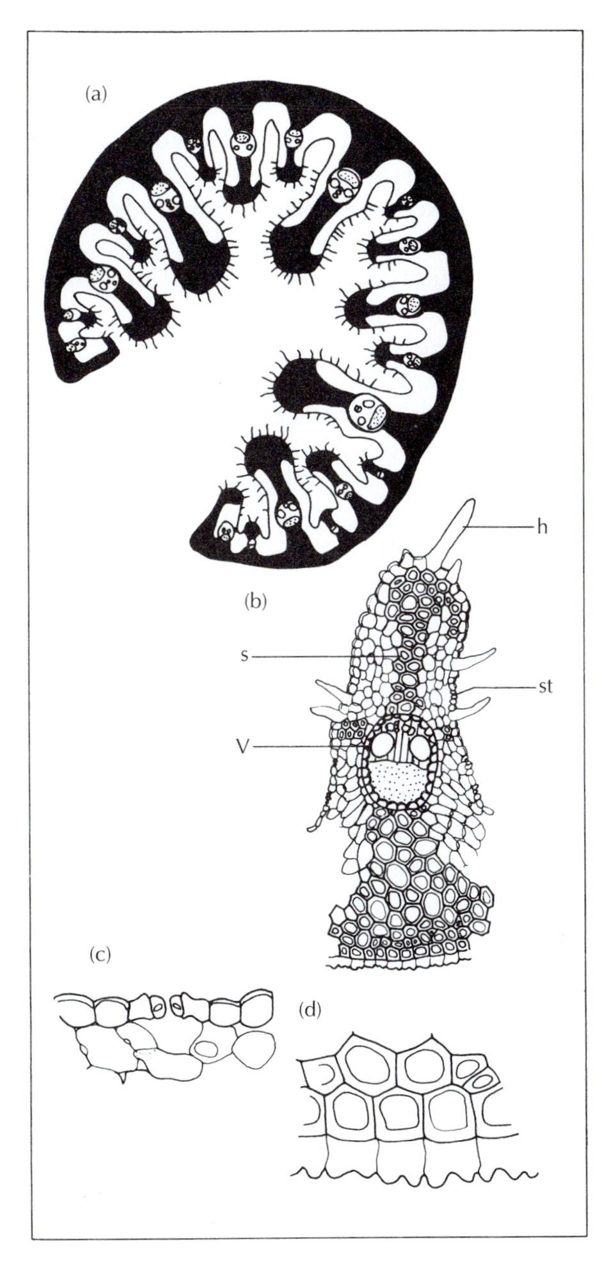

Fig. 8.2 *Ammophila arenaria.* (a) Low power, plan leaf TS. (b) Detail of rib (black areas represent thick walled cells). (c) Adaxial epidermis with stoma. ×300. (d) Abaxial epidermis with very thick cuticle. ×300. s, sclerenchyma; V, vascular bundles; h, hairs; st, stoma.

can be exposed to strong drying winds when it is probably too cold for the roots to deliver enough water to meet evaporation losses. The roots themselves often grow in waterlogged soils or standing water. Consequently, when conditions for transpiration and root action are satisfactory, the root adaptation is probably beneficial.

Internal adaptations in xerophytes may take one of two main forms: they can store water, and the plants are then described as being succulent, or they can provide structural rigidity, with the ability to resist collapse and tearing on drying, and then the plants might be described as sclerotic. Tearing and disruption of tissues is one of the main causes of permanent injury resulting from excessive desiccation. Chlorenchyma enclosed in rigid, lined channels is less likely to be torn than that which is in an unprotected, relatively unstiffened mesophyte leaf. In succulent plants there is very little, if any, mechanical tissue, and the xylem of the vascular system is usually not strongly thickened. Many crassulas (Fig. 8.3), aloes, etc. are of the succulent type, and *Hakea*, *Leptocarpus* (Fig. 8.4) and *Ulex*, gorse, are of the sclerotic type, as is *Ecdeiocolea* (Fig. 8.5).

Some *Haworthia* and *Lithops* species show only translucent leaf tips above the ground level. The rest of the leaf is buried, but contains chlorenchyma and water-storing mesophyll. These are often called 'window' plants. Light is able to penetrate to the photosynthetic tissue through the translucent cells which function like optical fibres.

The association of some other characters with xerophytes is much more dubious. Hairs have been supposed to help reduce surface wind speed and hence evaporation rates, but hairiness is often much more of a family character. There are many xerophytes belonging to families in which hairs are rare that manage quite adequately without them. Thin-walled hairs could easily increase water loss under some conditions, but most hairy xerophytes, for example *Gahnia* species, *Ammophila* and *Erica* species, have hairs with thickened walls and some also have thick cuticle. Extreme hairiness is found in the plants of high tropical mountains, subjected to extreme diurnal temperature fluctuation. Some plants, such as mesembryanthemums, have balloon-like hairs, which when fully turgid allow gas exchange to occur between the air and the stomata beneath them. If water is short, the hairs partly collapse, press against one another, and effectively block gas exchange and water vapour loss. Some hairs that do function as an insulating layer, reducing water loss in dry, windy conditions have a waterproof layer of suberin-like material built into their cell walls near the hair base (like a Casparian thickening). This effectively prevents water loss from the body of the leaf through the hairs which might otherwise leak out by a passive wicking effect. In the other extreme, hair cells might have thin walls, and thus be vulnerable to water loss – but equally capable of taking up water from mist or moist air. Hairs like these are common on epiphytes growing high in the canopy of trees in cool mist forests. The roots of epiphytes serve

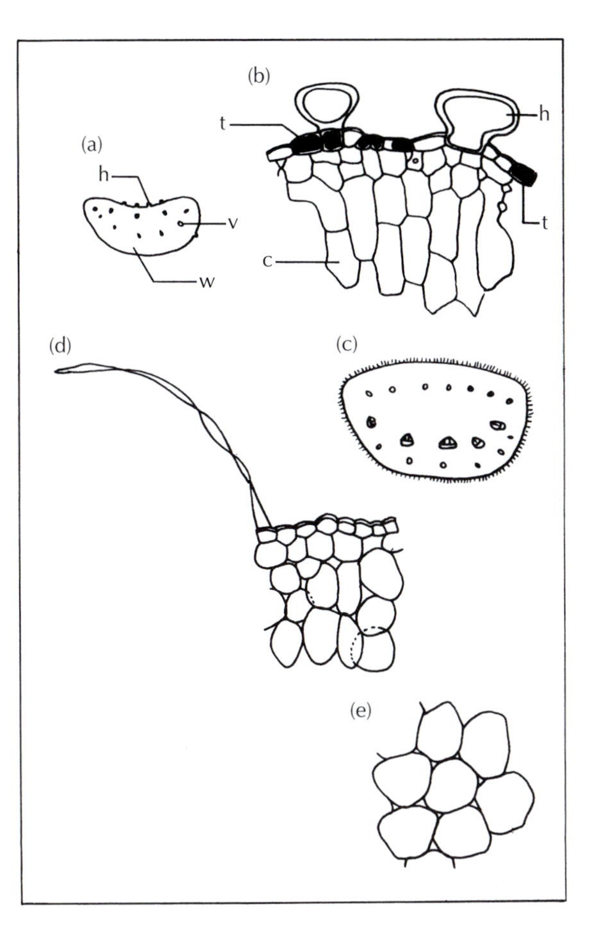

Fig. 8.3 (a,b) *Crassula* sp. (c–e) *Senecio scaposus*. (a,c) Plan TS leaf; mechanical tissue absent, central mesophyll cells store water. (b) Detail of outer part of (a). (d) Outer part of (c). (e) Central part of (c). (b,d,e) ×54. c, chlorenchyma; h, hair; t, tannin; v, vascular bundle; w, water storage tissue.

principally for anchorage; water is taken in by the leaves, aided by hairs like these when they are present.

There are numerous mesophytes with hairs.

The fine texture of the leaf or stem surface can have marked effects on the flow pattern of air above it. The boundary layer is the layer of air immediately next to the surface. It is thinner over a smooth than a rough surface. The boundary layer consists of relatively still air, but surface roughness can lead to turbulence in air flowing over it. Water vapour loss through stomata occurs more rapidly when the boundary layer is thin. Species habitually growing in humid environments, sheltered from the wind, tend to be smooth surfaced. In the more exposed environments, for example, hot, rocky hillsides, it is common to find leaves with quite rough surfaces, the roughness increased by prominent flakes of surface wax.

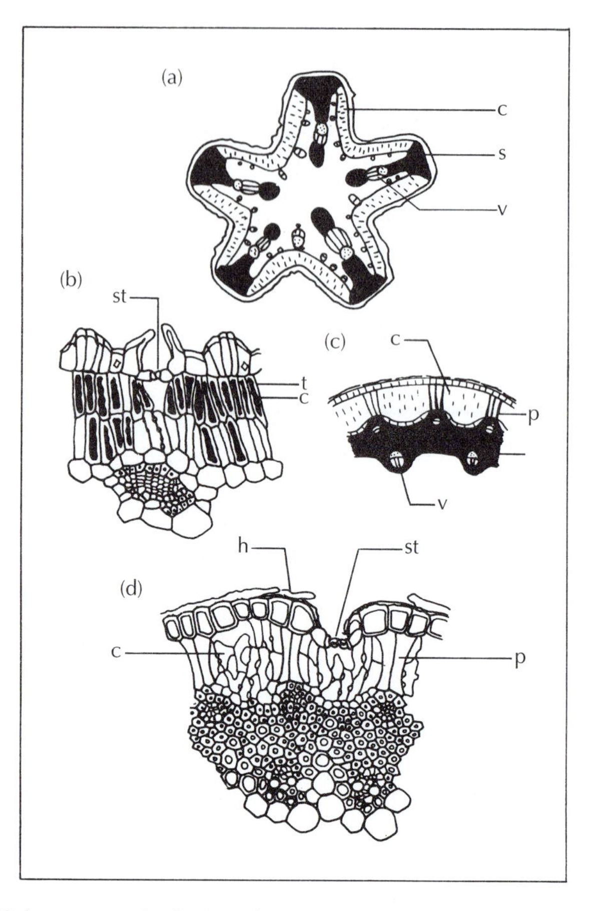

Fig. 8.4 (a,b) *Hakea scoparia*, leaf TS. (c,d) *Leptocarpus tenax*, stem TS. Note sunken stomata (st) in both and abundant strengthening sclerenchyma (s). Hairs (h) cover the *Leptocarpus* and tannin (t) is present in the chlorenchyma of *Hakea*. The pillar cells (p) in *Leptocarpus* divide the chlorenchyma into longitudinal channels. c, chlorenchyma; v, vascular bundle. (a,c) ×15; (b,d) ×120.

Increased surface roughness can be brought about by leaf form, prominent veins and corrugated laminas. Often it is contributed to by micro-characters as well. In smooth-leaved plants epidermal cells have flat outer periclinal walls. Even a slight doming of these walls can increase surface roughness, and if the walls are developed into papillae, they have a marked effect on the boundary layer. Of course hairs, discussed above, introduce a new dimension to surface roughness. The outer walls of epidermal cells may exhibit fine striations and micropapillae. The size of these features, and their occurrence, which is apparently correlated with plants exposed to wind and heat stress, suggests that they play a part in modifying the boundary layer. They may also increase light and heat scatter.

The edges of leaves, their margins, can be entire, or variously toothed (dentate). Dentations have marked effects on reducing turbulent air flow,

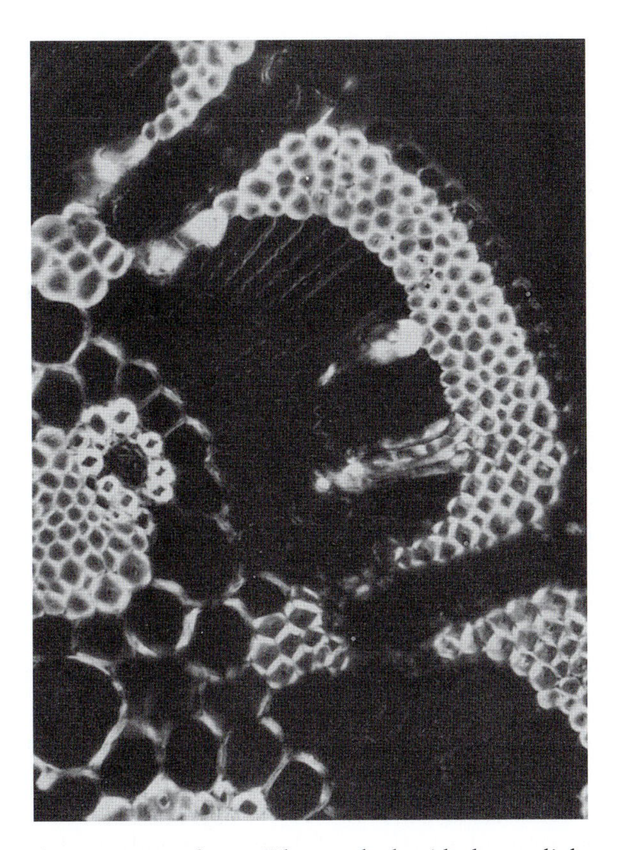

Fig. 8.5 *Ecdeiocolea*, outer part of stem. Fibres and sclereids show as lighter cells. This xerophyte has a deeply grooved stem, with stomata on the flanks of the grooves. Note the strong development of hypodermal fibres to the outer side of the thin-walled chlorenchyma. TS in polarized light, ×550.

and are thought to help maintain the integrity of the leaf margins in strong winds.

On the mountain tops, species in less dense air are subject to high insolation and have to contend with very high levels of damaging UV light. Chloroplasts can bleach and be rendered ineffective if they intercept too much UV light. It has been found that some species have evolved natural sun screens, often involving polyphenols, filtering the UV. These substances may be in the epidermal cells, or in a layer or two of otherwise colourless cells beneath the epidermis but above the chlorenchyma. While in the cells the substances may be translucent and colourless, but sometimes they are yellow to brown, and could have a direct blocking effect on light, reducing its intensity. There is considerable commercial interest in finding out how these chemicals work, and their structure.

Many xerophytes have a hypodermis, the cells of which are thick-walled (Fig. 8.5).

Compact mesophyll, with few air spaces, has also been thought of as a xeromorphic character (e.g. *Pinus* species, with plicate mesophyll cells), however, many succulent and sclerotic xerophytes have mesophyll or stem chlorenchyma that has abundant air spaces (e.g. *Laxmannia*, *Hypolaena*). Experiments are needed to determine the significance of high volumes of internal atmosphere which can be a feature of both xeromorphic and hydromorphic plants. Could the function be the same in both types of plant, because they could each be growing under conditions which permit only a low rate of flow in the transpiration stream?

There are certain montane parts of the world where little, if any, surface water can be seen for many months of the year, and where soil water is often frozen. The cushion plant is the characteristic life form in such places. From observation of the compact habit, reduced leaf surface area, short internodes, widely penetrating roots and slow growth rate, it might be thought that the anatomy of all such plants would conform to a xeromorphic type. However, this has not proved to be entirely true. Some species, such as *Pycnophyllum molle* and *P. micronatum* of the Caryophyllaceae, do show the expected adaptation. They have extreme reduction of the leaf surface area, leaves are closely appressed to the cylindrical stems, and sunken stomata are present only on the adaxial protected surface, amongst papillae. But other species, like *Oxalis exigua* (Oxalidaceae), have very little apparent xeromorphic modification. *Oxalis exigua* has hairs and papillae on the leaves, but the stomata are superficial. The chlorenchyma of the leaves is not compact. Its leaves are similar to those of the mesic members of the genus. The stem and roots exhibit an interesting modification, probably useful to a plant which has to penetrate the cracks between rocks. There is no interfascicular xylem or phloem produced during secondary growth and the vascular bundles remain separate. Cambium in the interfascicular regions produces parenchyma. Apparently, the roots and stems can twist and deform without undue compression of the vascular supply, as seen in many lianas.

Azorella compacta (Umbelliferae) shows anatomical features which appear to be related to the harsh high montane environment. The leaves are small and very shiny (thus reflecting ultraviolet light). Contractile roots are present which help to keep the plant firmly anchored, despite frost heave. The vascular bundles are separate, as in *Oxalis exigua*. Resin ducts, a family characteristic, are frequent. *Anthobryum triandrum* (Frankeniaceae) is also well adapted to cold, drying conditions. The leaves are furrowed, with stomata confined to the furrows. However, the vascular cylinder is compact, not composed of separate bundles. So, once more we have evidence that some 'family' characters can be conserved in much modified and reduced plants, and that various species with diverse anatomy can cope with a particular set of environmental conditions.

Halophytes often show succulence which is normally associated with dry conditions; they grow in saline areas where there is, in effect, a physiological drought. Although surrounded by water, the roots have to extract it from the soil against a considerable suction force. (This point has application in liquid feeding of plants through a gravel bed in glasshouses. The gravel has to be periodically flushed with salt-free water or the salt concentration becomes high enough to dehydrate the plants.)

Plants growing in soils which freeze for part of the year are also subjected to 'drought' conditions. Many conifers in such habitats have needle-like leaves and the ability to regulate water flow adequately, both in conditions of adequate and inadequate water supply.

The sap of plants which commonly flower and produce leaves before the snow and frost have gone is often of a mucilaginous nature, and acts as a kind of antifreeze.

Ribbed columnar cacti are perhaps the best example of adaptive changes in morphology and anatomy to meet functional needs in a xeric environment. Their shape as a fluted column allows the stem to contract and expand in response to water loss and uptake without damaging any cells. That is the distance between ridges changes with water content, the ribs are closer when water stress is happening and farther apart when water has been accumulated. This flexibility is facilitated by the presence of hypodermal collenchyma. Also the shape allows for direct sunlight for only a brief time each day, thereby avoiding overheating of the tissues while the ribs themselves maximize reflected light. The spines act as shade along the ribs. The thick layer of epicuticular waxes maximizes reflection of damaging ultraviolet light. The ribs act like radiators in dissipating heat, as do the spines which are modified leaves. During cool nights the ribs and spines are areas where moisture condenses and drips and/or runs to the roots at the base of the plants. So spines not only can prevent herbivory, but they can also provide shade, act as radiators and help collect moisture.

Mesophytes

Mesic conditions are suitable for broad-leaved plants with fairly soft, thin or somewhat sclerified coriaceous (tough) leaves. In temperate or tropical submontane zones, many mesophytes pass the winter months in a leafless form, either as deciduous trees or perennial herbs, and their buds have bud scales. The fringes of the mesic zones tend to have a higher proportion of evergreens with coriaceous leaves. There is a gradation from mesic to xeric conditions in many areas, and plants showing adaptations to both situations may grow side by side. Mesophytes tend to have anatomical variations that are related more to the family to which they belong than to the environment in which they grow. In many ways the anatomy of mesic plants is considered the norm from which xerophytes and hydrophytes depart. As a result the latter are usually compared to the former.

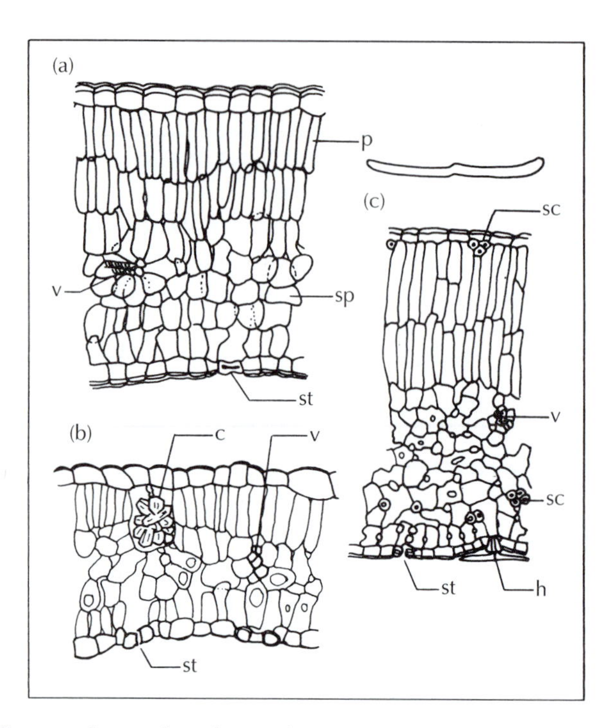

Fig. 8.6 Small parts of mesophyte leaves (lamina) in TS. (a) *Arbutus unedo*. ×109. (b) *Corylus avellana*. ×120 (c) *Olea europaea*. ×109. c, cluster crystal; h, hair; p, palisade; sc, sclereid; sp, spongy mesophyll; st, stoma; v, vascular bundle.

Consequently, it is difficult to generalize about the anatomy of mesophytes. The epidermal cells frequently have only moderately thickened outer walls and a thin or slightly thickened cuticle. The stomata, normally confined to the lower surface, are usually superficial. The mesophyll normally consists of one, two or more layers of closely packed palisade-like cells. Cells of the inner layers may be the least densely packed, and border the loosely arranged spongy mesophyll. Sclerenchymatous tissue is absent or sparse, and may be represented by a small number of sclereids. Sclerenchymatous sheaths to vascular bundles are rare, except in relation to larger primary veins, midrib or petiole. See Fig. 8.6 for examples of mesophyte leaves.

Tropical rain forests are adequately supplied with water. The dominant life form is the very tall tree. Leaves are often long-lived, because there are few seasonal stresses which might necessitate the adoption of a regular deciduous habit. Some rain forest trees continuously lose and replace leaves. Others may be deciduous every few years; these often flower before the new leaf growth (e.g. the Bignoniaceae). The leaves are often relatively tough (coriaceous) but have large surface areas. Many have an extended 'drip tip'. Bud scales are rarely developed.

The relative humidity inside the canopy of the rain forest is normally approaching 100%. Many epiphytes grow in the relative shade. Some, for example certain Bromeliaceae, may have a construction by which leaves funnel water to the centre of the plant where it is held in 'tanks' formed by leaf bases. The normal roots of these and other epiphytes are simply anchors, and do not extract nourishment from the plants upon which they grow. Many of the epiphytic Araceae and Orchidaceae have special aerial roots with modified enlarged epidermal and cortical tissues (velamen) which can absorb and retain atmospheric moisture (see Fig. 4.1). In the shelter of the large trees, understories of shorter trees flourish, often with frond-like leaves of great length.

Plants living on forest floors, in dense shade, have to survive in conditions almost diametrically opposite to those to which xerophytes are subjected. They have to be extremely efficient in light collection. The relatively humid, windless environment enables species with large, thin, relatively unprotected leaves to thrive. They are often very chlorophyll-rich, and appear dark green. A number of the epiphytic Gesneriaceae in the Old World tropics have a particularly interesting and as yet unexplained adaptation. The upper epidermis of the leaves is multilayered, and consists of colourless cells. In some species it may make up to two-thirds of the total thickness of the leaf. The chlorenchyma is relatively thin, with a conspicuous layer of widely spaced palisade-like cells and some spongy mesophyll (Fig. 8.7).

Shafts of sunlight may penetrate the leaf canopy above; their angle changes during the day. Light reaching a leaf surface at 90 degrees is less likely to be reflected than oblique light. Many of the species adapted to this environment have means of presenting to light a high proportion of their surface at 90 degrees. Some lower plants, such as *Selaginella* species, achieve this by having highly domed epidermal cells, containing chloroplasts. As the sun tracks, each cell can present part of its surface normal to the source. Other species, for example from among the begonias, have a surface in which the epidermal cells develop together to produce a series of multicellular domes with much the same effect. Undulating leaf surfaces are also common.

In general the leaves themselves are thin. This enables the chloroplasts to be presented in one to three cell layers into which the light can penetrate. Each layer effectively shades the next one in. Most interestingly, when viewed from the underside, many of these leaves adapted to low light interception appear purple. This is because they have an inbuilt sheet of anthocyanin-containing cells which act as reflectors. The anthocyanin may be in the lower epidermis, but more commonly it is in an inner layer, immediately below the chlorenchyma. The chlorenchyma above this layer is usually itself only one or two layered. This allows what little light there is to penetrate the chlorenchyma fully. This means that light striking the

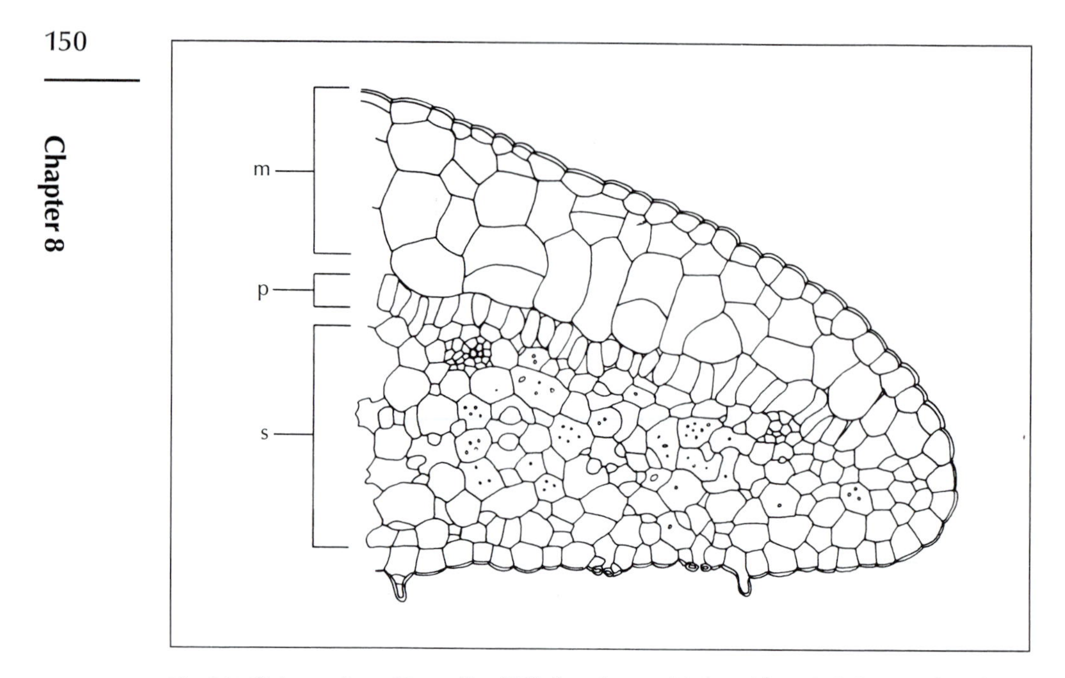

Fig. 8.7 *Codananthe* sp. Part of leaf TS showing multiple epidermis (m), a single palisade (p) and a large quantity of spongy mesophyll (s). ×102.

upper surface of the leaf, which would otherwise pass through the leaf, is reflected back for a return visit to the chlorenchyma. This parallels the heat/light reflecting membranes applied to windows in buildings, but makes efficient use of the reflected energy. Examples include members of the Commelinaceae, Marantaceae and Gesneriaceae.

Many of the epiphytic Bromeliaceae have numerous hairs and scales on their leaves, thought to be capable of absorbing water from the very humid atmosphere in which the plants grow.

Apart from the root modifications and the somewhat coriaceous leaves of the upper storey trees, rain forest angiosperms appear to have more anatomical characters relating to the families to which they belong than to the environment in which they grow.

Hydrophytes

Hydrophytes, plants growing immersed in water or with the leaves floating and, perhaps, aerial inflorescences, show many anatomical features that clearly relate to their habitat, and in some instances family characters are so reduced as to be difficult to define.

Most stems and leaves have large air spaces between layers of internal tissues. These assist in buoyancy and also gas exchange. Many aquatic plants

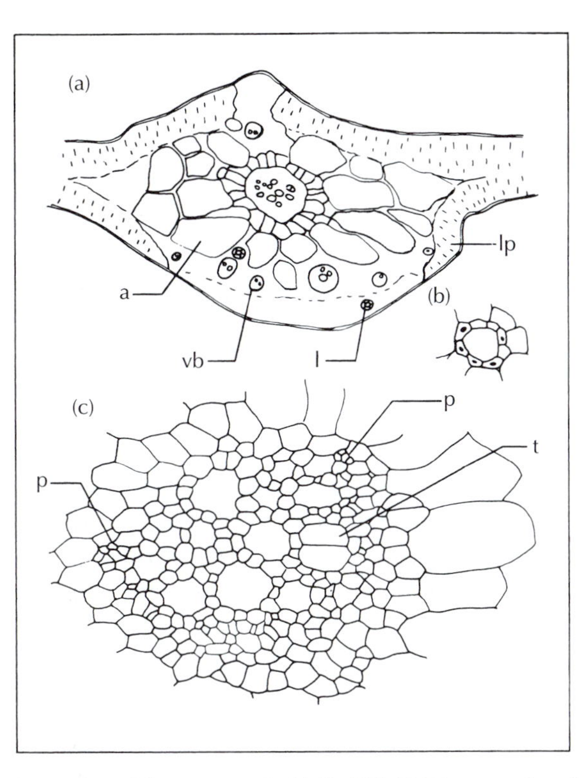

Fig. 8.8 *Limnophyton obtusifolium*, part of midrib, TS. (a) Diagram showing large air spaces around central vascular complex. ×15 (b) Laticifer. ×110. (c) Central vascular tissue. ×200. a, air space; l, laticifer; lp, loose palisade chlorenchyma; p, phloem; t, tracheary element.

contain such buoyancy tanks. The internal septa (walls bounding the air cavities), although multicellular, are often very thin. Individual cells in the septa are frequently star-like (stellate parenchyma), efficient in the use of materials, and allow gas flow from chamber to chamber. In bulrush, *Typha*, species, the upright leaves are divided into chambers similar in function to bulkheads in supertankers. They are an excellent example of economy of use of materials in producing a tall, mechanically strong structure. The upright, flattened blades are twisted. This adds to their strength, and improves wind spillage, which reduces the impact and potential damage of strong winds. Buoyancy in floating aquatics may be achieved by using trapped air round the surface of the leaves, rather than by having air-filled chambers. Water lettuce, *Pistia*, for example, has a layer of closely arranged, hydrophobic hairs on its upper surface making it difficult to wet or sink.

Cuticle is poorly developed or absent. Stomata are usually absent from submerged surfaces, and are often present on the upper surface of floating leaves. Vascular tissue, particularly xylem, is poorly developed and sclerenchyma is normally absent (Fig. 8.8).

Morphological adaptations include the reduction or absence of lamina or very linear leaf form in the submerged leaves of plants growing in running or tidal water, for example *Zostera, Posidonia*.

Plants growing in acid bogs have particular problems to overcome, particularly since mineral concentration is low in the water, and the availability of nitrogen may become a serious problem. A number of plants from different families have developed anatomical features that help them to survive such conditions. Among these, the animal trapping (so-called 'insectivorous') plants are of particular interest. All have specialized glandular hairs on the leaf surface, for example *Pinguicula, Drosera*. These hairs may be of two types, stalked ones secreting very sticky substances which trap the victim, and the other, sessile, secreting digestive enzymes. The leaf gradually rolls over to enclose the animal and opens out again when digestion and absorption is complete. Another, *Dionaea*, has sensitive trigger hairs on the lamina, three on either side of the midrib. The hairs are hinged at the base. They require two or three tactile stimuli to cause the leaf to fold shut vigorously. Marginal teeth mesh together, forming a prison from which the prey cannot escape. Reddish, glandular hairs then secrete digestive enzymes and absorption follows. Specialized hinge or motor cells are present along the midrib.

Applications

The application of information about anatomical modification in plants developed in response to various environments may at first sight seem obscure.

The morphology and anatomy of a plant can give horticulturalists a good guide as to the sort of growing conditions they should provide. Take for example an orchid with conspicuously swollen leaf bases, indicating the facility for water storage, and with aerial roots. It would be clear that the plant was an epiphyte which required support on a branch or log, and that it would need a very humid warm atmosphere. On the other hand, a plant with a rosette of thick, succulent, closely packed leaves with transparent tips would clearly be xeromorphic, would need to be planted deep in a quickly draining soil or compost so that the leaf tips were level with the surface. It would need bright light and probably a period of each year with little or no watering. It would need to be protected from frost, and may need additional heat.

The taxonomist also finds anatomical data of importance when dealing with plants from different families that have made a parallel response to a given environment, producing a similar morphology. A number of the monocotyledon families are like this. Because it is quite common for the anatomy to have retained some characters that are diagnostic for the family, these can be applied to solve problems of affinity. Both xerophytes and some hydrophytes are amenable to this type of study.

The plant breeder might find it worth looking at the anatomy of wild relatives of crop plants if he wishes to integrate some drought resistance, or extra structural rigidity, for example, into the crop.

It is clear, then, that some of the anatomical features that are apparent in plants are modified to an extent in relation to the environment in which the plants grow. No generalizations or sweeping statements can be made, though, and each species must be assessed on its own merits.

Several key aspects of adaptation that are quite commonly dealt with in undergraduate classes are explored in the CD-ROM, *The Virtual Plant*.

CHAPTER 9

Economic aspects of applied plant anatomy

Introduction

Many of the applied aspects of plant anatomy have been referred to in the previous chapters, but some do not fit well into descriptive text. We have therefore amplified some of these examples in this chapter and introduced new ones drawn from experience at our laboratories. In writing this chapter, we have had to be very selective – a whole book could be written on this subject alone – but we hope that the following interesting examples will serve to show the wide range of applications to which knowledge of plant anatomy can be put. Anatomy is particularly useful in taxonomically identifying disassociated plant parts whether those parts be leaves, roots, stems, fruits or seeds of living or fossil plants.

Identification and classification

It is not always appreciated how important it is to be able to give the correct name to a plant. Cytologists, geneticists, ecologists, plant breeders, chemists and anyone using plants for medicine, food, furniture, fabric or building material, or those conducting molecular research on plants, must be able to identify their source material or they may not be able to continue with their work. They would not know if further plant specimens or timbers were from the same species that they started with; their results and applications would be unpredictable; and the foundations for sound scientific botanical research would be undermined. Identification depends on a stable, logical, usable and basically sound system of classification. At present, many plants can be identified adequately if all organs, for example flowers, fruits, leaves, etc., are present. The traditional herbarium methods can then be applied. However, there are very large numbers of plants which have been classified using macromorphological features alone.

A more natural, accurate and reliable classification results from also taking into account features of morphology, anatomy, palynology, biochemistry, population studies and so on. This ideal can rarely be attained, but once the 'alpha' taxonomy of a family has been studied, the synthetic approach should be used for revisions, as has been done now for a considerable time. Should revisions based entirely on hand-lens studies of herbarium material be outlawed? Certainly not, but on the other hand once done the incorporation of anatomical and other data may well lead to better and easier identifications and classification.

Taxonomic application

Systematic anatomy has a long history. From the early days of microscopy it has fascinated people who saw first the wide range of variations in plant anatomy, then began to recognize patterns of similarity, and eventually realized that in many instances, plants sharing large numbers of anatomical characters in common were probably closely related. This led to a series of attempts to work through the plant kingdom in an orderly way, and record what was present. For the angiosperms, in particular, this process led to the production of series of books. The work was started in earnest on the dicotyledons, and when the first review was done the monocotyledons were started. In the present day, the work on dicotyledons is being revised, and completion of the first run through the monocotyledons is in progress. There is an enormous number of papers on angiosperm anatomy, and before setting out on new work it is sensible to find out what has already been published. In addition to the normal search engines, it is worth looking at the Plant Micromorphology Bibliographic Database on the website of the Royal Botanic Gardens, Kew. This is important, because it goes far back into the literature and is kept up to date. Quoting from the web site 'This is a unique bibliographic database maintained by the Micromorphology Group (Anatomy Section and Palynology Unit). It contains over 95,000 articles and is probably the most comprehensive computerized index to higher plant micromorphology in existence. It covers most work published on plant anatomy in the twentieth and twenty-first centuries, and is regularly updated with new literature. All aspects of angiosperm and gymnosperm plant structure are covered, together with vegetative anatomy of pteridophytes. Common subject areas include ontogeny, ultrastructure, techniques, palaeobotany, embryology and seed anatomy. Internet searches are available free of charge for scientific purposes. (If the searches are used in a publication, the "Micromorphology Group, Royal Botanic Gardens, Kew" should be acknowledged.) Contact: pa.database@kew.org.

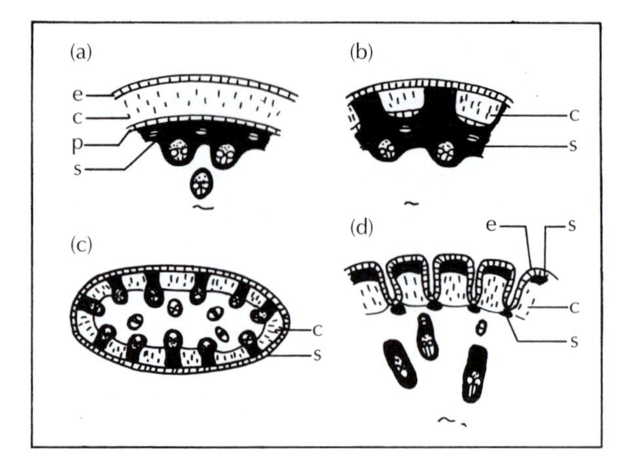

Fig. 9.1 Some differences between Restionaceae, Ecdeiocoleaceae and Anarthriaceae. (a,b) Restionaceae. Stem TS, most species have the general anatomy as shown in (a), with a continuous parenchymatous sheath; in some genera the sheath is interrupted by extensions from the sclerenchyma cylinder, as in (b). No vascular bundles occur in the chlorenchyma in all but one or two species. None of the species has hypodermal fibres or lacks a sclerenchyma cylinder as exhibited by Ecdeiocoleaceae (d). Anarthriaceae (c) differ in addition by having subepidermal fibre strands associated with vascular bundles; they may also have a sclerenchyma cylinder. Neither Anarthriaceae nor Ecdeiocoleacea has a parenchyma cylinder. c, chlorenchyma; e, epidermis; p, parenchyma cylinder (interrupted in (b)); s, sclerenchyma.

Anatomical data are easily applied to improving classifications and can often be used in making identifications. Take for example an instance where two new families were first recognized from their distinctive anatomy. The southwest Australian genera *Anarthria* and *Ecdeiocolea* were formerly treated as members of the Restionaceae. An extensive anatomical survey of the family Restionaceae showed the two genera to be misfits. Consultation with a classical taxonomist proved that there also were taxonomically valid distinctions at the macromorphological level. Co-operative research resulted in two new families being recognized, Anarthriaceae and Ecdeiocoleaceae. Figure 9.1 summarizes some of the main differences. These families have subsequently been shown to be chemically distinct, and molecular studies also support their separation. Thus, a hypothesis based upon macromorphology was tested using anatomy and other approaches and subsequently supported.

There are occasions when herbarium botanists find that it is difficult to ascribe a particular species or genus to a family, or where general affinities are suspected but there is insufficient evidence for them to place a taxon in a particular family. Here, additional anatomical evidence may be of help and

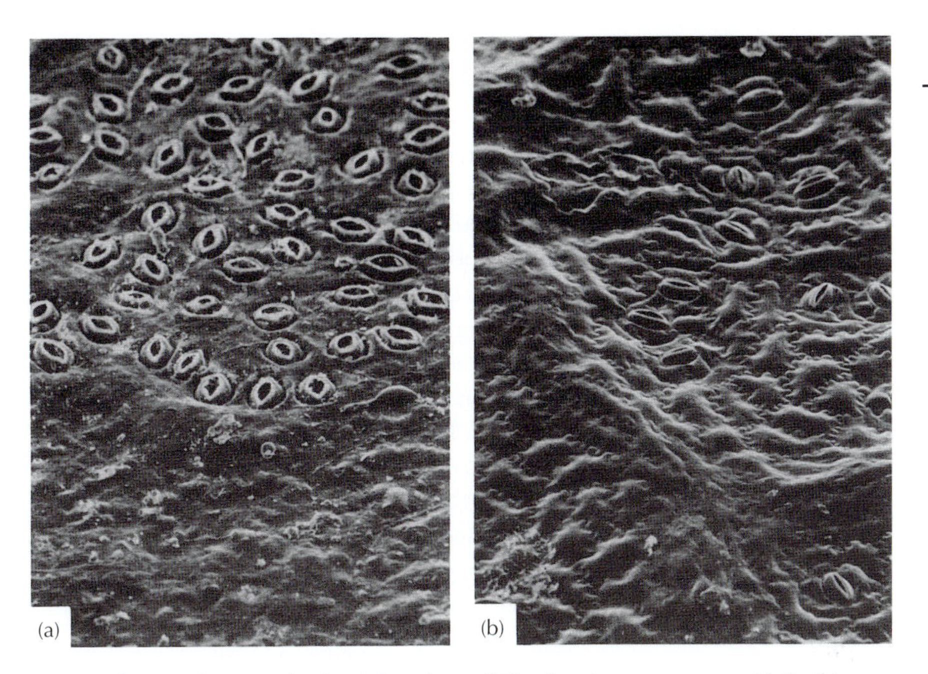

Fig. 9.2 Group of stomata in abaxial surface of *Eleutharrhena macrocarpa* (a). In (b), *Pycnarrhena pleniflora*, the stomata are scattered over the abaxial leaf surface. Both SEM, × 300.

there are many times when little extra helpful information comes from the anatomy. Recently, tree leaf material collected in China was examined anatomically; although there were no flowers or fruits on the herbarium sheet, the taxonomist thought that he knew the close relatives of the plant in question. The anatomy confirmed his views that the plant was *Pycnarrhena macrocarpa* Diels (Menispermaceae). A further study of species from this genus led to the discovery that two distinct genera were involved and a new genus *Eleutharrhena* was named by Forman to include *P. macrocarpa* using evidence from morphology, anatomy and palynology. In *Pycnarrhena*, stomata are scattered over the abaxial leaf surface; in the *Eleutharrhena* the stomata are in distinct clusters (Fig. 9.2).

Of course, the correct classification of plants is important, but it is often of more direct importance to know exactly to which species a specimen belongs. When flowers and fruits are absent, plant anatomists come into their own. Leaf fragments, wood and roots or twigs may have readily recognizable features which can be seen with a lens, but more often than not, identity has to be confirmed with the microscope. It is possible, for example, to check the identity of non-flowering aloes by looking at the leaf surfaces under the SEM. The appearance of the leaf surface in these plants can also indicate which subgroup they belong to.

Most of the drugs which are still extracted from plants come from leaves, bark, roots or rhizomes. Leaves often become fragmented and detached; bark, roots or rhizomes can be difficult to identify from their macroscopic appearance. The proper authentication of crude drug material is essential for standards of safety and quality to be maintained. For these purposes, accurate anatomical and morphological descriptions of the drugs have been published. The legal standards are found in such volumes as the British and European Pharmacopoeias and the British Pharmaceutical Codex, and those of other countries. In these books, the style of morphological and anatomical descriptions is very brief and to the point. Only those characters that will help to identify the material are given. Usually, these short monographs are carefully revised by a committee of experts. Herbalists are also aware of the need to have adequate control of the material they use, and work has been carried out to produce proper standards in reference works.

It is still often quicker to find out the identity of a crude drug (in the fragmented state) from its anatomy than from its chemistry. Importers of crude drugs are often experienced enough to know if they are buying pure material, or if adulterants are present. Sometimes samples will be sent for anatomical confirmation. For example Ipecacuanha, used in cough mixture can be adulterated with roots from alternative inferior species. Here microscopy can be used to give an indication of purity. The authentic source of the drug is *Cephaelis ipecacuanha* (Rubiaceae). Although rarely adulterated with other roots these days, there was a period when *Ionidium* (Violaceae) and other roots were regularly mixed in with the authentic material. Most of the adulterants have wide vessels in the xylem, whereas those in *Cephaelis* are narrow. The substitutes also lack characteristic starch granules, which are simple or, more usually, compound, with two or five or up to eight parts. The individual granules are oval, rounded or rounded and with one less curved facet, they rarely measure more than 15 μm in diameter. Sometimes *Cephaelis acuminata* is used as a substitute. This species is similar anatomically, but has starch granules up to 22 μm in diameter.

Sometimes closely similar substitutes are put on the market when the usual source of material is unavailable, for example, when Bolivian *Guarea* bark is difficult to obtain, and a substitute from Haiti is available. Microscopic study has shown that the substitute is from a different species, because the groups of phloem fibres are dissimilar but chemical tests prove it to be equally suitable for use. Occasionally the substitute may be poor and unsuitable. *Rheum officinale* root and rhizome is used medicinally, but *Rheum rhaponticum* is the vegetable. Fortunately, chemical and anatomical tests can be applied to detect which species is present. *Digitalis purpurea* and

D. lanata are used medicinally. They can be distinguished from one another on anatomical grounds, because the anticlinal walls of the abaxial epidermal cells are more beaded in *D. purpurea*.

Herbal remedies used as folk medicines from tropical parts of the world are often only available in fragmentary form. Those wishing to determine the identity of such fragments need to use anatomical methods.

Food adulterants and contaminants

Some herbs are used extensively as seasoning. These are often imported in the form of dried powdered plant parts, usually rhizomes, roots or leaves. Again it would be easy to introduce useless or sometimes even poisonous adulterants which would be difficult to detect with the naked eye. We have examined samples of dried mint, *Mentha* species, for purity, only to find considerable quantities of *Corylus* (hazel) leaf fragments included! *Ailanthus* leaf has also been used as a mint adulterant.

With the advent of the Trade Descriptions Act in the UK, manufacturers must state the contents of food products. It is essential for them to have adequate quality control, and to be able to identify all the materials they use.

Foreign bodies sometimes get into food by accident. Often these are small and fragmentary and can be identified only with the microscope. A splinter of wood in butter was found to come from a species of *Pinus*. The importer and packers hoped to be able to determine if the splinter could have come from the country of origin of the butter, or whether it might have been introduced during the packing stages. Buns and cakes containing sultanas periodically also contain other fruits which have become mixed with the sultanas during the drying process, when the sultanas are laid out in the sun. *Medicago* fruits are often involved. Some of these are prickly and unpleasant to eat! We have examined an object from a tin of baked beans which looked remarkably like a piece of a mouse. It turned out to be a piece of rhizome from the parent plant. It is often the case that odd-looking inclusions in food are only pieces of the parent plants.

Vitis, grape vine, stems have been found in currant buns, an *Avena* coleoptile, looking like a mouse tail, was present in a meat pie, and so on. Figure 9.3 shows an unsavoury looking shoot from a potato which occurred in a meat pie.

Starches from various plants have quite distinctive grain or granule features, so it is often possible to see if the stated materials have been used in a product unless the grains have become too hydrolysed (Fig. 9.4).

Animal feeds are made from the byproducts of other food manufacturing processes, or from seeds and fruits grown especially for the purpose.

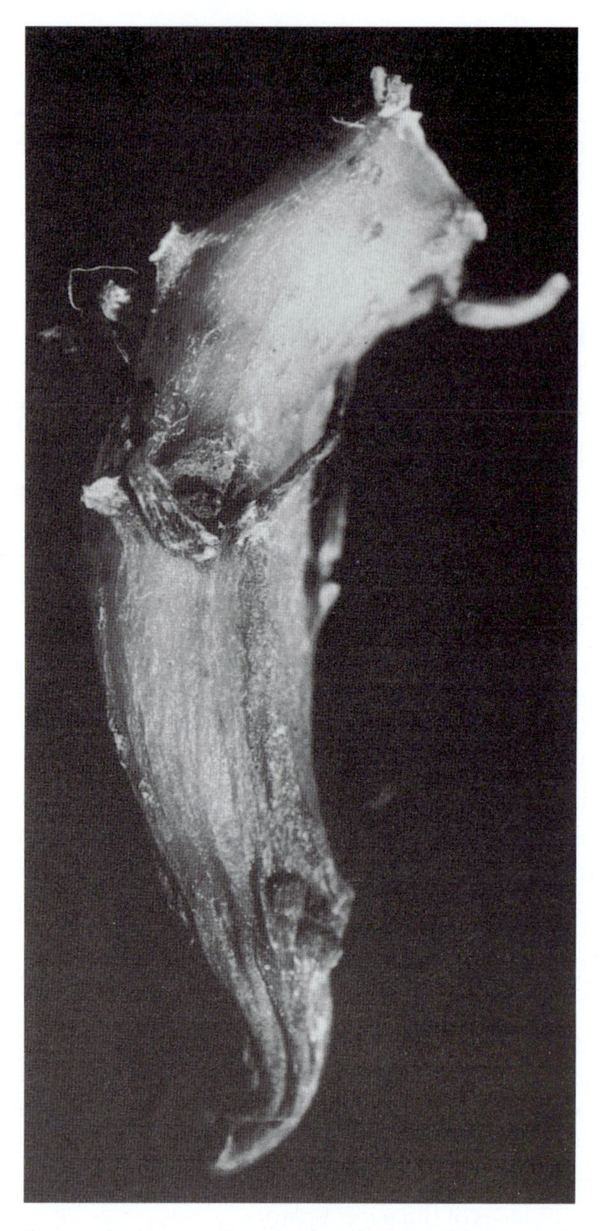

Fig. 9.3 Shoot of potato from meat pie, mistaken for something worse!

When ground as a powder the constituents are difficult to detect by methods other than microscopy. There is plenty of scope for adulteration in feeds, and careful microscopical quality control is essential.

There are also examples of marjoram being adulterated with *Cistus*. These impurities were readily spotted, because some hairs did correspond

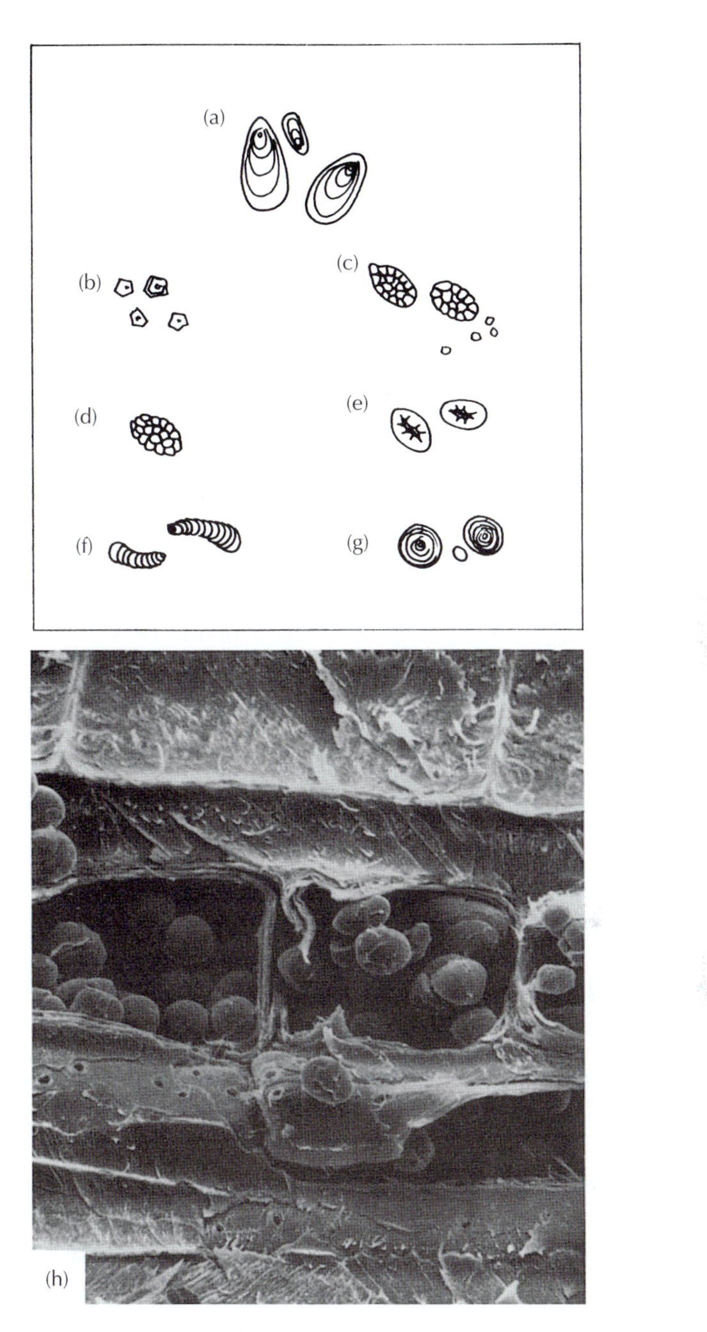

Fig. 9.4 Starch grains: (a) potato; (b) maize; (c) oat; (d) rice; (e) pea; (f) banana; (g) wheat. (h) Starch grains in xylem ray tissue of *Fabrisinapis*, SEM. (a–g) × 200; (h) × 3000.

to the species labelled while others corresponded to the adulterant, thus revealing a mixture.

Mucuna hairs, from the fruit pods, are very sharp and brittle, and contain an oil that is irritant. We came across them being used by a landlord who wished to evict a tenant. He had sprinkled them liberally in the blankets of a bed, causing the tenant to come out in a rash! A sample of the fine powder composed of hairs was sent to the New York Botanical Garden for identification because it was used as an irritant in mail to a judge. The sample turned out to be the specialized hairs, glochids, of the cactus genus *Opuntia*, which traditionally were used as itching powder. *Hedera helix* (ivy) hairs on a garment have been valuable in helping identify the scene of a crime in a recent murder case. In another murder case, fragments of partly decomposed oak (*Quercus*) leaves on shoes could be identified on a range of anatomical characters, including hairs. This with other evidence showed that a suspect had been near the place that a body had been found.

Tobacco (*Nicotiana*), together with other members of the Solanaceae, have rather characteristic glandular hairs. Some small cigars are enclosed in a paper made from macerated tobacco plant. In Great Britain, the law states that such cigars must be made entirely of tobacco. At Kew, we once looked at some so-called tobacco papers to ensure that only *Nicotiana* had been used. The presence of glandular hairs of the correct type was quickly established, and epidermal cells with sinuous walls were also found. However, we also discovered some hardwood vessel elements and softwood tracheids, and obviously other pulp had been added to strengthen the paper.

Animal feeding habits

Animal pests sometimes consume crop plants. It is often possible to find out what has been eaten by studying the composition of faeces, or stomach contents. A true estimate of potential losses can then be obtained. We have looked at faeces from rabbits, foxes, badgers, coypu, etc. and even millipedes! Of course the fragments of plant are very small when they have passed through an animal's digestive system. They are first fixed in FAA, and then washed in water. Then there is a sorting process, using a binocular microscope. Similar looking fragments are put into a petri dish, and the sample divided as far as possible into its components. Following this fragments from each dish are examined using temporary mounts under the light microscope. We always hope for good characters like silica bodies, hairs, stomatal types and so on. It is a big help to have a set of reference slides made from vegetation growing in the area from which the animal concerned was captured. It was suspected that some African cattle were being injured by eating grasses with sharp silica particles in them. The cattle only ate the grass concerned when other plants were unavailable. We examined the faeces and reported that there were silica bodies and sharp hairs present. Domestic animals occasionally eat poisonous plants, and we may be called upon to identify the fragments. The owner of the animals can then take precautionary measures against further livestock poisoning.

Wood: present day

Trunk wood

Most samples sent to Kew for anatomical identification consist wholly or mainly of wood. The samples are derived from many different sources and can be broadly divided into wood of recent origin and archaeological material. Furniture is made from woods carefully selected for their appearance and strength. Fashions have changed and it is common for certain species to have been selected for a period and then superseded by others. In addition, some woods were unavailable at certain periods. Consequently, by knowing which species were involved in the manufacture of antique furniture, it may be possible to date the piece, and occasionally the furniture expert may be able to get a good idea of who made it. Some craftsmen worked only with a carefully selected, characteristic range of woods. When repairs are necessary, it is also helpful to know which species should be used. The only way of being absolutely certain which woods were used is, in most instances, by making a microscopical study. Those who claim to be able to identify woods 'on sight' are either extremely experienced or over-bold, and many make errors.

The country of origin of carved wooden items can sometimes be established from the identity of the wood. Care must be taken because woods can be transported and then carved a long way away from their original sources. We have looked at items collected by Captain Cook on his voyages to try to determine where they could have come from, and this has proved successful. We once had for identification a wooden mask, carved in the likeness of a dog. This proved to be alder wood and its association with North American Indians was confirmed.

The Trade Descriptions Act has again provided problems for builders and manufacturers where woods are concerned. If they state that a particular wood has been used, this must be correct. The British Standards Institute has published a list of common names and the species from which the woods come, and this is the authoritative work which has to be followed in the UK. The only way to be certain that the correct wood has been used is to compare sections of it with those from a standard reference collection of microscopic slides. On one occasion a door said to be made of solid mahogany was brought to the laboratory. It turned out to be laminated, and no true mahogany was found in it – in fact the middle layer of veneer was birch.

Properties of woods related to structure have been mentioned in Chapter 3. We are occasionally asked to suggest substitute woods for some specialist purpose, when the supply of the normally used species has ceased. This can be difficult, but it is sometimes possible to suggest other species, which from their anatomical make-up might be expected to have similar properties.

Wood used as a backing for paintings, such as icons, is brought to the laboratory from time to time. The purpose in finding out the identity is often related to establishing the name of the artist, or the country of origin. We have examined the wood from a good many walking sticks; an amazingly wide range of species has been used for this purpose!

Preservation of wood is of considerable economic importance. A great deal of experimental anatomy is carried out in various parts of the world in order to establish the nature of the process of decay, the identity of the organisms involved and the prevention of their degrading activities. The 'sound' wood has to be very carefully examined and described. Close observations then have to be recorded on all stages of the decay processes and the action that the various organisms have on the wood.

Tree roots

Considerable damage, running into millions of pounds, is caused each year to buildings either directly or indirectly by the action of roots of trees or shrubs. There may be a number of different tree species near to the buildings concerned. All or some of them might have roots beneath the foundations. It would be excessively expensive to try and trace the roots back to their parent trees by excavation. Fortunately it is possible to identify most roots of trees growing in the British Isles from aspects of their root anatomy, largely from features of the wood (secondary xylem). In some instances it is possible to identify to the species level, but more often only the genus can be identified, for example *Quercus*, oak, or *Fraxinus*, ash, and *Acer*, maples and sycamore. In the Rosaceae, identifications can be made only to the subfamily level, for example Pomoideae and Prunoideae. Current research is aimed at finding additional characters in this family.

Sometimes it is not possible to get closer than the family, as for example in Salicaicae. In trunk wood, *Salix* and *Populus* can normally be separated because Salix usually has heterocellular rays and *Populus* homocellular rays. However, in the root wood this distinction does not hold. Indeed, root wood is often slightly dissimilar in its anatomy from trunk wood of the same species. This means that one cannot rely on the descriptions contained in reference works on wood anatomy for accurate identification of roots. Root anatomy is also quite variable within a species, so the only way to be sure of making the proper identification is to compare the root sections with reference microscope slides taken from a range of authenticated specimens. Figure 9.5 shows two roots of *Acer pseudoplatanus* (TS) grown under very different conditions and some normal trunk wood for comparison.

Wood: in archaeology

Wood or charcoal is often preserved in sites from antiquity. The best preservation occurs in localities which are either very dry or continuously wet. Fluctuating drying and wetting encourages the activity of microorganisms and/or insects and can lead to the rapid decay of wood.

Charcoal, usually in the form of fire ash or the burnt remains of structural posts in post holes, often retains even very delicate features of vessel element wall pitting and perforation plates. Figure 9.6 shows Romano-British *Alnus* charcoal. It can be difficult to see details of the anatomy on first examination of the surface of a piece of charcoal, because it is often damaged and dirty. After a period of drying in an oven at 50°C, the charcoal will fracture readily. If care is taken to snap it along the radial longitudinal, tangential longitudinal and transverse planes, good surfaces for study can be produced. The specimens are mounted in plasticine or blue-tack on a microscope slide, and examined under the epi-illuminating microscope.

We have tried embedding and sectioning charcoal (with a diamond saw), but so much material is lost in the process that it was found not to be worthwhile. The very small fragments can be examined in the SEM, after coating, but generally the light microscope is adequate.

It can be determined if the makers of the fire had selected particular woods for their burning properties, or if the remains merely represent what was growing locally and easily accessible. Moreover, an idea may be gathered about the composition of the vegetation of an area at particular times.

Some sites are very rich in waterlogged or dry, preserved wooden objects. The Sutton Hoo burial ship, for example, contained many wooden grave goods. Interesting examples from this site are some small pots with silver gilt rims. On excavation these were thought to be made from small gourds, fruit from the Cucurbitaceae. Microscopical study of thin sections showed the structure to be of walnut wood, probably from near the rootstock, where burr-wood could be obtained.

With improved techniques for recovering wooden wrecks and, subsequently, conserving them by special impregnation techniques, interest has

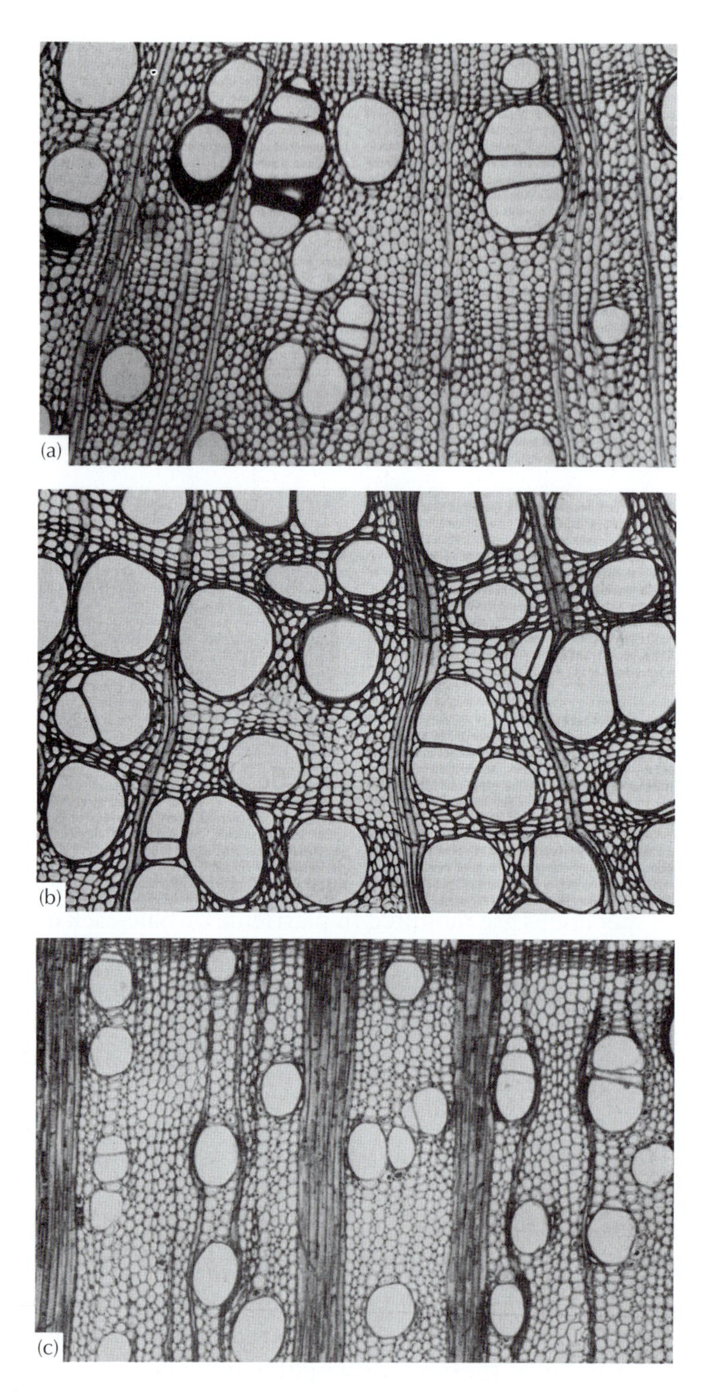

Fig. 9.5 *Acer pseudoplatanus* roots grown under different conditions (TS). (a) From normal and (b) from waterlogged soils. (c) Normal trunk wood. All × 130.

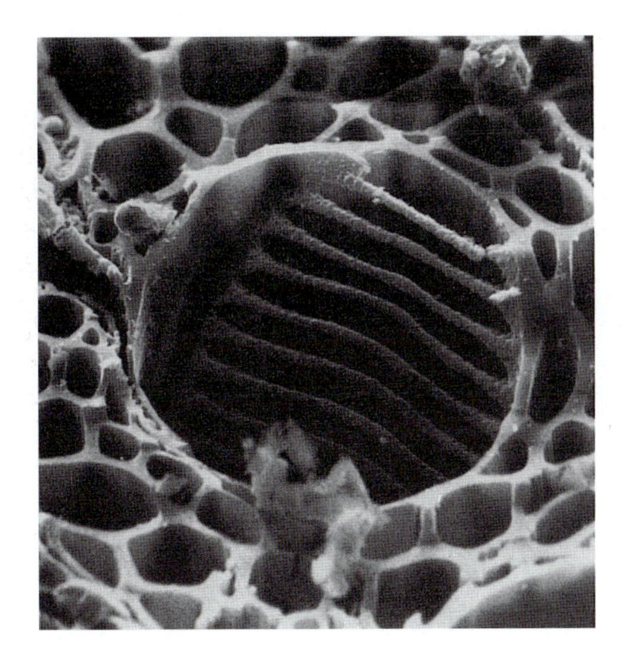

Fig. 9.6 Charcoal of *Alnus glutinosa* from Romano-British London. Details of the structure are well preserved, particularly the scalariform perforation plate.

increased in naval architecture. The timbers of a warship from the Punic wars were remarkably well preserved and were readily identified after many centuries in sea water. An oak Iron Age boat from Brigg in South Humberside also proved to be fascinating. No 'nails' were used to secure one timber to another, but the main logs were sewn together with twisted willow twigs passed through regularly pierced holes along the edges of the baulks of timbers In the Bronze Age, trackways were built across swampy ground in Somerset. The hazel faggots (*Corylus*) used in these were well preserved in the waterlogged conditions. We look at archaeological material from all sorts of wooden objects: spear shafts, shields, buckets, right through to structural timbers. Much of this work is very time-consuming. Often some details of the anatomy are lost, and very careful comparisons with reference materials need to be made before identifications are given Because of the potentially enormous quantity of fragments of wood that could come from even one fire, it is sensible to sort them visually into groups and limit the initial sampling to some examples from each group.

Wood products

Archaeological plant remains other than from wood can sometimes be remarkably well preserved. The sandal shown in Fig. 9.7 from ancient Egypt is such an example. *Cyperus papyrus* is a major constituent of the sandal, and

Fig. 9.7 An Egyptian sandal from antiquity, found to be made from papyrus (*Cyperus papyrus*) and palm species of *Borassus*.

some *Borassus* (palm) is also present). However, some of the samples are waterlogged and compressed. It is often possible to 'revive' such material. The secret is to section it in the compressed form and revive the sections, by floating them briefly in sodium hypochlorite solution or in chlor-zinc-iodine. Temporary mounts are best made in 50% glycerine.

The structural properties of wood are utilized in modern building methods by using not only solid timber, but also laminates, plywoods, chipboards, hardboards and the like. These materials are tested to destruction so that their properties can be properly evaluated. Microscopic examination of the failure areas can give a good guide to areas of weakness.

Forensic applications

Forensic work often involves the identification of small pieces of plant material other than wood, although in addition to safe ballast, wood splinters might come from such things as windows, doors and their frames, weapons and the like, and thus play an important part in police work. A wide range of particles of plants may become attached to clothing or footwear which relate to the scene of a crime. Plant fragments found on suspects may link them with the location of the crime. Clothing itself is made from a variety of fibres, a number of which may be of plant origin. Microscope slides of macerated textile fibres make up part of the reference collection of forensic laboratories. Drug plants, such as *Cannabis sativa*, frequently have diagnostic characters whereby quite small pieces may be identified microscopically.

An increasing number of plant species are being sold for consumption as drugs – some as adulterants, others as substitutes. It is a hard task to keep up with the introduction of additional species, particularly because the product is often in a very finely powdered form. Quite a lot of time and effort has to go into analysing such finely powdered drugs. Anatomical characters can be used with such confidence for identification that they may contribute part of the evidence given under oath in court.

Palaeobotany

Just as anatomical features can aid in the identification of archaeological materials, they can be employed in the identification of fossils and place them among extant plant families and genera. Fossils are usually composed of only some plant parts with the crucial identifying features associated with flowers and fruits being absent. Anatomical features, however, allow us to determine if the plant part is a root or stem or leaf simply by the presence of exarch or endarch protoxylem in the case of the first two. Further examination and comparison to a reference collection of extants may allow us to assign the fossil to an extant taxonomic group. Some fossils may be sectioned, others may be ground down to thin sections suitable for microscopy, and still others are fossilized in the form of charcoal which may be treated in the same way as recent charcoalified archaeological artefacts.

Postscript

In this chapter, we have seen the basic cellular structure of the vegetative parts of a number of common and less common plants described in simple terms. The evidence presented here shows that plant anatomy is not just an academic subject but has been drawn from a wide range of applications, many of economic importance, others of legal consequence and a number which simply served to answer intriguing questions.

CHAPTER 10

Practical microtechnique

Safety considerations

Local health and safety regulations should always be followed. Treat all chemicals as potentially dangerous, and always wear appropriate protective clothing and goggles whichever of the techniques you are using. Pay particular attention to the appropriate use of fume cupboards. It is most inadvisable to do practical work involving any chemicals or the use of sharp instruments on your own. The majority of the techniques given here require there to be at least two persons present, and some should be carried out only by trained technicians so that should any safety issues arise, they can be dealt with immediately. The authors cannot be held responsible for harm resulting from application of any of the techniques described below. Waste disposal must be done according to local regulations. Do not pour waste liquids down the sink.

Materials and methods

Any readily available plant can be used to teach or learn plant anatomy. Over the years a rather small number of species have become 'approved' as 'standard' plants for study. This has had an incredibly stultifying effect on the study of plant anatomy generally, with the result that many biology teachers have forgotten that there are many plants that can be used for study in the classroom. Indeed, many university botany departments, adhere to a generalized 'standard plant list' in teaching plant structure as well. The plants chosen are in many cases thought of as 'typical', but often they are quite atypical. Many botanists go through their lives thinking that *Zea maize* is a typical monocotyledon, but grasses in general are very specialized and represent a very restricted view of monocotyledons as a whole. We frequently adhere to peas, lettuce, maize and sunflower, barley, tobacco and beet in our physiological work, because botanists are often unaware that other plants, which can be grown with equal ease, are more varied and interesting in their structure. Scanning the literature will demonstrate just how often a species has been used.

Material is best if collected fresh. It can be examined fresh for cell contents, cytoplasmic movement and so forth. But when studies involve cell structure and histology, it is better to fix the fresh material by chemical means. Fixatives, when correctly formulated, will kill the plant material, preserve its general shape and size, and render the tissues suitable for sectioning and, depending on their potency, preserve the cellular detail. Dried herbarium material can be used for anatomical studies. Some plants revive easily, but others are unsatisfactory. If there is no fresh material available, then dried material can be fixed after boiling in water for about 15 minutes and after this allowing the material to cool. A few drops of detergent may be added to aid wetting the specimens.

Killing and preserving cell contents

This is one of the most critical steps in tissue processing. This process should be accomplished with the minimum disturbance of the protoplasmic organization within the cells, and a minimum disturbance and distortion of the cellular arrangement. In addition to killing the protoplasm, the process must be able to fix the undistorted structure and, further, be able to render the mass of material firm enough to withstand handling. Thus the requirements for a good preservation are as follows:

1 Kill the protoplast without distortion.
2 Preserve to fix the fine detail.
3 Harden the material.

Fixatives

Alcohol and formaldehyde-based fixatives

Seventy per cent alcohol is typically used as a plant fixative in schools and laboratories, because it has little effect on the user should they accidentally spill it over themselves provided it is immediately rinsed off with water. Alcohol by itself tends to harden the plant tissues and can cause changes in shape. It is to be avoided as a histological fixative, as no cell detail will be preserved at all.

Formaldehyde has to be treated with care and a great deal of respect. Do not breathe the vapour, and use a fume cupboard. Plant tissues require quite aggressive fixation techniques, and there is no reason why students should not use more potent liquids for fixing plants if suitable care is taken. For general histological purposes, the formalin–acetic acid–alcohol (FAA)

Table 10.1 The composition of formalin–acetic acid–alcohol (FAA).

850 ml 70% ethyl alcohol
100 ml 40% formaldehyde
50 ml glacial acetic acid

mixture given in Table 10.1 works reasonably well. Always use a fume cupboard.

Note: FAA is a corrosive liquid, and if it comes into contact with the skin **it should be washed off immediately**. The best preventive measure is to wear laboratory gloves. It is well worth the trouble and care to use material fixed in FAA because materials that are preserved in it section well and can be kept in the reagent indefinitely. However, be careful to store vials and bottles containing FAA in a well-ventilated space, as the fumes are harmful and should not be inhaled.

Irrespective of the fixation technique used, plant material to be fixed is normally cut into portions to enable rapid penetration of the fixative. Care should be taken to ensure that the portions of plant tissue are cut so that they can be readily identified and oriented. Bottles with wide mouths and polypropylene screw or push-on tops are ideal for storage, and can be obtained in a range of sizes. It is best to keep the plant in fixative for at least 72 hours before continuing on with the preparation process. Plant material may be kept in FAA and can be stored for as long as required, but the bottles should be inspected regularly for evaporation and topped up with 70% alcohol if necessary. This is the most volatile of the constituents.

Specimens to be sectioned are removed with forceps and washed in running tap water for 30 minuets to 1 hour. They can then be handled safely.

Non-coagulating fixatives

There are other fixation options available for detailed anatomical and cytological studies, which will preserve cytoplasmic details as well as prevent serious plasmolysis which is usually evident when using FAA and its associated dehydration procedures. Feder and O'Brien (1968) reviewed principles and methods used in *Plant Microtechnique*, which introduced the concept of using non-coagulant fixatives such as osmium tetroxide, acrolein glutaraldehyde and formaldehyde, coupled with the use of plastics such as glycol methacrylate polymer instead of wax. The advantages of using polymers, is that thin sections (1–3 μm) may be made, and these sections show excellent cell structural detail. Osmium tetroxide is particularly dangerous, and should be used only by competent, trained people in a fume hood, following all the safety precautions.

We recommend **acrolein** as an alternative fixative, especially where the researcher is intent upon preserving (and needs to resolve) cellular details which would not be preserved at all with FAA. Ten per cent acrolein is routinely used as a biological fixative. It may be made up in tap water or added to a suitable buffer solution, such as a phosphate buffer, not unlike that used in electron microscopy preparative techniques. Dehydration is a little more complicated than with FAA, but well within the scope of the average anatomy laboratory. **Caution: Wear protective gloves and use a fume hood when working with acrolein.**

Dehydration and infiltration

Pieces of thin leaf material are fixed and hardened in about 12 hours whereas thick leaves or small stems will require at least 24 hours. Woody twigs should be kept in FAA for about a week before continuing with processing. Note: Propionic acid may also be used, in which case the formula is designated as FPA. Other well-known killing and fixing fluids are chromo-acetic and Flemming's fixative, but these are potentially dangerous and should not be used by untrained people without proper supervision.

Dehydration is necessary in order to allow the material to be infiltrated with an appropriate supporting medium (paraffin wax; polyester wax or a polymer, such as glycol methacrylate) to provide adequate support for the material during sectioning. The fixing procedures involving the use of FAA and acrolein as the primary fixative differ slightly, and are described below.

This series of operations removes water from the fixed and hardened tissues. The removal of water is a necessary preliminary step to the infiltration of the specimen into a matrix that is not soluble in water. Complete removal of water ensures adhesion of the matrix to the external and internal surfaces of cells and tissues. The process consists of treating the tissue with a series of solutions which contain an increasing proportion of the dehydrating agent and progressively less water. Some anatomists regard tertiary butyl alcohol (TBA) as being an ideal dehydrating agent. The cost is high, but the results are well worth it!

Procedure for infiltrating through a TBA series

The procedure which we recommend for infiltrating plant tissues through a TBA series is outlined in Table 10.2.

Begin infiltration in 50% liquid paraffin/50% wax after step 9, for 24 hours, and then followed by three changes of wax.

When the tissue has been infiltrated with paraffin wax in an embedding oven, the wax effectively supports the tissues. Infiltration consists of dissolving the wax in the solvent containing the tissues, gradually increasing the concentration of the wax and decreasing the concentration of the solvent. This process is carried out in an embedding oven, usually at 50–70°C, depending on the melting point of the wax used. After infiltration with wax

Table 10.2 The tertiary butyl alcohol series.

TBA	95% ethyl alcohol	Absolute ethyl alcohol	TBA	Water	Paraffin oil
1	50	0	10	40	0
2	50	0	25	30	0
3	50	0	35	15	0
4	50	0	50	0	0
5	0	25	75	0	0
6	0	0	100	0	0
7	0	0	50	0	50
8	0	0	0	0	100

as stated, the wax matrix serves to support the tissues against the impact of the knife. It is therefore very important that the infiltration procedure is carried out properly and carefully!

Procedure for fixing in acrolein

Warning: When following this method, please ensure that all waste is disposed of into designated, labelled waste containers. Do not discard down sink. **Caution: Wear protective gloves and use a fume hood when working with acrolein.**

1 Cut small tissue segments into cold 10% acrolein – preferably in a Petri dish and on ice.

2 Put tissues in 10% acrolein into capped vials and leave at 0°C for 24 hours in a refrigerator.

3 Dehydrate in 2-methoxyethanol and return to refrigerator at 0°C for 12–24 hours.

4 Decant the 2-methoxyethanol carefully, and substitute with cold ethanol, n-propanol and n-butanol, in sequence, all at 0°C for 12–24 hours each.

5 Transfer specimen into 25 : 75 n-butanol–paraplast wax chips. Place in an embedding oven for 12–24 hours which is set at temperature that is a few degrees above the melting point of the wax. Do not cap the vials at this stage.

6 Decant mixture carefully, transfer to pure paraplast wax. Make three changes each of 12–24 hours each time.

When the tissue has been infiltrated with the paraplast in an embedding oven the wax effectively supports the tissues. Infiltration consists of dissolving the wax in the solvent containing the tissues, gradually increasing the concentration of the wax and decreasing the concentration of the solvent. This process is carried out in an embedding oven, usually at 50–70°C, depending on the melting point of the wax used. When infiltration with wax is complete, the wax matrix serves to support the tissues against the impact of the knife. It is therefore very important that the infiltration procedure is carried out properly and carefully!

Mould the blocks and cut sections.

Sectioning

The safety razor blade can be used to produce sections which are thin enough for study under magnifications of 100–400×, or sometimes more. However, practice is needed. Those with a steady hand will get better sections with the thinner, double-sided razor blade, but very thin blades are too flexible. Many anatomists bear the scars of early battles with tough plant material! It is advisable to cover one side of a double-edged blade (if you use these) with masking tape, to avoid cutting yourself, rather than the specimen.

Cutting freehand sections is a time-consuming process, and is not suited to the mass production of teaching material. However, for simple demonstrations, inquiries or identifications is quite suitable. For the production of large quantities of slides of the same specimen for class study, a more refined method of sectioning is needed. A rotary or rocking microtome should be used where wax-embedded specimens are going to be used to prepare thin sections A microtome is indispensable for sections under 10 μm thick, or for subjects where serial sections are needed of, for example, flower buds, where the various parts would separate and become disarranged when not embedded before attempting a sectioning procedure.

Sections of between 15 and 30 μm are required for histological studies. These can be cut with a sledge or sliding microtome. The sledge microtome allows the specimen which is to be to be sectioned to be firmly held in a universal clamp (to allow for correct orientation) and the knife is brought towards and over the specimen. Models in which the knife is fixed, and the specimen is made to move past the knife, are not so universally useful and somewhat more dangerous!

The moving knife sledge microtome can be used for the following types of material, using 50% alcohol for lubrication (applied to the knife blade with a small camel hair brush).

1 Tough materials, such as wood. The material should be cut into approximately 1 cm cubes, orientated as described in Fig. 10.7. The cubes are boiled in water until waterlogged, i.e. they sink when cold water is added to the container. The cubes are removed, cooled and clamped in place. If they are very hard, they may need to have a jet of steam directed on the surface to be cut (Fig. 10.1), but often they are soft enough to cut directly. Some woods contain silica, a substance so hard that it rapidly blunts the microtome knife. Silica can be removed by standing the wood for 12 or more hours in 10%

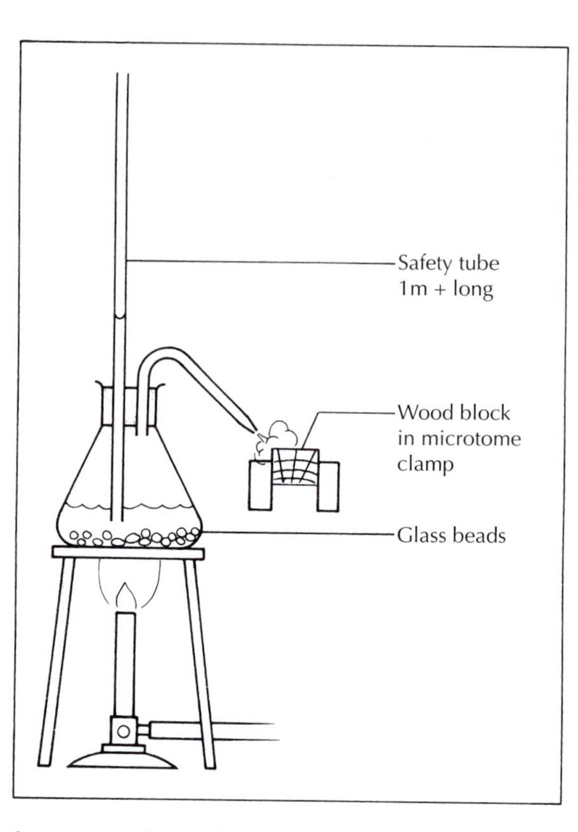

Fig. 10.1 A simple apparatus for producing a steam jet to soften wood prior to sectioning.

hydrofluoric acid, in plastic containers. **Very great caution** should be taken with this acid. It causes serious burns even in low concentration, and the burns heal very slowly. It is not recommended for class use, but could be used by trained technicians. After treatment, the wood must be washed in running water for several hours.

2 Twigs may need to be boiled before sectioning. Softer stems are best fixed and washed before sectioning. Many cylindrical objects need some support in the microtome clamp. Support is normally provided in the form of cork or pith. Pith tends to become soggy when wet, and cork can contain unexpected sclereids which blunt the knife, but on the whole we prefer cork. Some people now prefer Styrofoam. Suitable bottle corks with few lenticels should be selected (Fig. 10.2A–D). Circular slices about 34 mm thick are cut off, using a razor blade. The discs of cork are cut across the diameter and the two halves placed side by side.

For transverse sections of stems, a notch cut in one half of the cork will help to keep the specimen correctly oriented, without allowing it to be compressed excessively. Alternatively, the cork may be cut along its long axis, and the two halves used to mount the specimen to be sectioned.

For making longitudinal sections, the cork slice is cut across the diameter at an angle (Fig. 10.2E–H). When the two parts are placed side by side with the material to be sectioned at the bottom of the V, clamping causes the outer parts of the cork to roll outwards, but the material is retained. If the disc was cut to produce a flat surface before clamping, the outward curving would release the specimen, as in Fig. 10.3.

3 Leaves to be sectioned transversely are rarely just the right width for the clamp. Wider leaves can be folded once or several times so that they form a sandwich in the cork, Fig. 10.4. With narrow leaves, it is best to put several leaves between the cork slices there is more chance of getting some good sections.

Most mesophyte leaves section easily. Some succulent or very soft hydrophyte leaves and stems can pose problems. In these plants the cells are thin-walled and burst easily if compressed when turgid. A simple and usually effective remedy is to allow the specimen to go limp on the bench for half an hour or so. It can then be firmly clamped, cut and the sections put in water. If they do not return to their natural shape in water, 50% alcohol may be used, or even, for a second or so, immersion in undiluted bleach such as Domestos or Parozone (sodium hypochlorite) will cause them to return to the uncompressed form.

Certain leaves contain silica bodies (particularly those of grasses and sedges) which blunt the microtome knife and tear the section. So if a section appears torn, first examine it to see if silica bodies are present. Hydrofluoric acid (10%) can be used to remove the silica bodies, but must be treated with the **utmost caution** (see above).

As each section is cut it will slide onto the knife blade, lubricated by the 50% alcohol. It should be lightly transported on a paint brush to a Petri dish of 50% alcohol. The section can be examined temporarily in water, or will keep for several months in 50% glycerine solution on a slide (stored flat). Starch distribution can be studied in such sections, and chloroplasts and other larger cytoplasmic inclusions can be seen.

Clearing

Sometimes it is an advantage not to have cell contents obscuring the tissue distribution but, before 'clearing' the sections, some should be studied with their inclusions. Sections can be cleared by transferring them from the 50% alcohol to a dish of water with a brush or fine forceps. Then, using a mounted needle or fine forceps they are placed in a cavity block containing undiluted sodium hypochlorite household bleach.

The time needed for cell contents to dissolve away varies from subject to subject and can be determined by visual inspection. It usually takes about 5

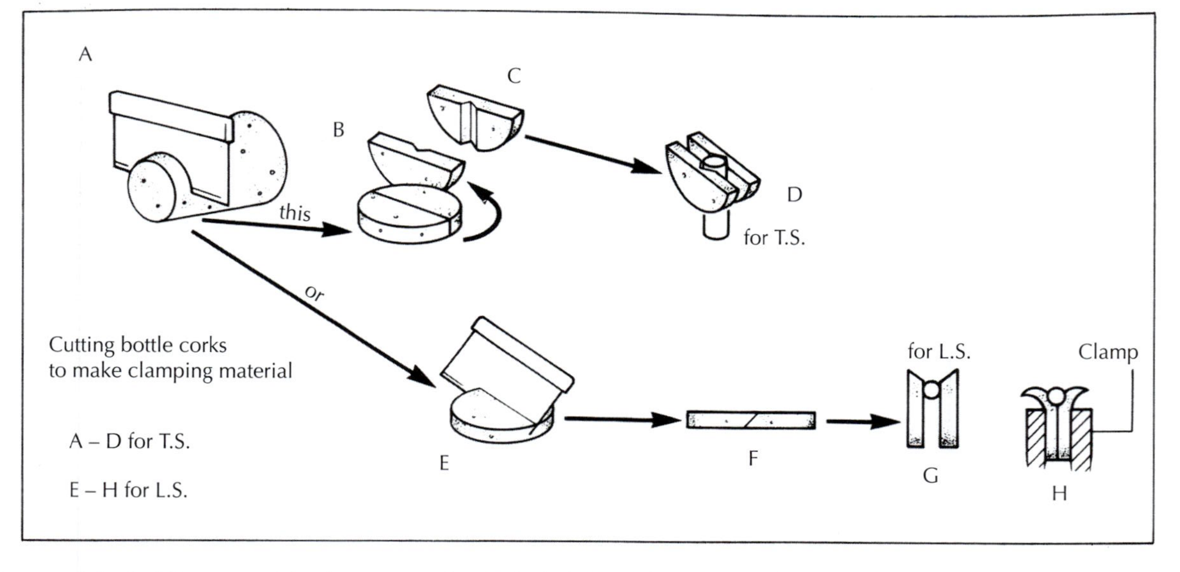

Fig. 10.2 Preparing a cork for holding material to be sectioned A–D for TS, E–H for LS; note that the oblique cut in cork E helps to prevent cylindrical stems from being released from the cork on clamping.

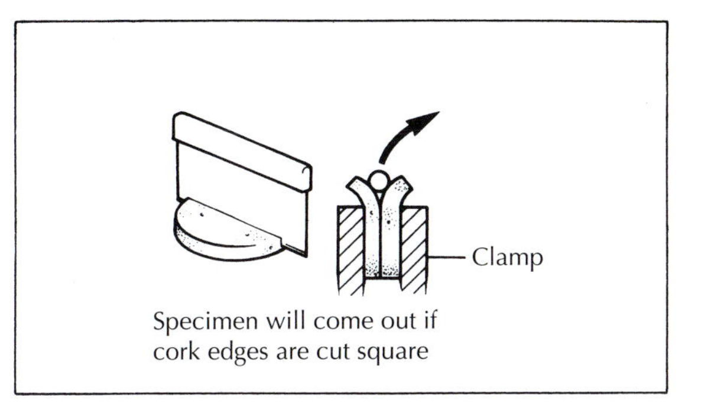

Fig. 10.3 The wrong way to cut cork for making LS of material. When clamped, the cork curls back and the specimen is released.

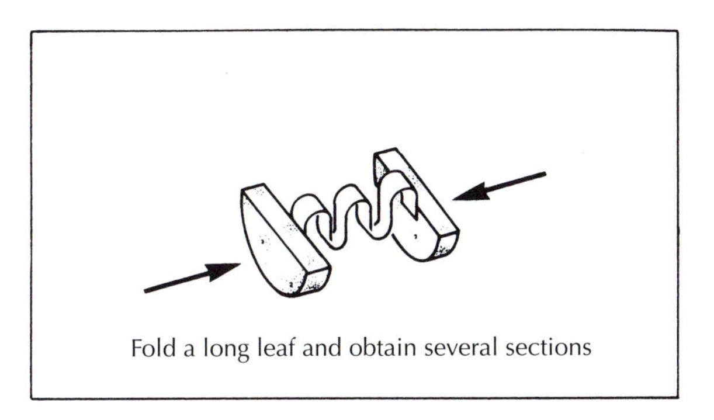

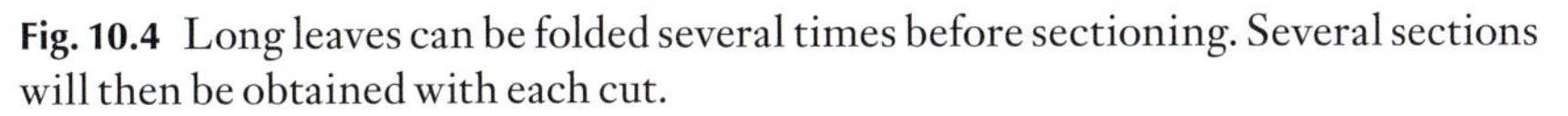

Fig. 10.4 Long leaves can be folded several times before sectioning. Several sections will then be obtained with each cut.

minutes. The whole section will dissolve if left long enough. After immersion in the bleach, the sections are thoroughly washed in water. Take care not to get the brush in the bleach as its bristles will dissolve.

Staining

Whilst observation in the natural state is important, it may be difficult for the student to differentiate, for example, between unlignified and lignified tissues and it is for this reason that we recommend staining sections. There are several staining combinations that may be used to enhance details within sections. Sections of freshly cut material should be washed gently, to remove cell debris which will obscure details once the section has been stained.

Two main types of stain can be used: (i) those which are temporary, whose colour fades, or which gradually damage the section; and (ii) those which are regarded as permanent. Even permanent stains may lose their colour if exposed to sunlight, so be careful and store your collection in the dark.

With care, stains can be selected to give the maximum contrast between the various cell and tissue types in the plant. They might be selected because they colour particular parts of the cell wall structure and indicate its chemical composition. The stains and the protocol described below are in daily use in many laboratories throughout the world, and not just those associated with the authors. Those who want comprehensive lists of stains, procedures and protocols, should see the books by Gurr (1965), Foster (1950) or Peacock revised by Bradbury (1973) to mention but three of the many guides to micro technique.

We have included other techniques which will be useful to the student on the CD-ROM. These can be accessed quite simply by following the 'Techniques' links associated with each of the exercises.

Temporary stains

1 1% aqueous methylene blue. All cell walls turn blue, except cutin or cutinized walls which remain unstained; cell walls take up a degree of intensity of blue depending upon their chemical composition and physical structure; various wall layers frequently stain differently. The stain may be mixed with 50% glycerine, about 10 ml of a 1% aqueous stain, to 90 ml of 50% glycerine, and sections mounted directly into this medium. This mixture is also useful for staining macerated tissues which are difficult to handle. A drop of washed macerate in water is mixed with a drop of the mixture on a slide, and the coverslip put on.

2 Chlorzinciodine solution (CZI, Schulte's solution). This solution consists of: 30 g zinc chloride, 5 g potassium iodide, 1 g iodine and 140 ml distilled water. Cellulose walls turn blue, starch turns blue-black, lignin and suberin turn yellow and moderately lignified walls turn green-blue. Sections are placed on the slide, and a drop or two of CZI added. This can be drawn off and replaced by 50% glycerine after 2–4 minutes, but satisfactory results can be obtained by adding 50% glycerine directly, and mounting in the mixture. This stain swells the walls and eventually dissolves them. Consequently, care must be taken when describing **wall thickness**.

4 Chlorazol black saturated solution in 70% alcohol. Stains walls black or grey and is particularly good for showing pitting.

5 Saturated carbolic acid solution. Sections are mounted directly in the solution (which should be kept off the hands). Silica bodies usually turn pink; this helps to distinguish them from crystals which remain colourless.

6 Phloroglucin/HCl. The phloroglucin is added to the section, and then the dilute HCl. Lignin turns red. **Caution:** HCl is highly corrosive of skin, clothes and the microscope!

7 Sudan IV. Sections can be mounted directly in the stain. Stains fats, cuticles turn orange.

8 Ruthenium red. Mucilage and some gums turn pink. Sections can be mounted directly in the stain.

A simple double stain

Fuchsin, aniline blue and iodine in lactophenol (FABIL) is a most useful stain and mountant for all types of plant material. Although not available commercially it is easily prepared from the following stock solutions: lactophenol : phenol (crystals), glycerol, lactic acid and distilled water in equal parts by weight.

Aniline blue, 0.5% in lactophenol. A
Basic Fuchsin, 0.5% in lactophenol. B
Iodine, KI, 0.6 g : lactophenol, 1 litre. C

The stain is made up by mixing the stock solutions in the proportions of A,4 : B,6 : C,5. This is allowed to stand overnight and after filtering it is ready for use and will keep indefinitely.

FABIL is superior to other commonly used temporary reagents such as aniline double stain, phloroglucinol or zinc chloride, the particular advantage being that it incorporates a differential stain, a clearing agent and a semi-permanent mountant. Sections cut from fresh or alcohol-stored material are transferred directly to the stain, enclosed by a coverslip and examined immediately. If desired, the stain may be replaced after about 10 minutes by plain lactophenol or 25% aqueous glycerol, but this is by no means essential. The solution is only slightly volatile and mounts can be kept for several months without drying up, although additions from time to time at the edge of the coverslip will prevent air bubbles from being drawn in. However, if it is necessary to store the mounts for a very long time the coverslip can be sealed with melted beeswax or nail varnish. The cytoplasmic cell contents, including nuclei and callose plugs of sieve tubes, are stained blue, Cellulose walls stain a paler blue and lignified tissue becomes a bright yellow, orange or pink, depending on the nature of the specimen. Staining is rapid but improves with time and overstaining is not possible, even after immersion for several weeks.

Much of the success of this reagent is due not simply to staining but also to differential clearing, so that in practice the tissues are rendered more distinct than the above-mentioned colour reactions would suggest. Moreover, because of the extremely good light-transmitting properties of the solution, cellular detail can be studied even in thick sections. FABIL may also be used for mounting fungal or algal material, including seaweeds, and causes very little distortion. Alternatively the aniline blue solution A, or fuchsin solution B, may be used alone. With fungi, gentle heating of the mount improves the absorption of the stain.

Safranin (1% in 50% alcohol) and Delafield's haematoxylin is a very useful combination. In cell walls cellulose turns dark blue; lignin turns red and cellulose walls with some lignins turn purple.

Freshly mix the prepared safranin with matured Delafield's haematoxylin in the proportions of 1 : 4; filter. The stock mixture can be used for up to about 1 week, but should be filtered before use each day.

Sections should be transferred from water (after washing all bleach away) into a suitable dish containing the stain, and covered. Most sections take up the stain in 24 hours, others will need less time. The sections should then transferred to a Petri dish in which there is 50% acidified alcohol (use a few drops of conc. HCI). This solution removes the stain, acting on the safranin first. Whilst the object is to obtain a satisfactory colour balance, only experience will tell when this has been reached.

Sections should be removed when they still appear to be slightly dark or overstained for best results – the colours look less intense under the microscope. The decolorizing action is halted by placing the sections into a Petri dish containing 95% alcohol. After about 5 minutes they can be transferred to absolute alcohol in a covered Petri dish. Five minutes later they can be transferred either to a 50–50 mixture of absolute alcohol xylene in a covered dish, or this step may be eliminated and they may be transferred directly into xylene. **Xylene fumes should not be inhaled**; use a fume hood. After 10 minutes in the xylene, the sections can be mounted in Canada balsam on the microscope slide. Any milkiness in the section at this stage means that water is still present, and the section should be taken back through xylene, then fresh absolute alcohol, fresh absolute alcohol xylene and fresh xylene before remounting. Sections which curl up or roll up should be straightened out in 50% alcohol. As they progressively dehydrate in purer alcohol they become more brittle and cannot be unrolled without breaking. Curled wood sections can be flattened by drawing them over the edge of a slide partly immersed in 50% alcohol (Fig. 10.5), a process needing three hands! Alternatively a section lifter can be used. Once on the slide they can be 'set' using a few drops of 95% alcohol.

If it is more convenient to stain overnight, the safranin-haematoxylin mixture can be used in the proportions 94 : 6. Although fast green can be used as a counter stain for safranin, we have found that haematoxylin produces a colour which photographs better on normal panchromatic film. Alcian blue may be used as an alternative to the haematoxylin; a 1% aqueous solution is satisfactory. It is easier to use and gives blue colours where the haematoxylin would have stained purple.

Fast Green can be used on its own as a stain for macerated material. The macerate is dehydrated by decanting alcohols of 50, 70, 90, 95% and abso-

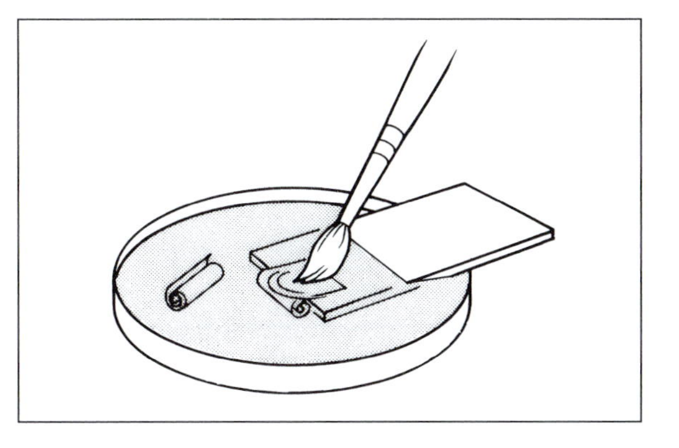

Fig. 10.5 Drawing a curled section onto a microscope slide.

Table 10.3 Dehydration of fresh material for permanent mounting using the stains Safranin and Fast Green.

Step	Time
Cut sections, and place in watch glass, in 40% ethanol	30–60 s
Rinse with 50% ethanol	30–60 s
Immerse in Safranin solution (in 50% ethanol)	30–60 s
Rinse in 95% ethanol	60 s
Immerse in Fast Green (95% alcoholic)	30–60 s
Wash in 95% alcohol	30 s
Rinse in 100% alcohol, two changes	30–60 s each
Immerse in xylene, two changes	1 min each
Mount in DPX or Euparal or Canada Balsam	

lute in turn from a tube containing the macerate. It is a help if a small hand centrifuge can be used to settle the cells at each stage. Finally, the cells are transferred to a slide bearing 23 drops of Euparal containing 23 drops of Fast Green per 10 ml. The coverslip is applied.

Preparing permanent slides

Freehand sections can be turned into permanent preparations quite easily. Once you have cut your sections, take unstained sections and follow the procedure outlined in Table 10.3. If the sections are thin enough, valuable information may be obtained using this relatively simple technique. **Xylene should not be inhaled.** It is much safer to use Euparal essence in its place, and mount in Euparal rather than Canada Balsam.

Whilst there are many commercial mounting media available, neutral Canada Balsam is preferred, as it is not likely to remove safranin from stained sections. Some modern substitutes, whilst being colourless, may be too acid and might gradually cause the safranin to leach out or contract markedly if over-dried, but with due care they may be preferable because they are safer to use.

Euparal is used either where it is undesirable to pass very delicate sections through xylene after absolute alcohol because they may distort, or for use when mounting stained macerated material, or for safety considerations.

Once the material has been mounted in mounting medium, and cover-slips placed over the specimens, they should be baked to flatten the sections in an oven set at about 58°C for 10–14 days, to thoroughly harden the mounting medium. Sections mounted in Canada Balsam are firmly set by this stage, and the slides can be stored upright. Those mounted in Euparal may be firm round the edges of the coverslip only, and great care should be exercised when handling them. Permanency can only be achieved by further baking.

Infiltrating procedures

In order to obtain relatively thin (10–15 μm) serial sections, it is imperative that the material to be sectioned is supported by a suitable medium during the sectioning process. There are several alternative support mediums, each of which has their problems and difficulties associated with their use. Wax infiltration is the easiest method. Paraffin wax (as detailed above) is a common embedding medium that is relatively easy to use. The tissue is infiltrated into the paraffin wax mixture (via a series of ethyl-butyl alcohol mixtures), liquid paraffin and then into the wax medium. Various suitable waxes are commercially available. The trend nowadays is to use a monomer wax such as Paraplast, which is a compound of purified paraffin and plastic polymers, in which is incorporated dimethyl sulfoxide (DMSO) which promotes rapid tissue infiltration. The material is then sectioned with a microtome, sections collected on glass slides and then processed further. Steps involved in this process are described below.

After collection and fixation, the material has to be infiltrated through a series of alcohols – this involves using a graded series of ethyl alcohol: In FAA-fixed material, a sequence using tertiary butyl alcohol (TBA) and the substitution of TBA by liquid paraffin before infiltration into wax is most commonly used. After a suitable infiltration period, the specimens are poured into moulds, which, when set, are trimmed, orientated and affixed to microtome chucks before sectioning at the required thickness.

Trim top and bottom faces of the square to be parallel to each other and orientate the block, so that when you section the material, a ribbon is formed. **Be very careful of the knife edge at all times**.

The operation of a rotary microtome can be learned best by watching the procedure adopted by an experienced worker (Table 10.4). This will be demonstrated to you individually and in groups as and when you are ready to cut your first sections.

If you run into a problem, do no hesitate to ask the instructor or demonstrator for help – they have experience and will rectify your problem.

Attaching paraffin sections to the slide

Now that you have obtained a ribbon you will need to expand these in a water bath, which has been set to a temperature 4–5°C below the melting point of the wax or expand directly on a wet slide on a warm tray. Carefully cut your ribbon into segments (short enough that they will fit under a coverslip after expansion) and pick these up; using a damp fine camel hair paint brush gently place these on the water surface. You will need to practise this step, to ensure that you do not inadvertently fold the sections over. Once the ribbons have expanded, dip one of your adhesive-coated microscope slides in under the ribbon, and gently guide the ribbon onto the microscope slide. Remove the slide, ensuring that the ribbon is placed as you want it on the slide, and place in a warming oven (set to a temperature about 5–10°C below the melting point of the wax), and allow the ribbons to dry and set for about 5–7 days. Once dried down, you may proceed to stain the sections yourself.

It is important to note that paraffin sections, either in the form of a ribbon or as single sections must be fastened to the microscope slide with an adhesive prior to staining, else the sections will simply fall off during the staining procedure. When applying adhesive to the slides, apply very little and spread it evenly over the slide. If you use excessive adhesive, this will result in an undesirable messy background stain that will appear after your sections have been processed.

Adhesion of the sections to the slide (and therefore their quality) is influenced by several factors, the most important of which are:

Table 10.4 Important DOs and DON'Ts which should be observed during microtoming.

DO	DO NOT
Work cleanly. Clean up after you have finished, and leave everything neat and tidy – as you would expect to find it	Leave the apparatus in a mess
Handle the knife with care, nasty accidents may be caused by carelessness	Forget to make notes as to which sections that you have finished cutting are to be found on a particular slide
Label all material!	

1 The slides must be perfectly clean – wash in alcohol, air dry and wipe with a paper tissue.
2 The adhesive must be suitable for the particular material.
3 The sections must be properly flattened by heating.
4 The adhesive must be left to harden completely, thus making it insoluble in the reagents used during the staining procedure.

Staining paraffin sections

Paraffin sections affixed to slides are stained and processed by immersion in reagents in staining jars. For our purpose, it is a safe practical assumption that the staining of cellular structures is based on the specific affinity between certain dyes and particular cell structures. As is the case during microtomy, staining requires that you work cleanly and carefully.

Sloppy workmanship will result in poorly stained slides, and further, will result in pollution of the stain sequence for all those students using the sequence after you. So work carefully. Staining requires that the sections be dewaxed, stained and mounted as permanent preparations that can be examined with a microscope. All the sequences to be followed during staining are very time dependent, and should be followed carefully – varying the time that the sections are in a particular stage of the staining procedure will give varying results.

Some of the more common stains and useful botanical stains are listed in Table 10.5. It is wise to study this information, before attempting to set up or use an existing staining procedure.

Table 10.5 Examples of staining procedures.

Material and stains	Solvent
1. Cellulose cell walls	
Haematoxylin (self-mordanting type)	50% ETOH
Fast Green FCF	95% ETOH
Bismarck Brown Y	70% ETOH
Acid Fuchsin	70% ETOH
Astra Blue	70% ETOH
2. Lignified cell walls	
Safranin	50% ETOH
Crystal Violet	50% ETOH
3. Cutinized cell walls	
Safranin	50% ETOH
Crystal Violet	50% ETOH
Erythrosine	95% ETOH
4. Cytoplasm	
Fast Green FCF	70% ETOH
Orange G or Gold Orange	100% ETOH
Astra Blue	70% ETOH

Effective staining of sections is achieved by using protocols and procedures that have been worked out over time. Do not try to cut corners, do not leave anything out. Your preparations will simply be a dismal failure and complete waste of time, effort and chemicals. We have included a basic staining procedure, as well as a few alternatives to this in Table 10.6a and b.

Many variations of the staining procedure in Table 10.6a exist. For example, it can be modified to a triple stain procedure. The choice of the third stain will depend on what feature of the section you wish to enhance, or clarify by using a specific counterstain. For example we may wish to counterstain with crystal violet. The process is modified at step 11 (Table 10.6b). Note: DO NOT remove all the slides from the last xylol step. Only take one at a time, apply the mounting medium and coverslip to it and then place the slide on a drying rack before removing another slide from the xylol.

Another commonly used staining procedure is a triple stain method. Flemming's Triple Stain, which has its application in cytological research. Details of this procedure are given in Table 10.7. De-wax and dehydrate as in the staining schedule in Table 10.6, then follow the procedure detailed in Table 10.7.

NOTE: You should always work carefully, and be aware that many of the chemicals that you are exposing yourself to could be dangerous. Use a fume hood for xylene.

Table 10.6a Safranin Fast Green or Astra Blue staining chart from wax-embedded sections.

Step	Stain	Duration
1	Xylol 1	5 min
2	Xylol 2	5 min
3	Xylol/abs. ETOH	3min
4	Abs. ETOH	3 min
5	95% ETOH	3 min
6	70% ETOH	2 min
7	Safranin O	Minimum 1 hour
8	95% ETOH	1 min
9	Ammoniacal 95% ETOH	Minimum 10 s, maximum 1 min
10	70% ETOH rinse	30 s
11	70% ETOH rinse	30 s
12	Fast Green or Astra Blue	1 min maximum
13	Absolute ETOH	1 min maximum
14	Clove Oil/xylol (1:1 v/v)	2 min
15	Xylene 3	2 min
16	Xylene 4	2 min
17	Xylene 5	Remove from here, as coverslips are applied

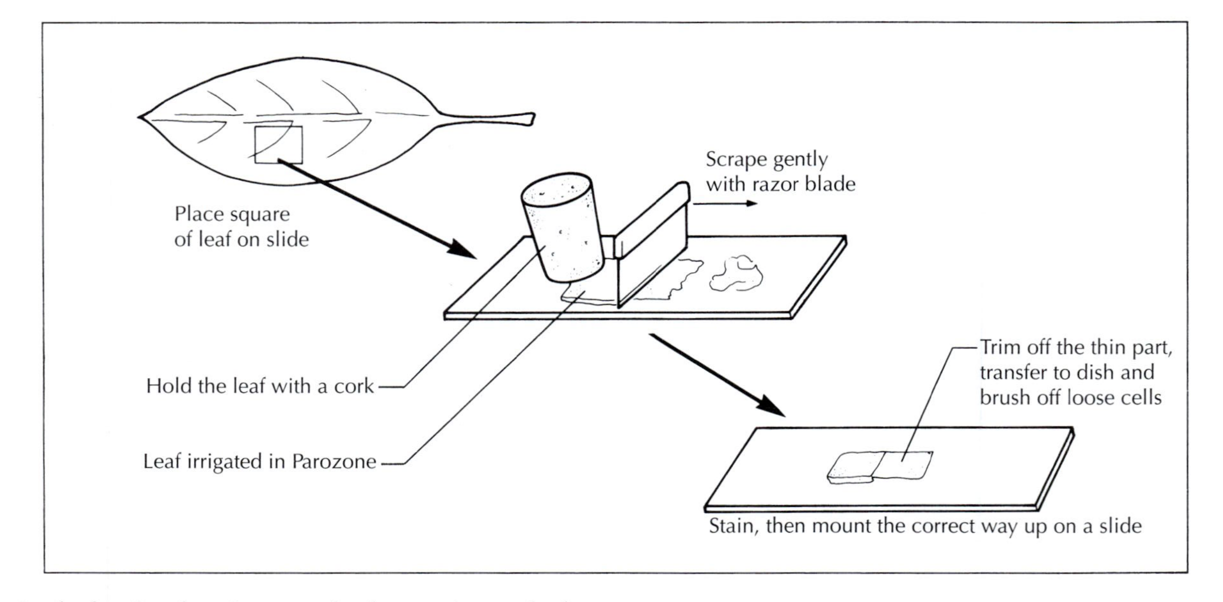

Fig. 10.6 Preparing leaf surface for microscopy by the scraping method.

Table 10.6b Modifications to staining chart 10.6a.

Step	Stain	Duration
11a	Crystal Violet	1 min maximum
11b	70% ETOH	rinse 30 sec
12	Fast Green	1 min maximum

Continue to step 13 in Table 10.6a

Preparation of surfaces

Many of the the staining methods applied to sections can equally well be used for surface preparations.

Leaf surfaces

The epidermis of most leaves can be readily removed by the scraping method. Only those leaves with very prominent veins, or large, numerous hairs pose problems, and these will demand a great deal of patience.

Material may be fresh, or washed after fixing. A suitable piece is cut from the leaf (Fig. 10.6). The surface which you wish to study is placed face down on a glazed tile or on a glass plate. It is irrigated with a few drops of sodium hypochlorite. One end may be held securely with a cork, and the other end scraped lightly with a safety razor blade. With practice a double-edged blade can be used, but it is better to start off by using a single-edged blade. The blade is held at 90 degrees to the leaf (Fig. 10.6) and the gentle scraping is continued, adding extra hypochlorite as necessary, keeping the leaf well irrigated. If the leaf is not severed by a forceful scrape, you will end up with a thin, clear area, which can be cut off, placed in a cavity block for a few minutes with sodium hypochlorite, and then washed in a petri dish containing tap water. Loosely adhering cells can then be brushed off with a fine camel hair paintbrush.

The preparation can then be viewed in water under the microscope. It is easy to see if enough has been scraped away. Experience will soon enable you to judge when the end point is reached in scraping. Make sure the surface is placed the right way up before the final mounting.

With some material it is unnecessary to scrape the leaf, since the epidermis can be removed by peeling it off the fresh leaf. This is done by folding the leaf to break the surface, and either stripping directly, pulling one part of the leaf downwards relative to the other, or by holding as thin a layer of the surface as possible between forceps and peeling it back.

Table 10.7 Flemming's triple stain procedure.

Step	Stain and procedure
1	Immerse slide in tap water 5 min
2	Stain: 0.25% aqueous Crystal Violet
3	Wash in clean tap water 30 s
4	Stain: 1% iodine + 1% KI in 70% ETOH 30 s
5	50% ETOH 30 s maximum
6	70% ETOH 30 s maximum
7	1% picric acid in 70% ETOH few seconds
8	Ammoniacal 95% ETOH few seconds
9	Abs. ETOH few seconds
10	1% Orange G in Clove Oil few seconds
11	Clove Oil few seconds
12	Xylene 3 – rinse (if too red return to abs. ETOH and proceed: if too purple, return to Clove Oil and proceed)
13	Xylene 4
14	Xylene 5: remove from here **and mount under coverslips**

Stem surfaces

A thin strip of stem surface can be obtained by making the first longitudinal section cut on the microtome just pass through the surface layers. This requires careful microtome adjustment, but is quite satisfactory.

Surface replicas

Sometimes it is not possible or desirable to remove the epidermis itself – a plant may be rare, or there may be little material readily available. A good imprint of the surface can be obtained with a film of cellulose acetate (nail varnish). It may be necessary to wipe the leaf surface with acetone to clean it. Then clear nail varnish is brushed on. More than one layer may be required, and it is advisable to make the individual coats quite thin. Once dry, the nail varnish can be removed as a film and then mounted under a coverslip, preferably in a medium with a different refractive index (immersion oil will work well) or little will be seen.

Although replicas can be made easily using a variety of materials (latex, for example), the epidermis itself is most useful in determining the shape of the epidermis, and its associated subsidiary and guard cells. Replicas can tell us little about cell contents.

Cuticular preparations

Various chemical treatments can be used to cause the cuticle to separate from the leaf. The cuticle is in many respects better than the replica, but can be very delicate.

One method is to digest away the leaf tissues using **nitric acid (care!)**. The cuticles will frequently float to the surface, or, when the leaf has been fully washed, can be teased off the dissolving tissues.

Clearing material

Whole thin leaves or stems or flowers can be made transparent by soaking them in chloral hydrate, washing and soaking in sodium hydroxide solution, alternately, for several changes, with several hours at each stage.

After the final washing, the organ can be carefully stained in safranin (1% in 50% alcohol) dehydrated through a very gradual alcohol series and mounted. Veins and sclereids show up well.

Alternatively, the preparation may be left unsigned for examination by various optical methods.

Standard levels

When plants are being examined for comparative purposes, either for identification or for taxonomic reasons, it is important that similar parts of the organs are looked at. The leaf, for example, is normally looked at in transverse section across its broadest region, or halfway along the length of the lamina. The surface of the leaf is studied near the central region of the lamina. The margin may also be examined.

Petioles should be examined in transverse section just where the lamina begins, halfway down its length and also near the base. Stems are normally sectioned in the middle of the internode, or, in addition, at the node (Fig. 10.7).

Roots are normally sectioned at a convenient level, since accurately defined positions are harder to delimit. For very detailed studies, of course, sections are required from many other levels, and sometimes serial sections are needed. These are particularly useful in the study of nodes and shoot apices.

Microscopy

Electron microscopy

This book is not intended for people who are using electron microscopes, but some space must be given to a brief description of their uses. There are two main types of electron microscope, the transmission microscope (TEM) and the scanning electron microscope (SEM). Clearly, the TEM and the SEM are now taken as being standard tools of the trade and essential adjuncts to the light microscope in many anatomical investigations.

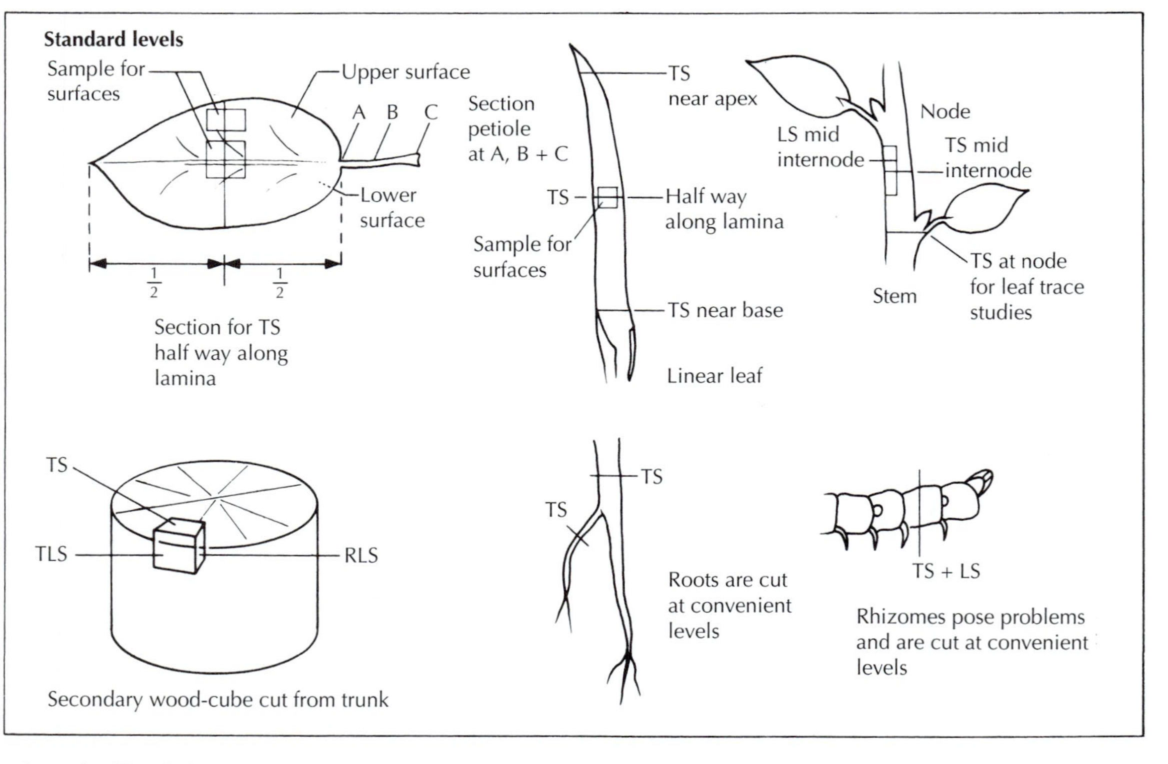

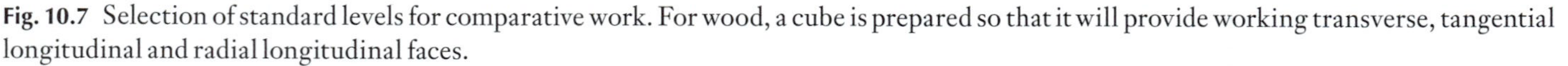

Fig. 10.7 Selection of standard levels for comparative work. For wood, a cube is prepared so that it will provide working transverse, tangential longitudinal and radial longitudinal faces.

Thin sections are examined in the TEM, or carbon replicas produced from specimens can be used where surface features are to be studied. Electrons are made to pass like a focused light beam through the section. Some parts of the specimen are electron-dense or are prepared by stains and fixatives to be electron-dense, whereas other parts are electron-opaque, permitting electrons to pass and form an image either on a special fluorescent screen, or directly on a photographic plate. The TEM uses electromagnetic lenses to focus the electron beam, which has a very much higher resolving power than a beam of light. That is, it can make distinct points which are very close together on the object. The largest commonly available microscopes can resolve between points about 20 nm apart, and have the ability to magnify above 500,000×. Naturally, at such high magnifications, only very small areas can be seen at a time. By comparison, the best light microscope using green light can give a maximum real magnification of about 1200×. Using ultraviolet light, slightly higher magnifications can be obtained.

Methods of preparation of the specimen for TEMs may be complex or straightforward, depending on what the investigator wishes to gain from the instrument. However, a very skilled operator is needed to produce good results. Fixation of the specimen is very critical, for example.

The SEM is used most commonly to examine the surface of specimens. Some specimens can be examined fresh for a brief time, but most are dehydrated carefully to minimize shrinkage and distortion, then coated with a very thin layer of metal, usually gold or gold/palladium alloy. This gives a better image and prevents contamination of the microscope by water.

Because of its relative ease of use, and because quite low (10×) magnifications as well as those of up to or above 180,000× can be obtained on it, the instrument has a very wide use in applied plant anatomy. Currently, resolution of better than 700 nm can be obtained as a matter of routine.

The specimen is bombarded with a focused ray of electrons. The electrons are made to scan in parallel lines over a rectangular area. Secondary electrons are emitted by the object, are collected by a series of electronic devices, and a synchronous image is displayed on a small cathode ray tube. Most tubes are about 10 cm square, and have about 1000 lines. The screen itself is photographed to produce a permanent record. Provided the coated specimen is kept clean and dry, it can often be used many times.

A great depth of field can be obtained with this instrument, about 500 times that of a light microscope. Many surface patterns of leaves, seeds and fruits, spores, etc., are being seen and understood properly for the first time of course. The cost will make it impossible for many people to own or even use an SEM, but some characters can be seen and illustrated or described, using other methods; it is astonishing how many can be seen with a good dark field, illuminating optical microscope, for example. It is only specimens which have to be magnified above about 1200× times that cannot be interpreted with the conventional light microscope.

Extending the use of the student's microscope

It is unusual for the student to have access to a high quality light microscope.

One simple way to make the microscope more versatile is to make Polaroid attachments. A disc of Polaroid material mounted over the eyepiece, and another fitted in the filter carrier or in a holder between the mirror (or light source) and the microscope slide, will allow you to turn the instrument into a simple polarizing microscope. Crystals become more readily discernible, as do starch grains and details of cell wall structure (Fig. 8.5 shows part of a stem section in polarized light.)

Thin cellophane placed in one layer over the lower Polarized sheet (the analyser) will cause the light beam to become elliptically polarized. This phenomenon gives a coloured background against which crystals, etc. will appear a different colour. Rotation of the polarized disc over the eyepiece (polarizer) will cause changes in the colours. This technique is useful for examining unstained, macerated material, sclereids in cleared material or for looking at hairs or surface details in preparations where staining would make the subject too dense.

Other optical techniques

Other higher-resolution optical techniques include phase contrast, anoptral contrast (see Fig. 6.4), dark ground, fluorescence, interference and, increasingly, laser scanning confocal microscopy.

The use of very thin (ultra-microtome) sections allows the light microscopist to see details of cell walls otherwise obscured, and the study of the wall structure of transfer cells is a case in point where the thin section allows observation of very fine details in such sections.

APPENDIX 1

Selected study material

Introduction

The majority of the readily available plant anatomy texts were originally written with northern hemisphere readers in mind, and used examples of plants mainly from north temperate or mediterranean climatic regions. Most of theses species will not grow readily in the tropics or other parts of the world. Consequently many people would have had problems in obtaining specimens of the plants for study. This is particularly true for those in the tropics and temperate areas of the southern hemisphere. Even if the species could be obtained, many excellent examples on their doorsteps would have been much more appropriate for study.

In this book, the problem is addressed by including the following lists and notes. They are organized as lists of families showing interesting anatomical characters in leaf and stem, followed by some selected thumbnail accounts. Then secondary xylem is covered, first with a list of anatomical characters found in particular gymnosperm woods, and finally a few thumbnail accounts of angiosperm woods.

Not all members of the families listed will show necessarily the features mentioned, but the features regularly occur where indicated. There are, of course, many other examples of families where features in these lists also occur. We have tried to include families that have widely cultivated members in order to make the list more useful on any continent. You will also find many additional examples on the Virtual Plant CD.

Local floras will help you locate members of the families listed, and you can see if particular species grow near to you.

Leaf

Anisocytic stomata: Brassicaeae, Plumbaginaceae.

Anomocytic stomata: Berberidaceae, Capparaceae, Liliaceae, Polygonaceae, Ranunculaceae.

Branched or dendritic hairs: Zamiaceae, Melastomataceae, Solanaceae, Piperaceae.

Calcified hairs: Boraginaceae, Loasaceae.

Capitate glands: Convolvulaceae, Lamiaceae, Sapindaceae.

Diacytic stomata: Acanthaceae, Caryophyllaceae.

Hairs of varied types: Asteraceae, Lamiaceae, Polygonaceae. Note: not all members of the families named above Malvaceae, Solanaceae. Hairs functioning as hydathodes: *Hygrophylla*. Glandular hairs, secreting mucilage: *Drosera*, *Drosophyllum*.

Hydathodes: Campanulaceae, Piperaceae, Primulaceae.

Hypodermis: Lauraceae, Monimiaceae, Piperaceae.

Latex-containing cells: Apocynaceae, Convolvulaceae, Papaveraceae.

Laticifers, articulated: Papaveraceae, *Hevea* and other Euphorbiaceae.

Laticifers, non-articulated: Apocynaceae, Asclepiadaceae.

Mucilaginous epidermis: Elaeocarpaceae, Malvaceae, Rhamnaceae, Salicaceae.

Papillose lower epidermis: Berberidaceae, Lauraceae, Papilionaceae, Rhamnaceae.

Papillose upper epidermis: Begoniaceae, Melastomataceae.

Paracytic stomata: Juncaceae, Magnoliaceae, Poaceae, Rubiaceae.

Peltate hairs: Bombacaceae, Elaeagnaceae, Oleaceae.

Salt glands: Frankeniaceae, Tamaricaceae.

Scales: Bromeliaceae.

Sclereids: Margraviaceae, Oleaceae, Theaceae, Trochodendraceae.

Silicified hairs: Poaceae.

Stinging hairs: Euphorbiaceae, Loasaceae, Urticaceae.

Tufted hairs: Bixaceae, Fagaceae, Hamamelidaceae.

T-shaped hairs: Malpighiaceae, Sapotaceae, Zamiaceae.

Stem

Cluster crystals: Bombacaceae, Cactaceae, Chenopodiaceae, Malvaceae, Rutaceae, Tiliaceae, Urticaceae.

Cortical bundles: Araliaceae, Cactaceae, Cucurbitaceae, Melastomataceae, Proteaceae.

Cystoliths: Cannabinaceae, Moraceae, Urticaceae.

Deep-seated cork: Bignoniaceae, Casuarinaceae, Hypericaceae, Rosaceae, Theaceae.

Intraxylary phloem: Apocynaceae, Asclepidaceae, Convolvulaceae, Cucurbitaceae, Lythraceae.

Medullary bundles: Apiaceae, Begoniaceae, Asteraceaee, Nyctaginaceae, Papilionaceae, Piperaceae, Saxifragaceae.

Primary medullary rays, narrow: Asclepiadaceae, Brassicaceae, Ericaceae, Meliaceae, Oliniaceae, Rubiaceae, Sapotaceae.

Primary medullary rays, wide: Asteraceae, Begoniaceae, Cucurbitaceae, Ficoideae, Nyctaginaceae, Papilionaceae.

Raphides: Balsaminaceae, Dilleniaceae, Liliaceae, Margraviaceae, Rubiaceae.

Secondary thickening from multiple cambia: Amaranthaceae, Chenopodiaceae, Menispermaceae, Nyctaginaceae.

Solitary crystals: Flacourtiaceae, Mimosaceae, Papilionaceae, Rutaceae, Tamaricaceae.

Superficial cork: Apiaceae Asteraceae, Corylaceae, Fagaceae, Labiatae, Meliaceae, Proteaceae.

A few examples

The short notes which follow mention only a few of the interesting characters that may be seen in each species.

Abrus precatorius (Papillonaceae) – Stem: superficial cork, fibres in cortex and phloem, rhombic crystals, phloem wide, with inflated rays, broad intrusions of phloem into xylem, wide vessels solitary and in long radial chains, narrow vessels and tracheids in clusters, rays heterocellular and 16 cells wide, parenchyma of xylem aliform and in tangential bands, pith sclerified.

Aerva lanata (Amaranthaceae) – Leaf: various hair types, stomata anomocytic and present on both surfaces, cluster crystals. Stem: phloem fibres, large crystalliferous cells in cortex, anomalous vascular tissue with succession of collateral bundles from cambial tissue, vessel elements with simple perforation plates and alternate intervascular pitting.

Aesculus hippocastanum (Hippocastanaceae) – Leaf: hairs unicellular and short uniseriate with warty walls, stomata anomocytic, petiole vasculature composed of cylinder enclosing amphivasal bundles, tanniniferous cells, cluster crystals.

Ageratum conyzoides (Asteraceae) – Stem: hairs, rounded cells of chlorenchyma, endodermoid sheath, fibre strands at phloem poles; vessels narrow in radial multiples with simple perforation plates.

Arbutus unedo (Ericaceae) – Leaf: cuticle thick, stomata anomocytic, palisade chlorenchyma both adaxially and abaxially, sclerenchyma caps to vascular bundles, rhombic and other crystals, tannin in some abaxial epidermal cells.

Averrhoa carambola (Oxalidaceae) – Stem: hairs thick-walled, unicellular; epidermis and hypodermis thick-walled, cortical fibres, vessel element perforation plates simple and oblique, tannin abundant, rhombic and cubic crystals abundant in cortex phloem and xylem.

Bidens pilosa (Asteraceae) – Stem: polygonal in TS, hairs, well developed endodermoid sheath, fibre caps to phloem poles, vessels narrow, in radial multiples, perforation plates simple.

Bougainvillaea sp. (Nyctaginaceae) – Stem: hairs short uniseriate, hypodermis, fibres at phloem cortex boundary, vascular system anomalous, outer bundles embedded in thick-walled prosenchymatous tissue, inner bundles in parenchyma, rhaphide sacs in cortex and pith.

Briza maxima (Poaceae) – Leaf: prickle hairs, silica bodies rectangular in epidermal cells with sinuous walls, stomata paracytic, sclerenchymatous girders opposite vascular bundles both abaxially and adaxially, bundle sheaths, inner sclerenchymatous, outer parenchymatous, chlorenchyma radiate.

Catalpa bignonioides (Bignoniaceae) – Leaf: hairs peltate and uniseriate, stomata anomocytic and superficial, epidermal cells with sinuous walls.

Cistus salviifolius (Cistaceae) – Leaf: hairs, non-glandular tufted and raised on mounds and glandular and capitate, stomata anomocytic, cluster crystals.

Coffea arabica (Rubiaceae) – Stem: phloem fibres, vessels solitary with simple perforation plates, rays narrow, rhombic crystals and crystal sand.

Coldenia procumbens (Boraginaceae) – Leaf: warty hairs with rosette of basal cells, stomata anomocytic, palisade adaxial, cluster crystals. Stem: collenchymatous outer cortex, vessels with simple perforation plates, rays narrow, pith of parenchymatous cells with conspicuous pits.

Cyperus papyrus (Cyperaceae) – Stem: outline triangular, stomata paracytic, conical silica bodies in epidermal cells above hypodermal fibre strands, network of parenchyma with large air spaces, vascular bundles scattered and embedded in parenchyma.

Elaeis guineensis (Arecaceae) – Petiole rachis: vascular bundles in very thick sclerenchymatous bundle sheaths embedded in parenchymatous matrix. Lamina: hairs, expansion cells above and below midrib, hypodermis, spherical silica bodies.

Epaeris impressa (Epacridaceae) – Leaf: epidermal cells axially elongated with sinuous anticlinal walls, stomata anomocytic and superficial on abaxial surface only, vascular bundles with sclerenchyma caps at phloem pole.

Euphorbia hirta (Euphorbiaceae) – Leaf: hairs, abaxial epidermal cells papillose, stomata anisocytic or anomocytic, laticifers, bundle sheaths with contents staining red in safranin, arm cells of spongy mesophyll clearly visible in paradermal preparations. Stem: vessels solitary or in radial multiples, perforation plates simple.

Fagus sylvatica (Fagaceae) – Leaf: cuticle thin except over petiole, epidermal cells with sinuous anticlinal walls, hairs, stomata anomocytic and superficial on abaxial surface only, bundle sheaths with paired crystals, tannin abundant in cells of petiole. Stem: cork arising in outer cortex, phloem fibres, vessels diffuse porous and solitary or in pairs, perforation plates simple (scalariform in some narrow elements), rays uniseriate to multiseriate and heterocellular, xylem parenchyma scattered.

Gloriosa superba (Colchicaceae) – Leaf: epidermal cells over veins elongated with straight anticlinal walls, epidermal cells between veins with sinuous walls, stomata anomocytic abaxial, vascular bundle sheaths parenchymatous, spongy mesophyll composed of arm cells.

Hamamelis mollis (Hamamelidaceae) – Leaf: hairs tufted consisting of 48 thick-walled radiating pointed cells sometimes raised on mounds, stomata superficial and anomocytic or tending paracytic, sclereids in mesophyll, large mucilage cells, cluster crystals, rhombic crystals, tannin cells.

Heteropogon contortus (Poaceae) – Leaf: adaxial epidermal cells larger than abaxial, stomata paracytic, prickle hairs, silica bodies square to oblong to saddle-shaped, sclerenchyma in margins and as abaxial and adaxial girders to main vascular bundles, parenchyma bundle sheaths, chlorenchyma radiate. Stem: sclerified hypodermis, cylinder of fibres to inner side of cortex.

Hyphaene sp. (Arecaceae) – Leaf: stomata appearing tetracytic, hypodermis, sclerenchyma bundle sheath extensions, fibre strands.

Lantana camara (Verbenaceae) – Stem: hairs both glandular and non-glandular, phloem fibres, vessels with simple perforation plates, intervascular pitting alternate, rays narrow and heterocellular, xylem parenchyma abundant.

Mangifera indica (Anacardiaceae) – Stem: cuticle thick, cortex with rhombic and prismatic and cluster crystals, tannin cells and cells with granular inclusions, phloem fibres, vessels angular and thin walled both solitary and in short radial multiples, perforation plates simple or a few scalariform intervascular pitting coarse and alternate, rays one or two cells wide and heterocellular, axial secretory ducts lined with thin-walled epithelial cells in phloem and pith.

Nerium oleander (Apocynaceae) – Leaf: cuticle thick, stomata and hairs in pits on abaxial surface, hypodermis, cluster and prismatic crystals, laticiferous canals near to veins. Stems: internal and external phloem.

Oxalis corniculata (Oxalidaceae) – Stem: some epidermal cells containing tannin, complete cylinder of cortical fibres, vessel element perforation plates simple.

Pittosporum crassifolium (Pittosporaceae) – Leaf: cuticle very thick, hairs, stomata paracytic with massive cuticular rim, sunken, hypodermis abaxially and adaxially, phloem poles to bundles disproportionately large, secretory canals of various diameters, cluster crystals.

Plantago media (Plantaginaceae) – Leaf: hairs uniseriate and short, with bicellular head, stomata anomocytic and superficial, epidermal cells with sinuous anticlinal walls.

Plumbago zeylanica (Plumbaginaceae) – Leaf: glandular hairs, stomata, anisocytic, enlarged tracheids at vein ends.

Polemonium coeruleum (Polemoniaceae) – Stem: hairs, stomata slightly raised, outer cortex of rounded chlorenchyma cells, inner cortex collenchymatous, phloem with transverse sieve plates, vessels solitary or paired and diffuse and angular, perforation plates simple and oblique, intervascular pits fine and rounded, rays narrow.

Rubus sp. (Rosaceae) – Stem: cork arising in middle cortex, suberized alternating with unsuberized layers, phloem fibres, primary rays broad, secondary rays 12 cells wide and heterocellular, vessels wide, in radial or tangential multiples, perforation plates simple, intervascular pitting alternate, pith composed of large and small parenchyma cells, cluster and rhombic crystals present in cortex and pith.

Salvadora persica (Salvadoraceae) – Stem: epidermal cells of uneven heights and some raised into mounds, phloem fibres, xylem with included phloem, vessel element perforation plates simple, intervascular pitting alternate.

Sphenoclea zeylanica (Sphenocleaceae) – Leaf: epidermal cells papillate, adaxial palisade chlorenchyma, parenchymatous bundle sheaths, cluster crystals. Stem: cortex with air spaces, phloem fibres, vessel elements with simple perforation plates, intervascular pitting alternate.

Tamarix gallica (Tamaricaceae) – Stem: cork superficial with large cells, phloem fibres, vessels solitary or in small radial multiples, perforation plates simple, rays 13 seriate and conspicuous composed of wide cells, crystal sand and irregular crystals abundant.

Tecoma capensis (Bignoniaceae) – Stem: cuticle thick, hairs unicellular, cork superficial and to outer side of chlorenchyma, cortical fibre caps and strands of phloem fibres alternating with soft tissue, innermost fibres forming interrupted ring, phloem appearing storied, xylem with narrow vessels in solitary or in short radial multiples, vessel walls thick, perforation plates simple and oblique, intervascular pitting coarse and alternate.

Theobroma cacao (Sterculiaceae) – Stem: hairs unicellular and thick-walled, mucilage cavities (canals) in cortex, phloem fibres, vessel elements with simple perforation plates, intervascular pits coarse and alternate.

Some selected softwoods (gymnosperms) in which particular features can be found

Axial parenchyma: *Sequoia*, *Taxodium*.
Bars of Sanio: *Sequoia sempervirens*, *Podocarpus*, *Taxus*.
Helical thickening in mature tracheids: *Juniperus*, *Taxus*.
Pitting (of tracheid to tracheid walls): alternate biseriate – *Agathis palmerstonii*; multiseriate – *Taxodium distichum*; multiseriate alternate – *Araucaria angustifolia*; opposite biseriate – *Sequoia sempervirens*.
Rays, tall (approximately 30 cells): *Abies alba*.
Rays, low (most less than 10 cells): *Juniperus*.
Ray tracheids: *Picea*, *Pinus*, *Larix*.
Resin ducts (axial): *Picea*, *Pinus*.
Resin ducts (radial): *Picea*, *Pseudotsuga*.
Torus margin (irregular): *Tsuga heterophylla*.
Torus margin (scalloped): *Cedrus*.
Window pits: *Pinus sylvestris*.

Some characters in the secondary xylem of selected hardwoods

The descriptions given here supplement the main text, and give examples of some of the features mentioned there. They are not intended to be complete.

Azadirachta indica, Meliaceae: Vessels solitary and in radial multiples, perforation plates simple, intervascular pitting fine; rays 1–4 cells wide, heterocellular; parenchyma vasicentric and in narrow tangential bands; crystals rhombic, chambered, abundant; gum in some vessels.

Buxus sempervirens, Buxaceae: Vessels narrow, mostly solitary, perforation plates scalariform with many bars, oblique; rays l–2 cells wide, heterocellular, marginal cells upright, central cells procumbent; parenchyma diffuse.

Ceiba pentandra, Bombacaceae: Vessels mostly solitary, perforation plates simple; rays up to about 8–15 cells wide, heterocellular; parenchyma vasicentric and in narrow tangential bands alternating with narrow bands of fibres; tannin or resin in many cells, crystals present.

Dipterocarpus alatus, Dipterocarpaceae: Vessels wide, mostly solitary, perforation plates simple, transverse; tyloses present; rays 1–4 or 5 cells wide, heterocellular; parenchyma vasicentric and apotracheal, scattered and in tangential bands; fibres thick-walled; vertical canals with thin-walled epithelial cells set in broad bands of tangential parenchyma .

Dombeya mastersii, Sterculiaceae: Vessels solitary and in short radial multiples, perforation plates simple, intervascular pitting alternate, pits circular; rays 1–4 cells wide, heterocellular; parenchyma aliform to aliform confluent; fibres thick walled.

Eucalyptus marginata, Myrtaceae: Vessels solitary and in radial and oblique multiples, perforation plates simple, tyloses present; rays 1–2 cells wide, heterocellular; parenchyma mostly amphivasal, fibres thick-walled, dense, septate.

Liriodendron tulipifera, Magnoliaceae: Vessels wide, thin walled, in radial, tangential and oblique multiples, occupying most of volume of wood, perforation plates scalariform, oblique, intervascular pits wide, opposite; tyloses present; rays mostly 2–3 cells wide, expanded at growth rings, heterocellular.

Pistacia lentiscus, Anacardiaceae: Vessels in long, radial multiples, some elements much wider than others, perforation plates simple; tyloses present; rays mostly 1–2 seriate, heterocellular, some with secretory canals; some fibres septate.

Pittosporum rhombifolium, Pittosporaceae: Vessels rounded, angular, solitary, or in radial to oblique groups, perforation plates simple, vessels with tails and very fine spirals; rays mainly 3–4 cells wide, heterocellular, some only 1 cell.

Rhododendron sp., Ericaceae: Vessels angular, solitary or in small groups, perforation plates scalariform, with many bars, some spirals present; rays 1–4 cells wide, some only 1 cell.

Robinia pseudacacia, Papilionaceae: Ring porous; wide vessels solitary or in short radial multiples, narrow vessels in clusters, perforation plates simple, transverse, intervascular pits vestured; tyloses present; rays storied, most 4–5 cells wide, more or less homocellular; parenchyma aliform confluent, storied.

Sparmannia africana, Bignoniaceae: Vessels angular, solitary or in short multiples, or in clusters, perforation plates transverse, simple, intervascular pits large, with narrow borders; rays 1–8 or more cells wide, composed of wide cells, heterocellular; rays making up large proportion of wood; fibres sparse.

Tectona grandis, Verbenaceae: Growth rings conspicuous; vessels solitary, in pairs or radial multiples, perforation plates simple, intervascular pitting fine, alternate; tyloses; rays mostly 1–3 cells wide, heterocellular; parenchyma initial and a little vasicentric; fibres septate; deposits in some vessels.

Practical exercises

Introduction

Practical plant anatomy does not present many problems in larger plant sciences departments. However, the trend is to have fewer and fewer 'classical' practicals. This is, we believe, leading to the situation where fewer plant biologists are able to comfortably recognize cell and tissue types in the cortex or stele of a stem from a plant such as the common bean, *Phaseolus vulgaris*, or for that matter even recognize the difference between a cortex or a stele. We believe that identification of the structural components of stem, root and leaf are critical to a number of plant-sciences-related disciplines. To overcome the resistance that the purchase of expensive microscopes always seems to generate, because they tend to be underutilized in departments, we have developed a series of virtual practicals which, together with other teaching and learning materials, are provided on the accompanying CD-ROM, *The Virtual Plant*.

We are convinced that theoretical anatomy remains useless unless a sound strategy exists in which a well-structured practical component is run in tandem with lectures. Sadly the trend today is less emphasis on observation through the microscope, and when this is undertaken, instructors who are ill qualified to do so often do it. The ability to see, observe, compare, reference and reflect on structure underscores theoretical knowledge and makes a great deal of sense. We believe that the combination of the book and the CD-ROM do that. *Plant Anatomy: an applied approach* taken together with *The Virtual Plant* is a powerful learning and teaching tool. In *The Virtual Plant*, we present a series of self-paced virtual laboratory exercises, which, if used in conjunction with additional theoretical information supplied in this text, means that the student will be able to work through plant structure more easily than is possible in the normal laboratory environment. These are introduced here, and the frustration of using hard copy versus interactive electronic media will become immediately apparent.

So in this chapter we show a glimpse of the contents of the interactive practicals. We necessarily repeat information given in previous chapters of this book (this saves you having to refer back to the relevant chapters, and shows you what is actually on the CD): however, there is also additional information. The intention is that the following simply provides a glimpse of the way in which the CD amplifies what we have written here, and goes considerably beyond it. In the CD, you can click on highlighted unfamiliar terms, and be taken directly to the inbuilt glossary. Here, of course, you will have to thumb through the glossary. On the CD you can move rapidly from the practicals to the main body of the text if you need to.

In the previous chapter there are examples of families and plants that have particular anatomical characters. Here you will find some more. In addition, there are examples from ferns and cycads.

Figure A2.1. Shows a screen dump showing the contents page of *The Virtual Plant.* Ten topics cover the use of the microscope, introduce the student to stem, leaf and root structure, before dealing with topics such as secondary and anomalous growth in plants, before highlighting examples of anatomy that illustrate the evolution of the vascular system, the structure of secondary wood, ending with a brief introduction to structural adaptiveness.

The Virtual Plant is richly illustrated with material encompassing more than 215 documents. The well-illustrated information on the CD-ROM exceeds 550 Mb in size. It includes an online glossary, which contains more than 500 definitions, often illustrated with an image which makes the meaning of the definition very clear to the reader. These are presented as 'pop-up' windows within the exercises, but may also be accessed as a stand-alone glossary from the menu as well, by simply accessing the **Illustrated Glossary** link on the introductory page. *The Virtual Plant* is a very useful adjunct to the laboratory. Students are often not able to see structures properly in the labortaory setting, because there are insufficient preparations available, or possibly few microscopes may be available for their use. We believe that the great advantage of *The Virtual Plant* is that the exercises and information may be accessed in any order – there is no need to work through these from the beginning in order to understand what is being presented. The material that we have chosen is that which is typically used in the normal teaching laboratory environment. We use a variety of plant examples, many of them have become 'standard' plants, but others introduce less familiar species that can be used just as well, and may be more readily available. Importantly, the information and photomicrographs that accompany the practical sessions will ensure that students from less well equipped institutions can, through *The Virtual Plant*, have as many advantages in their practicals as those enjoyed by students in better equipped academic environments.

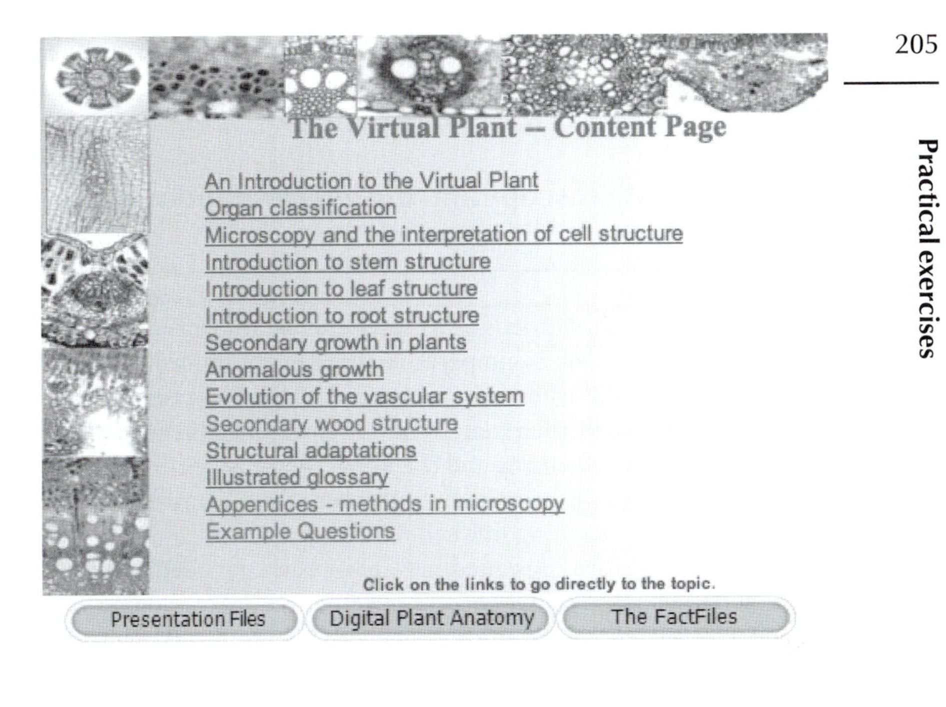

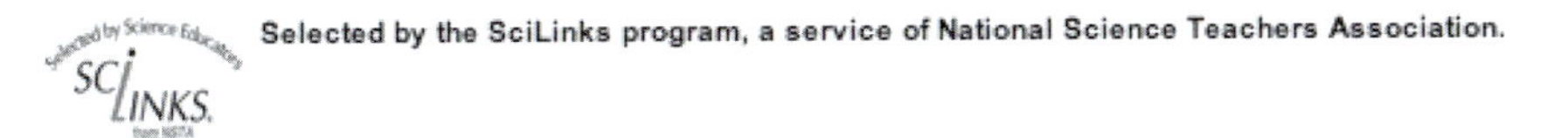

Fig. A2.1 A screen dump showing the contents page of *The Virtual Plant*. Note the Presentation Files, Digital Plant Anatomy and The FactFiles links.

The Virtual Plant contains 11 PowerPoint presentations, which we believe are useful starting points for instructors wishing to present an overview of practical sessions, using many of the specimens that we have used to illustrate *The Virtual Plant*. These PowerPoint presentations are provided 'as is' – they may be modified as necessary to enable the instructor to make best use of the material for preparation of the practical sessions. These can also be used widely as pre-practical sessions, during which students are introduced to some of the material and concepts in understanding anatomy and identification of cells and tissues.

In addition to the PowerPoint presentations, we have included a number of high-resolution digital images which may be freely used to illustrate and reinforce the learning process. No additional information is provided with these images other than the species name, and whether the image is of cells, from a root, a stem or a leaf.

The content of the virtual exercises are outlined below, using images and text extracts from the exercises themselves.

Exercise 1: Microscopy and the interpretation of cell structures

In this short session, we focus on the use a typical student compound microscope. We believe that it is imperative that correct procedures are adopted from the beginning when using the microscope, otherwise students will become frustrated by their inability to see small objects and detail. Pay particular attention to focusing and the use of the substage diaphragm, which aligns and focuses the beam of light through the objective, thus clarifying structural detail including fine cellular details.

The tasks that are presented here have been included to help develop basic skills needed to enable correct and effective use of a compound microscope. Correct use of the microscope will lead to less frustration during practical sessions. This will result in a more stimulating learning experience, as you use your microscope to discover more about the microscopic structure of plant cells and their interrelationships with each other.

If a microscope is available, we deal with removing it from its case, and examining it on the bench, then identifying its components. We have illustrated a Zeiss Standard microscope for the purposes of the exercise. Moving the mouse over the image of the microscope – say over the eyepieces – will, if you click, will bring up a new image, detailing the eyepieces and providing additional information to the reader as well. In many cases, we have adopted the technique of requiring the student to click on something – part of an image, or on a link, to access new more relevant information, where this is appropriate to the learning module.

Structure is best demonstrated by using simple examples. We have chosen to start with pollen grains. Pollen grains are small and to the naked eye will have no visible surface detail. However, place them on a slide in water, under a coverslip, and you may well see a unique surface structure. We have chosen to use *Hibiscus* pollen in this exercise. The bright yellow *Hibiscus* pollen is easy to see and scatter on a microscope slide. And they have some interesting surface detail which is very easy to see with the compound microscope.

Determination of scale and relative size is something that causes a great deal of difficulty in the laboratory environment. It is an exercise that can take a lot of explaining, as it is often quite difficult for the student to grasp the difference between **actual** and **virtual** size. In reality how big is the object that is being observed using the microscope? We have developed a relatively simple method for determining size, making use of a simple

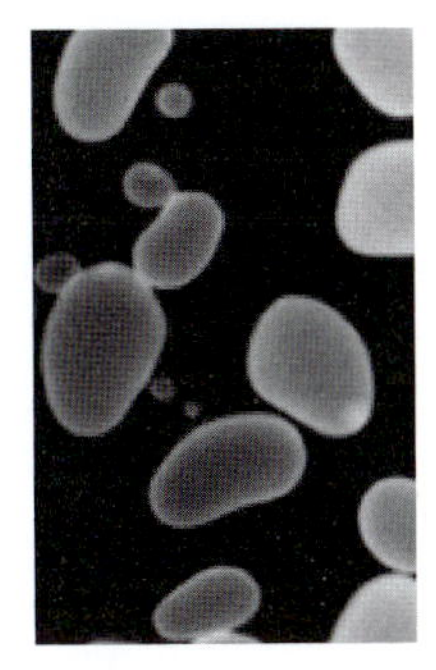

Fig. A2.2 Starch grains, iodine-potassium iodide negative image.

micrometer slide which can be constructed using an electron microscope specimen grid, which is mounted on a slide. Given the known thickness of the bars, and the approximate size (length from one bar to the next in microns), one is able to work out useful measurements such as the diameter of the field of view, and the area of the field of view seen through the eye-piece and using all available objectives. The concept of virtual magnification is thus easier to comprehend as a result of this exercise.

We have chosen to use potato starch grains, as these have substructure which is visible using most average student compound microscopes. Trying to resolve the substructure is a useful exercise, as this will demonstrate to the students just how the **iris diaphragm** which is built into the **substage condenser**, affects the image that one sees. The addition of iodine as a stain introduces the concept of **cell histochemistry** as well at an early stage during the learning experience. Histochemical stains are very important as these impart colour to the specimens, and the colour that one sees not only helps identify the structures, but also helps to determine the composition of the wall itself (Fig. A2.2).

By using simple structures, the microscopy exercise will introduce the student to easy observation, help understand size, staining procedure, the process of irrigating a specimen and how to mount a specimen on a slide, under a coverslip.

Exercise 2: Introduction to stem structure

All stems require some mechanical support, which varies from plant to plant. Structural modifications may also vary depending on the environment that the plant is exposed to. Many stems are divided into cortex and stele, with the division occurring just outside the vascular bundles. Staining a fresh section in iodine, for example, may reveal a high proportion of starch in this boundary layer, which is appropriately called the starch sheath.

The exercise introduces the student to primary stem structure in the major seed plant groups – monocotyledons, dicotyledons and the gymnospermous plants. We introduce the concept of structural (mechanical) and functional (water-transport, carbohydrate-translocating) tissues and the cells which comprise these complex tissues.

The stem has a more complex structure than the root. In the first place, it represents the aerial part of the plant, which bears leaves, and is divided into nodes and internodes. The consequences of this are in the form of the stem apex modified to accommodate production of new leaf and stem tissue because the development of the stem apex is intimately bound up with the formation of leaf primordia.

Even simple stems may be structurally complicated. For example, the presence of a branching vascular supply to the leaves results in the internal structure at the node being much more complicated than that of the internode, which usually bears no appendages. Another example is that the cortex of the stem usually contains more tissue types than the corresponding region in the root, a feature that may be related to the aerial habit of the stem and its photosynthetic capacity, as well as its mechanical function. Thus, we may find stomata in the epidermis and chloroplasts in the cortical parenchyma. We explore the distribution and structure of mechanical tissue such as collenchyma and sclerenchyma which is usually abundant in stems.

All young stems are not the same, yet there are common features in their structure. A principal aim of this exercise is to demonstrate common and distinctly stem-like features in vascular plants in general.

The exercise starts with a 'road map' to the specimens that have been selected as potential study specimens to be examined (Fig. A2.3). In this exercise, we have chosen a number of species that we feel are representative, and universally available.

The longitudinal section of the *Coleus* shoot apex serves as a good introduction to the structure of the shoot apex. This section, and the detail images associated with it, serves to illustrate the relative simplicity on one hand as well as the complexity on the other, of a structure such as a shoot apex. A core objective of the image and the associated reading is to help students to recognize the principal components of the shoot apex.

The shoot apex is a complex structure. At first glance, the system seems to contain very similar cells to other regions of a plant. However, the actual zones that are involved in cell division are very limited. The most important can be recognized in the apical dome, between the two young developing leaf primordia in the *Coleus* shoot apex shown here. The meristematic zone is divided into two layers usually, an outer layer that is responsible for formation of the outer regions of the plant, including the first-formed protective layer, called the protoderm and, beneath this, a region of thin-walled

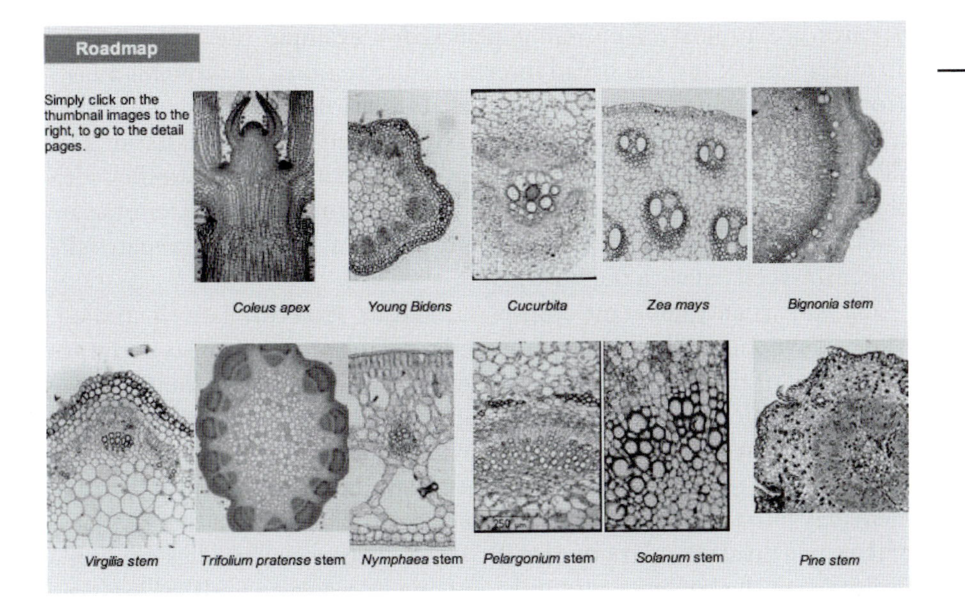

Fig. A2.3 A typical *Virtual Plant* roadmap, from which elements of the exercise may be selected in any order. It illustrates the specimens which have been included in this exercise. Clicking on these will short cut the reader to the specimen concerned. So, if you want to look at the *Trifolium* stem, simply double click on the image . All images are shown as 24-bit colour thumbnails for web viewing.

cells, which will form the cortex. Look carefully, and you should be able to recognize a region of meristematic tissue called the procambium, which consists of a group of narrow, elongated, densely cytoplasmic cells. The procambium is responsible for the formation of the first vascular tissues – the protoxylem and protophloem – which occupy the inner region of this young developing stem. This region is called the stele, which forms the central core of the stem. So, whilst the shoot apex appears at first glance to be a simple structure, it is actually a very complex and important functional part of the plant, responsible for forming the many cell types found in the aerial parts of the plant.

The anatomy of young stems is illustrated using 10 examples of dicotyledonous, monocotyledonous and gymnospermous plants. *Bidens pilosa* has been included, as it is an example of a species that undergoes limited secondary growth, and thus this species is ideally suited to demonstrating all stages of primary as well as some secondary differentiation. This is limited to the vascular bundles only. It is also a suitable example to use, to demonstrate the development and differentiation of the mechanical support tissues associated with the cortex, as well as the vascular bundles.

The Cucurbit stem is an example of an herbaceous dicotyledon, but the stem structure differs in many respects from that of *Bidens*. Some of the features are typical of climbing plants, for example the rather wide-diameter vessels and the broad interfascicular regions. *Cucurbita* is a fast-growing herb, which has **hollow stems and petioles**, so not much carbon resource is allocated to the production of lignified mechanical tissue. It is also a good example of a plant which has bicollateral vascular bundles with phloem outside (external) as well as inside (internal) to the xylem tissue. Look for sieve plates within the phloem, as well as protein aggregates (P-protein) associated with the sieve plates.

Zea mays is an important crop plant, and its structure is well studied. Maize is a monocotyledonous plant, and resembles other grasses in the arrangement of tissues in the stem leaf and root. The stems of monocotyledons are characterized by usually having a single ring of vascular bundles immediately beneath the epidermis, under which a spiralling series of vascular bundles occurs. The peripheral vascular bundles are those that immediately join the leaf traces, and, as such, any differences in the structure of the superficial vascular bundles and the deeper-seated bundles might reflect some of the known structural components associate with the leaf vascular bundles. The vascular bundles are very similar to each other and are representative of the 'typical' construction within the monocotyledons. The phloem consists of large-diameter sieve tubes and smaller companion cells. Phloem parenchyma, usually found at the periphery of the phloem strand, is also located between the large metaxylem vessels (**MX**), which are connected to several tracheids. The protoxylem in monocots is limited to one or two protoxylem vessels, which may be destroyed, leaving a protoxylem lacuna behind, during the rapid elongation growth usually associated with monocotyledons.

Nymphaea petiole is a plant of contrasts. The rooting system and petioles are submerged and functional leaves float on the surface of the water. The adaxial leaf surface is exposed to sunlight and, given that the abaxial leaf surface is in direct contact with water, whole leaf surface temperatures are unlikely to become too stressful. Because of its continual exposure to direct sunlight, this hydrophyte will possibly have a high rate of photosynthesis.

The vasculature is 'typical' of a hydrophyte, in that the xylem is much reduced, and remains functional to traffic transpiration water as well as nutrients. The phloem, on the other hand, occupies a significant proportion of the cross-sectional area of the vascular bundle and contains many large sieve tubes and associated companion cells. *Nymphaea* illustrates several key structures commonly associated with submerged aquatics. For example, large intercellular spaces are found in the petiole. These play a significant role in gas exchange, and provide an internal pathway for this process. These airspaces are surrounded and delineated by long finger-like columns of parenchyma cells. The columnar cells contain chloroplasts and are presumably photosynthetic. Also, the stomata are on the upper leaf surface to facilitate gas exchange with the atmosphere. In many cases, limited mechanical supportive tissue is evident, for example, large astroscleríeds are often associated with the parenchymatous cells delimiting the internal airspaces, adding some mechanical strength to an otherwise fragile structure.

Pelargonium stem: *Pelargonium* is an example of an herbaceous plant that demonstrates secondary growth within the stem. It therefore makes an ideal study plant because many features such as the development of a periderm, perivascular fibres and interfascicular as well as fascicular cambial activity can be demonstrated using *Pelargonium* species. The micrograph illustrating stem structure shows a cross-section of a stem which has completed its primary growth phase, and is at the onset of secondary development.

Solanum (potato) stem: The potato (*Solanum tuberosum*) is a member of one of few families that have phloem which is located, as is normal, external to the metaxylem (and to the fascicular cambium when this is formed) as well as internal to the inner side of the protoxylem. If you do not look carefully, you may miss the phloem, which is separated from the protoxylem by a few rows of parenchyma. The internal phloem is formed by procambial strands although it is not clear if these are remnants of the procambial strand that would have differentiated into the more 'normal' collateral component of this vascular bundle, or if they are reformed prior to differentiation of the internal phloem.

Pinus (the pine) stem: The gymnosperms are an important study group, as they are of great economic importance. Structurally, gymnosperm stems differ from the dicotyledonous and monocotyledonous examples shown earlier, in that the vascular tissues are less specialized. The phloem, for example, lacks sieve tubes and companion cells, and more primitive sieve cells manage the conduction of carbohydrate. The sieve cells are accompanied by albuminous cells and associated parenchyma cells. In contrast to the dicotyledons, gymnosperm wood contains tracheids, and no vessels. All gymnosperms contain resins of some kind, which are transported in resin ducts or resin canals that are usually associated with the cortical tissues in the stem.

Exercise 3: Introduction to leaf structure

This exercise covers a wide range of leaves, and again, we have selected what we believe to be useful examples in order to make understanding and comprehension of leaf structure clearer, and more interesting.

Leaves are the principle sites of photosynthesis in plants. Sunlight is utilized by the chloroplasts within the mesophyll cells within the leaves, and carbon dioxide becomes incorporated into complex carbohydrate molecules. Leaves are also the principal site of water loss, through a process called transpiration during which water vapour is lost to the atmosphere through stomata. The evapotranspirational water loss can help lower the leaf temperature several degrees below that of the ambient air temperature. Nature's 'air-conditioning systems' serve an important function not only in temperature regulation, but also in ensuring a constant 'pull' of water from the soil through the roots to the leaves, thus ensuring a continuous supply of nutrient during this process.

When studying the anatomy of various organs, it is important to keep the functions that these organs perform in mind. This is particularly important in the case of the leaf. An understanding of leaf anatomy is impossible unless it is correlated with some knowledge of leaf function. For example, the main functions of the leaf are photosynthesis and transpiration, both of which involve gaseous exchange between living cells and the atmosphere. It is necessary therefore to consider such features as:

1 The total absorbing surface, both for gas exchange and sunlight utilization.
2 The permeability of the epidermis to gases.
3 The extent of the total intercellular space.
4 The nature of, and distribution of, vascular tissues.
5 Climate and habitat.

Suggested specimens

We have suggested specimens that may be available in the field as well, so that you could compare what is available with what is presented in *The Virtual Plant*. We have taken text from the exercise, in order to illustrate how this topic is dealt with.

Ligustrum leaf: An example of a mesomorphic dicotyledonous foliage leaf. The midrib contains a single, large collateral vein. Here and on the lamina the upper epidermis has a thick cuticle, while the lower epidermis has a much thinner cuticle. In this mesomorphic leaf, the mesophyll is organized into a palisade (upper) and a spongy (lower) photosynthetic zone. Palisade cells are arranged vertically, with small intercellular spaces between them, whilst spongy mesophyll is much more loosely and randomly arranged to allow for rapid and efficient gas exchange. Intercellular spaces occur below the stomata. The vertical arrangement of the palisade cells means that many of the chloroplasts within the mesophyll cells are shaded from direct sunlight – thus reducing the number of mole quanta of light reaching them, and thus limiting light damage to the sensitive photosynthetic machinery within the chloroplasts themselves. The two veins that you can see in this micrograph are embedded between the palisade and spongy mesophyll. They are therefore classified as minor veins.

The midrib of *Ligustrum* contains a single, large collateral vein. The large vein is completely surrounded by parenchyma cells, which form a bundle sheath, which separates the vascular tissue from the non-vascular tissue. A well-developed cambium can often be seen which separates the upper (adaxial) xylem from the lower (abaxial) phloem. Most of the xylem and phloem is primary. Some of the parenchyma cells above the lower epidermis are collenchymatous and thickened. The midrib of *Ligustrum* contains a single, large collateral vein. Here and on the lamina the upper epidermis has a thick cuticle. The large vein is completely surrounded by parenchyma cells, which form a bundle sheath. A well-developed cambium separates the upper (adaxial) xylem from the lower (abaxial) phloem. Most of the xylem and phloem is primary. Some of the parenchyma cells above the lower epidermis are collenchymatous and thickened. The lamina or blade of the leaf contains many different vein orders. Try to recognize as many as you can. If provided, look at the lower leaf surface to see the reticulate or netted venation pattern which is typical of most dicotyledonous foliage leaves.

Amaranthus leaf: At first glance, there are many similarities between the *Amaranthus* leaf and that of *Zea mays*. What do you see first? One cannot help but notice the large bundle sheath cells that contain conspicuous large chloroplasts. Look at the surrounding mesophyll. It is described as radiate and is referred to as the Kranz mesophyll. These Kranz cells contain smaller chloroplasts than the underlying bundle sheath cell chloroplasts. *Amaranthus* is an example of a C_4 dicotyledonous plant. The mesophyll and stomata (**S**) occur on adaxial and abaxial leaf surfaces. Look carefully at *Zea mays* and compare the structures. What you can see that is similar includes the arrangement of the bundle sheaths and their surrounding radiate mesophyll as well as the chloroplast dimorphism with larger chloroplasts in the bundle sheath cells, and smaller ones in the radiate Kranz mesophyll.

Nerium oleander leaf: *Nerium* is an example of a xeromorphic leaf. If you examined the oleander leaf using a dissecting microscope, you would be able to see numerous spots on the leaf surface. At higher magnification, you would be able to see that these are not spots, but holes in the leaf surface, and that the holes contain numerous hairs. Examining the leaf section using a compound microscope would reveal the structural detail more clearly. There are many invaginations in the lower epidermis. These invaginations of the epidermis form stomatal crypts, which contain many trichomes as well as a large number of stomata. Note the distribution and size of the veins. Microscopic examination should make use of both compound and dissecting microscopes where appropriate.

Nymphaea leaf: The floating leaf of the waterlily is an excellent example of a hydromorphic leaf. *Nymphaea* shows the reduction of the xylem tissue, but not of the phloem tissue which is common in hydrophytes. Sclereids are scattered throughout the parenchymatous cortex – these offer some mechanical support. The abaxial leaf surface lacks a cuticle and the stomata are found in the adaxial surface. Metaxylem vessels are large and few in number.

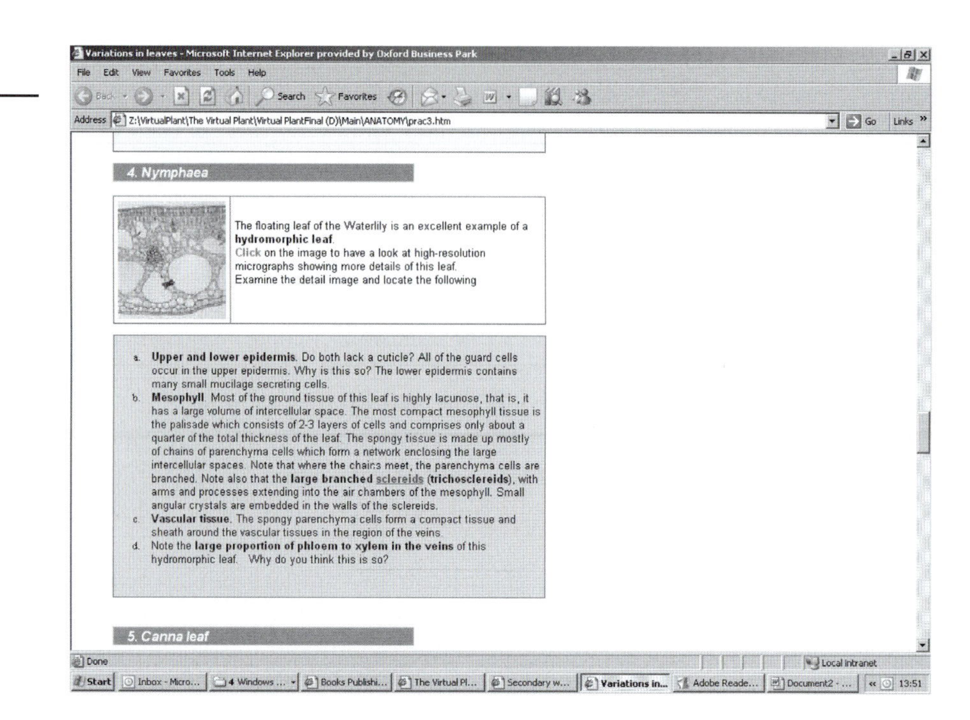

Fig. A2.4 Screenshot of the *Nymphaea* exercise.

The associated narrower-diameter protoxylem elements are much narrower in diameter. Sieve elements contain darkly-staining material (P-protein). The intercellular spaces beneath the palisade are supported by columnar parenchyma cells (all containing chloroplasts and therefore assumed to be photosynthetic). Vascular bundles occur centrally in the lamina and are supported by the adaxial palisade and abaxial spongy mesophyll. Intercellular spaces associated with the abaxial side of the leaf are large and again support is maintained through **columnar** mesophyll cells interspersed with astrosclereids. Spongy mesophyll does not occur in the leaf, as there is no direct need for this tissue due to the presence of the large intercellular spaces.

The three images in Figs A2.4–A2.6 illustrate the 'drill-down' principle employed to provide easier access to information in smaller units than are commonly found in texts.

In most cases, each image is linked to several other pages where additional information may be obtained. In the case of the waterlily in Fig. A2.4, clicking on the image accesses a page with a labelled image and plan of the tissue distribution in the petiole.

Clicking on the link at the bottom of the page in Fig. A2.6 provides yet more additional information. This 'drill-down' concept is used throughout *The Virtual Plant*, where this is appropriate.

Canna is an example of a monocotyledonous plant. It is a widespread genus, often growing in wetlands where it can choke waterways. It is

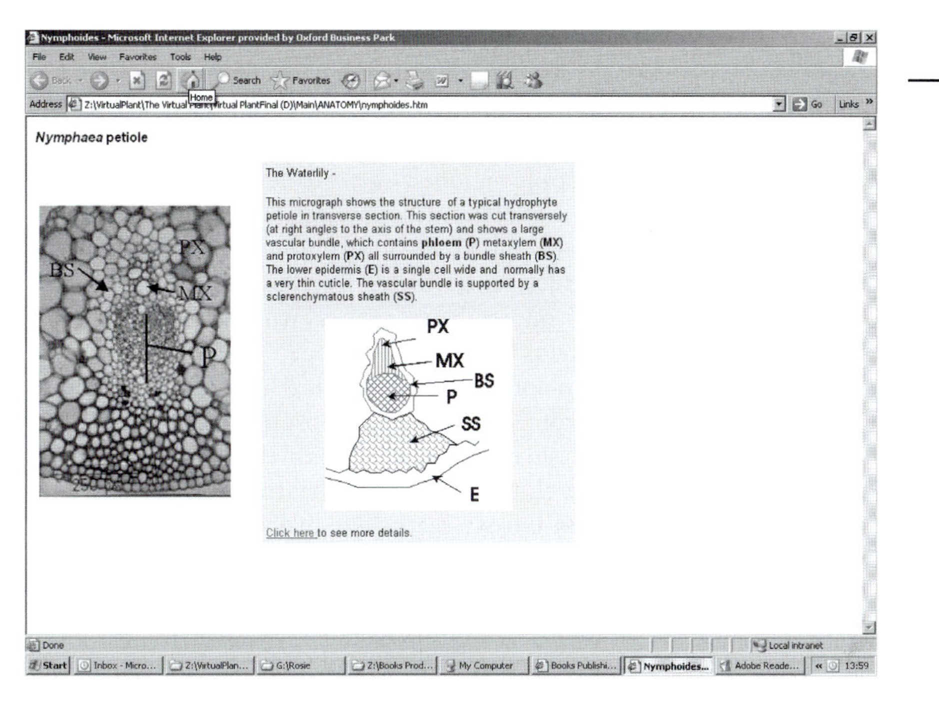

Fig. A2.5 Screenshot of the *Nymphaea* exercise, detailing the anatomy of the waterlily.

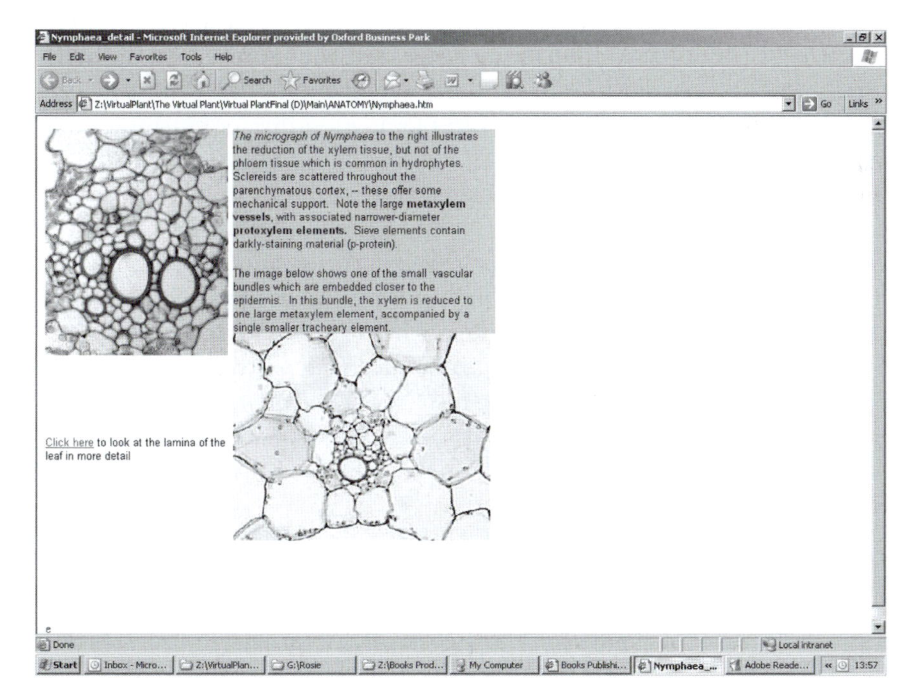

Fig. A2.6 Screenshot of the *Nymphaea* exercise.

popular because of its showy flowers, and there are many hybrids. The petiole of *Canna* contains large conspicuous airspaces, which are demarcated by aerenchyma cells with thin, finger-like projections that touch neighbouring cells with which they share a common wall. In these examples, we have cut sections through the petiole region of *Canna* to demonstrate the aerenchyma cells which make up a lot of the mesophyll associated with the midrib region. These cells are thin-walled and translucent (colourless). These cells have many arms, or projections, which effectively forms a large three-dimensional network of intercellular spaces within the leaf This airspace facilitates buoyancy as well as gas exchange.

Zea is an example of a C_4 plant. Several anatomical features are commonly used to distinguish C_4 from C_3 plants. The most notable of these is the presence of a bundle-sheath surrounding the veins in the leaf, which usually contains large, conspicuous chloroplasts. Note that the mesophyll radiates from the bundle sheath – it looks wreath-like, hence its German-origin name, the Kranz mesophyll. Another diagnostic feature which can be used to separate monocot from dicot foliage leaf is the fact that when monocot leaves are viewed in transverse section (i.e. the section that you can see in the sections presented in this exercise), all the veins are cut in transverse plane, whereas typical transverse sections of dicots will have veins which are mostly cut obliquely. Many xerophytic monocot leaves have groups of large, swollen epidermal cells, interspersed with smaller epidermal cells. These large cells are called bulliform cells. These bulliform cells have a significant role in the life of the plant, in that in times of water stress, the leaf is able to roll up, due to the rapid loss of water from the cytoplasm of the bulliform cells. This results in a smaller portion of the leaf being exposed to the atmosphere and, as a result, a lower rate of water loss due to transpiration or evaporation.

Phormium (Agavaceae), or New Zealand flax, is grown extensively in many countries as a source of high quality fibre. Examine a cross-section of the leaf and identify characteristics that confirm this as another example of a monocotyledonous foliage leaf. Fibre caps are conspicuous, and occur in association with the vascular bundles. These are long sclerenchymatous fibres. These fibres impart a great deal of mechanical strength to the leaf, and are obviously of commercial interest. New Zealand flax was used in the past quite extensively for medicinal purposes. The sap, which is quite sticky, is said to be a mild disinfectant, as was the sap from the roots. It finds use today in the soap and cosmetic industry.

The *Pinus* leaf is an example of a Conifer (Gymnosperm). Conifer leaves vary from needle-like to rather flattened structures. The pine leaf is of course an example of a needle-like leaf. The epidermal cells have very thick walls and a prominent cuticle. The guard cells are sunken and attached to prominent subsidiary cells. The outer cells of the ground tissue are differentiated as a hypodermis and are sclerenchymatous. Many of the remaining cells (the mesophyll) of the ground tissue have internal ridges projecting into the cell lumina. In transverse sections the mesophyll appears compact but longitudinal sections would reveal the presence of intercellular spaces. Most pine leaves contain two or more resin ducts, and the leaf you are examining contains several such ducts. The pine leaf contains two vascular bundles, which are surrounded by transfusion tissue. Transfusion tissue is composed of tracheids and parenchyma cells. The transfusion tissue is surrounded, in turn, by a conspicuous endodermis, the innermost layer of ground tissue. The endodermal cells may have somewhat thickened walls, especially the outer tangential wall, which contains conspicuous simple pits. Endodermal cells may contain a well-developed Casparian strip. Each vascular bundle contains a vascular cambium. The micrographs show that these needles may therefore produce some secondary xylem and secondary phloem.

Exercise 4: Introduction to root structure

Roots serve two important functions – the first is purely mechanical, as they are responsible for anchoring the plant firmly within the ground. The second important function is ensuring that an adequate supply of water and nutrient reaches the aerial parts of the plant, via the xylem. The roots also contain phloem tissue which is the pathway through which assimilated carbohydrate moves from the points of synthesis (source) to the sites of storage (sinks).

The *Ranunculus* root is often used as it contains a very prominent pericycle. The pericycle cells are lignified, and have conspicuously thickened radial and inner tangential walls. The pericycle contains a few thin-walled passage cells, which, as their name implies, allows the passage of water and nutrient from the cortex to the xylem within the stele. Passage cells are also the route taken by carbohydrate to the living cells within the cortex.

During this exercise, you will be able to take a close look at, and study the structure of, roots.

The *Ranunculus* root serves to illustrate the relatively simple structure of a young dicotyledonous root. The main reason for the relative simplicity in root anatomy is the fact that there are no leaves (and therefore no associated leaf traces entering the root vascular system). The primary vascular core (stele) remains relatively undisturbed, and becomes more complex with the introduction of secondary vascular tissues which are found only in roots of dicotyledons and gymnosperms. The cortex lacks secondarily –

modified or lignified – mechanical support tissue. Why? Obviously, the structure is supported by the soil within which it grows. The root has a fairly simple structure. The epidermis has a fairly thick cuticle, with numerous epidermal hairs (trichomes) that are extensions of these cells. The **exodermis** is formed in the outermost layer of cortical cells and consists of a layer of parenchyma cells which take over the function of the epidermis when the epidermis is shed. The cortex is composed of parenchyma. You may see large airspaces in the cortex, which are bounded by parenchyma cells arranged in radiating plates. This condition is usually found in waterlogged soil. Otherwise the cortical parenchyma is associated with small intercellular spaces. Cortical cells will contain starch grains. Xylem – seen in transverse section, as a lignified four- or five-rayed star. The points of the star are the protoxylem and the rest is the metaxylem, which becomes differentiated from the outside towards the inside of the root (i.e. it differentiates centripetally). This primary xylem is composed of xylem vessels, the last formed and widest of which lie at the centre of the stele. The arrangement of xylem, with the narrow-diameter protoxylem vessels lying towards the outside, is referred to as an exarch condition. The phloem occupies patches and is composed of small cells, with dense contents, lying between the protoxylem points. The phloem is composed of sieve tubes and companion cells. The companion cells are much narrower diameter cells than the sieve tubes. The xylem and phloem alternate in strands with one another, with the protoxylem occurring just beneath the pericycle.

Zea mays root: The monocotyledon root of *Z. mays* is also divided into a cortical and stelar zone. Like all primary roots, an endodermis forms the boundary between the cortex and the stele. Internal to the endodermis is the pericycle, which is the outermost layer of the stele. Note again the regular alternation of xylem and phloem, with strands alternating as in the *Ranunculus* root. However, monocotyledonous roots contain many more xylem and phloem strands than do dicotyledonous or gymnospermous roots.

Iris root, TS: Again, this is another example of a monocotyledonous root, but this time one with a very prominent endodermis – seen here as the layer of cells with the prominent thickening of the radial and inner tangential walls. The endodermis is the innermost layer of the cortex. The wall thickening forces water and other molecules to take a symplasmic route from the cortex to the stele, and vice versa, through the unthickened passage cells.

Helianthus root: The example we have chosen to highlight is of a mature root, which illustrates the extensive secondary growth that can take place in a sunflower during one growing season. Secondary roots of *Helianthus* may contain large secondary xylem vessels that are arranged radially in files and these are usually interspersed with narrower vessels and tracheids. Variable width parenchyma rays separate the xylem.

Secondary growth is usually more apparent in the stem and root systems of dicotyledonous and gymnospermous plants. It is characterized by the development of secondary tissues which include new xylem and phloem conduits (within the axial system), as well as the development of other secondary tissues, including fibres as well as a new protective layer, called the periderm, which replaces the primary epidermis as that tissue becomes damaged during secondary stem and root growth. Secondary growth commences in the vascular bundles and will spear to the interfascicular region between the bundles. In some plants, such as the *Helianthus* stem at left, secondary growth is limited but will result in the formation of a ring of secondary xylem and phloem, which will fill in the interfascicular regions between the vascular bundles.

There are three phases involved in the development of the plant – the embryonic, the primary and the secondary. The primary plant body is composed of the division and differentiation products of the apical meristems. This includes the procambium, which differentiates to form the primary vascular tissue. The secondary plant body is composed of the division and differentiation components of the fascicular and interfascicular cambia. In this exercise we explore the various stages of development in the plant body, using a number of different examples, each of which illustrates various aspects of development of the primary and secondary plant body.

Bidens pilosa is a weed species. As such, the prime object for this species is rapid growth, niche occupancy and flowering. In order to do this some energy has to be expended on mechanical support structures in order for the plant to be able to successfully invade space and occupy a niche. As is typical of many of the weedy species, this means that the plant does not invest heavily in lignified tissue, but uses other strategies (such as partial lignification only and hollow stems) to provide enough mechanical support. Have a look at the illustrations of *Bidens* which have anatomy that is is in many respects 'typical' of a young dicotyledonous stem. Cambial activity spreads laterally beyond the confines of the vascular bundle, so the fascicular (within the bundle) and the interfascicular (between the bundles) cambia join to form the vascular cambium.

The sunflower undergoes limited secondary growth as well, but, like *Bidens*, secondary growth within the sunflower stem is confined to the vascular bundles. In other words, the interfascicular cambium does not develop much beyond production of a few scattered secondary phloem elements.

Chrysanthemum is another example of a dicotyledonous stem in which limited secondary growth takes place. Although the cambial zone is complete, the interfascicular region does not produce significant secondary vascular tissue in this species either.

Quercus, the oak, is a tree genus which undergoes considerable secondary growth. The outer layers in the example have been replaced by a periderm, and a ring of very thick-walled sclerenchymatous perivascular fibres interspersed with sclereids lies exarch to the secondary phloem. There are detailed images that show that the cambial zone (shown in the leftmost thumbnail for orientation purposes) consists of several layers of cells, indicating that the sections were made from actively-growing material. Click on the left image above, to see details of secondary growth and the right image, to see a detail of the lenticel which is involved in gas exchange.

Prunus species generally undergo considerable secondary growth. As a result growth rings will be evident in the xylem and the periderm will be continually replaced, usually from parenchymatous cells within the secondary phloem that form a new cork cambium at regular (seasonal) times.

Helianthus secondary root development: When seen in cross-section, dicotyledonous roots that have undergone secondary growth have a somewhat confusing anatomy. If one carefully examines the central region of an old root, it may be possible to distinguish the primary from the early secondary xylem. It should be possible to determine how many primary xylem strands occurred in the young seedling's root. The micrographs in this exercise show that there were four primary xylem strands in this old root cross-section.

The beetroot (*Beta vulgaris*) exhibits some evidence of anomalous secondary growth patterns. This is evidenced by the formation of successive cambia, each of which gives rise to a ring of vascular bundles and wide zones of parenchymatous tissue between these bundles. Note that the successive bundles are arranged more or less along the same radius as preceding ones.

The herbaceous pea stem is much like any other herbaceous dicotyledon stem, in that its vascular bundles show limited secondary growth, with some evidence of a fascicular cambium forming a limited amount of secondary xylem and secondary phloem. Like many other herbaceous plants, the pea wastes little energy in the production of lignified secondary xylem, and mechanical support for this stem is provided by the primary phloem fibres as well as the functional xylem elements.

Exercise 6: Anomalous growth

The word anomalous means to deviate from the general or common order or type. Thus, the term anomalous growth reflects a growth condition which is not commonly seen, and which is present in a limited number of families or genera. This exercise explores a few examples of anomalous growth, but bear in mind there are many to choose from! The examples that we have chosen illustrate aspects that are common – and include multiple cambia, included vascular bundles, and multiple vascular cylinders. This exercise illustrates that the development of the stem, root or leaf of higher plants does not always follow a recognizable pattern in all cases. This phloem is part of a primary bundle embedded in xylem. The xylem part of the primary bundle is some distance away. In *Bougainvillea* the stem also produces **included phloem tissue**, which is buried either side of secondary xylem. In sugar beet, **supernumerary cambia** may be produced which result in a significant increase in secondary vascular tissue.

In the sweet potato, additional cambia produce some tracheary elements as well as a few phloem elements. These secondary cambia mostly produce additional storage parenchyma **exarch** (towards the periphery of the structure) and **endarch** (towards the centre) of the tuber.

Anomalous growth can therefore be defined as a growth form which does not follow recognizable patterns that occur commonly in the majority of vascular plants. In this exercise, we explore a number of plants which exhibit varying degrees of anomalous growth structure. In all cases, if one can learn to recognize the tissues that are formed, then one can work out the relationships of the tissues to one another.

The root of *Beta* shows anomalous secondary growth patterns. This is shown by the formation of successive **supernumerary cambia** each of which gives rise to a ring of vascular bundles and wide zones of parenchymatous tissue between these bundles. The illustrations show that successive bundles are arranged more or less along the same radius as preceding ones.

Carrot roots undergo limited secondary thickening. As can be seen in the accompanying photomicrographs, this secondary growth is unlike that seen in typical (normal) secondary growth in roots. The carrot, like beetroot, forms **successive cambia**, and **multiple rings** of vascular bundles.

Dracaena is an example of a palm-like tree. They are monocots that grow quite tall and thick, yet they lack 'normal' secondary growth. *Dracaena* is not a true palm; in common with *Cordyline* it has a **peripheral secondary thickening meristem**, a structure absent from true palms. This meristem produces both new vascular bundles and ground tissue (parenchyma). *Dracaena* is an unusual plant, in that the vascular bundles are surrounded by very **prominent fibre bundles**. In this sense, *Dracaena* is not anomalous. However, the stems undergo a specialized secondary growth which manifests itself in the production of additional parenchymatous elements. Their later growth pattern is termed **diffuse secondary growth**, and consists mostly of a proliferation of ground parenchyma cells and additional vascular bundles near the periphery.

Bougainvillea is a member of the Nyctaginaceae and is an example of a dicotyledonous stem which displays **anomalous secondary growth**. In this transverse section, near the centre of the stem, you will see some

primary vascular bundles embedded in lignified pith parenchyma. Move the slide towards the outer regions, and you will notice that there has been fairly extensive production of secondary vascular tissue. Look for the **vascular cambium**. Secondary phloem and secondary xylem lie on either side of it. The secondary xylem is composed of **tracheids, fibres** and **narrow-diameter vessels**. Interspersed with the secondary xylem you should recognize small pockets of phloem and what look like large-diameter **metaxylem** vessels. These are reminiscent of the primary bundles towards the centre of the stem. These are in fact primary vascular bundles embedded within the secondary xylem, hence the use of the term **anomalous growth** in this instance. The phloem is described as being **included phloem**, which by definition is phloem tissue which lies between regions of secondary xylem. Whilst the physiological advantage of the formation of included phloem has not yet been studied, one could speculate that in this instance, the included phloem would be well protected from predators and pests and, of course, be well-supplied with water and nutrient. The **anomalous growth** results from differential cambial activity. Newly produced vascular cambia result in the outer lateral meristem becoming quiescent, and this cambium returns to activity only when the internal vascular cambium (which produce the individual embedded bundles) becomes less active. Vascular cambia are said to not produce rays in Nyctaginaceae (lateral meristems do), but do produce vessels and associated axial parenchyma and sometimes fibres to the inside and variable secondary phloem to the outside.

Beta stem The beetroot stem is anomalous, in that there are several layers of **primary vascular bundles** visible towards the centre of the stem. An active **vascular cambium** is capped by **secondary phloem strands** interspersed with a few secondary phloem elements, which are derived from the cambial ring.

Campsis radicans (Bignoniaceae) has been included in the anomalous growth exercise, as it is a climbing vine. At first glance, this stem looks like a typical dicot stem which is undergoing secondary growth. The illustration in this exercise shows a broad band of secondary xylem on the outside of the primary xylem. A vascular cambium is visible as is some secondary phloem. The vascular cambium has formed some secondary phloem. Strands of primary phloem fibres are present. So, what's different here? This looks just like a typical young woody stem. Not so. Look carefully at the tissue nearer the centre of the stem internal to the protoxylem. A cambial zone is evident, and internal phloem has developed. This internal cambium in Bignoniaceae forms what are termed inverted medullary bundles.

Boerhaavia is a member of the Nyctaginaceae and has been described as having C_4 and C_3 physiology and mixed anatomy, with some species showing $C_{4,}$ others C_3 type anatomy and related physiology. The stem in *Boerhaavia* undergoes well-defined anomalous secondary growth, which is characterized by the presence of successive rings of xylem and phloem. The cambium is composed of fusiform initials only, which give rise to rayless secondary vascular tissues. The cambium is described as being storied when cell division ceases. Each successive ring of cambium is originated from the outermost phloem parenchyma cells.

The cambial ring is functionally segmented into fascicular and interfascicular regions which produce mostly conducting elements of the xylem and phloem with some parenchyma, the latter to parenchyma cells. The xylem parenchyma cells develop into conjunctive tissue following thickening and lignification of cell walls. Alternate bands of lignified and parenchymatous bands are distinct in the stem.

Dicranopteris is a pteridophyte (fern). The structure of the vascular tissue or stele in rhizomes has often been used to separate certain groups of pteridophytes. The simplest form of vascular structure is the protostele, in which there is a solid vascular core or strand of tissue which contains xylem towards the centre of the stele, and external to this, a strand of phloem. In other instances the central protostele may contain non-vascular parenchyma cells and this condition is termed a medullated protostele or an ectophloic siphonostele. In this definition, a siphonostele is any uninterrupted stele with an undifferentiated centre. Where external as well as internal phloem coexists, the structure is known as an amphiphloic siphonostele or sometimes, equivalently, a solenostele.

Dicranopteris is known to contain xylem vessels in which the end walls are clearly perforate, compared with the lateral wall pits which are associated with a pit membrane.

Serjania stem: The liana *Serjania* a member of the Sapindaceae. This stem is anomalous because it consists of several vascular cylinders enclosed in a common periderm. Each bundle has separate pith. Metcalfe and Chalk describe this as a compound xylem mass and the structure is also referred to as extra-stelar bundles. The secondary thickening develops from a conventional cambial ring, or may be anomalous as in this case, as it forms concentric cambia. These features are common in the Sapindaceae.

Exercise 7: Evolution of the vascular systems

This exercise deals with the process of vascularization in plants. Whilst there are many specimens that we could have chosen to include, we have deliberately focused on those that we feel illustrate the process of vascular evolution in a way that the student will understand and, we hope, will encourage the student to study further. Vascularization in higher plants is dependent upon the need for supply of water, minerals and other nutrient via the xylem conduits, as well as the need to provide an efficient pathway through which assimilated carbohydrates can be transported. In primary stems as well as in leaves, these tissues always occur on the same radius in vascular bundles, whereas they do not follow this pattern in the root.

Vascularization was an essential step in the land migration, as well as the development of more efficient (but not necessarily larger) plants. This exercise examines some aspects of the evolution and development of the vascular systems in plants. We will look at some examples of hydrophytes, as well as some 'typical' land plants, which show variable structural features. Water plants (hydrophytes) generally have a very different structure to those that occupy the land niche. For example, hydrophytes do not need (nor have they) much in the way of mechanical supporting tissues. Hydrophytes also have reduced vascular supply – this is evidenced in reduction of the water-conducting tissues within the xylem. Xerophytes, on the other hand, have much-reduced diameter of tracheids and vessels, have fewer intercellular spaces and may have other physiologically important structures to regulate water and assimilate transport. Xylem vessels and tracheids connect to surrounding parenchymatous elements, through pit membranes, thus ensuring passageway for water and other dissolved organics and inorganics that need to be transported throughout the plant body.

Our first example is *Lycopodium*. The Lycopodiales include the living genera – *Lycopodium s.l.* and *Phyllglossum*. There are a large number of fossil members, which were at their height of diversity during the Devonian and Carboniferous (some 408–360 million years ago). These plants are commonly known as the club mosses. What is interesting with respect to transport of water and carbohydrates is the great variability that we can see in the arrangement of the vascular tissues in the stems. *Lycopodium claviatum* shows an example of a dissected protostele. This is recognized as primitive. *Lycopodium saururus* contains a plectostele in which plate-like xylem strands are interspersed with the xylem. The phloem consists of sieve cells and albuminous cells. Interestingly, the plectostele is found to this day in all young dicotyledonous and monocotyledonous roots.

Selaginella is our next choice. There are about 700 species within the genus *Selaginella*. The herbaceous sporophyte of *Selaginella* is morphologically very similar to *Lycopodium*, but the stem anatomy differs markedly between these genera. This micrograph shows the vascular tissue, which is suspended centrally within a hollow stem by elongated cortical cells called trabeculae.

Rumohra is a fairly common fern, often found in botanical gardens. The rachis shows reduced vascular tissue, which is composed of a dorsiventrally flattened central xylem plate, with the phloem tissue on either side of it. The vascular cylinder is separated from the outlying cortical region by a well-defined endodermis and pericycle.

Encephalartos belongs to the Cycadophyta within the gymnosperms. They are palm-like plants that appeared about 285 million years ago, near the start of the Permian period during the age of the dinosaurs. They are truly living fossils. There are 11 genera that contain about 300 species in all. The petiole is photosynthetic, and contains numerous sunken stomata. The petiole and leaf is very fibrous and has large strands of thick-walled fibres just beneath the epidermis. The vascular bundles contain large and small xylem elements and are separated from one another by an endodermis. Cycads exhibit true secondary growth, which is generally described as very slow to 'sluggish'. The xylem consists of tracheids and fibre tracheids, and the phloem of sieve cells, albuminous cells and parenchyma cells.

Araucaria is a member of the gymnosperms. The wood is similar to that of *Podocarpus* and is composed entirely of long, tapering tracheids, (the files of narrow-diameter pink-red cells) interspersed with which are parenchymatous rays. The phloem (not shown in this image) is composed of sieve cells and albuminous cells. The phloem is not organized into longitudinal conduits as in the angiosperms. The micrograph shows a number of primary rays, which traverse the secondary xylem starting in the pith to the left, and running all the way out to the secondary phloem.

Our next choice is *Cucurbita*. This is an example of an herbaceous dicotyledon. This plant belongs to the Cucurbitaceae, which are almost all exploited by man as a source of food. The pumpkins have interesting anatomy, in that most of their stems are hollow-cantered and the vascular bundles contain two distinct groups of phloem. The one, the internal phloem, is found on the inside of the water-conducting xylem. It is outlined in blue in the micrograph to the right. The other is called the external phloem, and is found on the outside of the xylem. It is outlined in red in the micrograph. The xylem is composed of narrow tracheids and wider-diameter **vessels**. The phloem within the pumpkin family is composed of companion cells and sieve tubes, which contain sieve plates.

The monocotyledons are an important well-represented group of plants. *Zea mays* is a member of the grasses, and is an important food crop worldwide. This monocotyledonous plant contains vascular bundles that are classified as atactostelic. An atactostele is considered to be the most evolved vascular bundle in vascular plants. The vascular system is composed of individual primary bundles, which are separated by broad zones of parenchymatous tissue. The xylem is reduced in each bundle to usually two large metaxylem vessels, and the largest bundles have protoxylem elements which are usually short-lived, and destroyed during stem elongation. They are reduced to lacunae (cavities) in vascular bundles of mature stems. The phloem is reduced to a few, large-diameter sieve tubes, surrounded by companion cells and associated vascular parenchyma cells. There are many vascular bundles in a 'typical' *Z. mays* stem and nearly all are surrounded by parenchymatous cells.

Our next specimen is *Nymphoides* (a waterlily) which is a hydrophyte – there are many distinct features which point to this plant's aquatic habitat.

The petiole lacks a visible cuticular layer, has a very reduced cortex, immediately beneath which are the vascular bundles, which contain reduced xylem. Much of the central region of this plant is composed of airspaces, which contain thin unicellular strands of specialized parenchyma, termed aerenchyma separating the individual airspaces within the stem. Vascular tissue occupies the central region of this specimen. Note that there are usually two vascular strands visible. What is unusual is that the phloem is opposite the protoxylem. The vascular tissue is surrounded by an endodermis (Casparian strips are clearly visible using the 40× objective). Immediately beneath this, is a pericycle. The specimen has thus many features of a 'normal' root.

Leaves show important trends in vascularization. The *Ligustrum* (privet) leaf is a good example. This leaf can be described as a 'typical' mesomorphic leaf. The image illustrates the division of the photosynthetic tissue into an upper part which contains long columnar palisade cells and a lower (abaxial) part, which contains more rounded spongy mesophyll. The vascular tissue which forms vascular bundles for the most part is sited in the mid region, and sits on top of the spongy mesophyll. The smaller bundles are thus embedded in the mesophyll, but each is surrounded by a bundle sheath. Note the position of the xylem and phloem within the bundle. Some of the bundles are seen in transverse section, some in longitudinal section, and some are oblique – why is this?

Amongst the gymnosperms, the pines are a well-studied group, mainly because of their economic importance. *Pinus* is an example of a conifer (Gymnosperm) leaf. Conifer leaves vary from needle-like to rather flattened structures. The pine leaf forms part of a needle-like fascicle, in which the base of the needles are wrapped in scale leaves. Morphologically, the fascicles are short shoots within which apical development has been arrested (stopped) – it is thus a shoot of determinate or restricted growth. The thick-walled fibres occur immediately under the epidermis. This layer is called the hypodermis. The mesophyll is composed of parenchymatous cells with an uneven outline. The mesophyll is separated from the vascular bundles by a conspicuous endodermis. Beneath the endodermis, the vascular bundles (two in this case) are surrounded by transfusion tissue which is involved in water transport from the tracheids in the bundles to the mesophyll, as well as serving an important role in the uptake pathway of assimilates from the mesophyll to the sieve cells within the phloem.

Monocotyledonous foliage leaves are remarkably different to either dicotyledonous leaves or the gymnosperms. Our next choice, the *Zea mays* leaf, illustrates the key issues of the leaf anatomy, which, as a C4 species, is dominated by its compartmentalized photosynthetic machinery. An obvious feature of the monocot leaf are the vascular bundles. There are three sizes or orders of vascular bundle in the maize leaf – large, intermediate and small bundles occur within the lamina (blade) of the leaf. This is typical for many of the monocots. Notice that the mesophyll contains small red-coloured objects. These are chloroplasts that carry out the first stages of photosynthesis. Carbon dioxide is trapped by malic acid within the Kranz mesophyll and is transferred to the bundle sheath via plasmodesmata, where the malate is released, and taken up into the Calvin cycle where the CO_2 is incorporated into sugars. Because the first product formed during photosynthesis is a C_4 acid, this cycle is called C_4 photosynthesis. C_4 photosynthesis is much more efficient than C_3 photosynthesis. Several anatomical features are commonly used to distinguish C_4 from C_3 plants. The most notable of these features is the presence of a bundle sheath surrounding the veins in the leaf, which usually contain large, conspicuous chloroplasts. Note that the mesophyll radiates round the bundle sheath – it looks like a wreath – hence its German-origin name, the Kranz mesophyll.

Exercise 8: Secondary wood structure

It is generally accepted that secondary xylem has undergone a long evolutionary history. The main trends can be seen because the various stages are often related to other 'marker' characters in flowers and fruits of the plants concerned. There are instances where habitat has seemingly reversed some of these trends in various species, but overall, their 'direction' can be fairly safely defined.

Taken in its simplest form, the evidence to hand indicates that the tracheid is a dual-purpose cell, combining properties of both mechanical support and water conduction in evolving groups of plants, which gave rise to fibres with simple mechanical function and to perforate cells, the vessel elements which are concerned with the conduction of water and dissolved salts. This division of labour is seen as a specialization, or advance.

The primitive vessel element shows much similarity to the tracheid as it is axially elongated, with oblique end walls in which are grouped perforations making up the scalariform, reticulate, or otherwise compound perforation plate. The lateral walls bear bordered pits, often in an opposite arrangement. The advanced vessel element is seen as a broad short cell with large, simple transversely arranged perforation plates at either end and alternating bordered pits on the lateral walls. Between these extremes is a variety of forms. In this exercise, we will focus on the structure of wood in stems that have undergone secondary growth. We illustrate the exercise with examples of tropical and subtropical origin, highlighting key features in these examples.

The species illustrated here are few. It is not possible, nor is it desirable, to include too many examples. Again the principle applied in this exercise is to illustrate different examples and, hopefully, to stimulate those who have an interest in wood anatomy to look further. However, the illustrations that we have chosen serve their purpose well as they illustrate the structure of secondary xylem in transverse, radial longitudinal and tangential longitudinal sections. All three planes are required when researching wood, or for that matter, cambium or phloem structure, not just TS. We have included dicotyledons as well as gymnosperms in this exercise.

Figure A2.7 shows how we have chosen to illustrate secondary wood structure in this *Virtual Plant* exercise. The images show transverse, radial longitudinal and tangential views of the wood of *Alnus nepalensis*. Note the large-diameter vessels, interspersed with narrow tracheids and parenchymatic elements. Rays are short and two to four cells wide.

Wood anatomy is best studied using transverse, radial and tangential planes of section. In this way, a picture can be built up of what the wood really looks like. A vanishing perspective point image reconstruction helps us understand just how vessels, tracheids, fibres and parenchyma intersperse to form secondary wood. In many cases, pop-up detailed information is available where we think this provides useful additional information as illustrated in Fig. A2.8.

Figure 14.9 shows a vanishing point reconstruction in 3-D. The reconstruction is not entirely accurate, as it was not made from registered matching sections of *Swietenia mahogani*. Drawing a 3-D diagram does

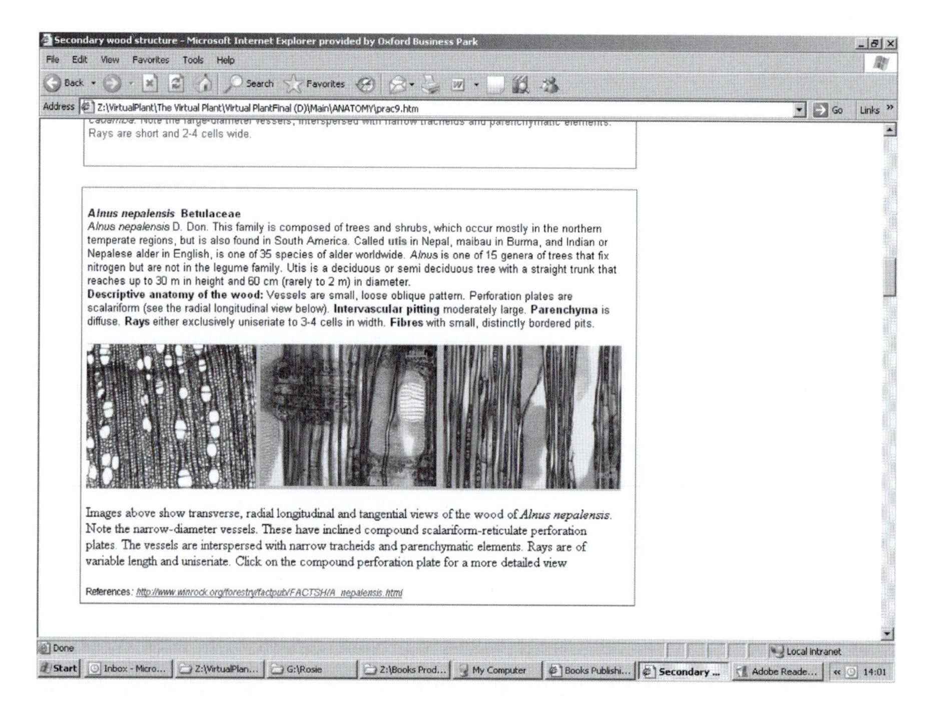

Fig. A2.7 Illustration of secondary wood structure in this *Virtual Plant* exercise.

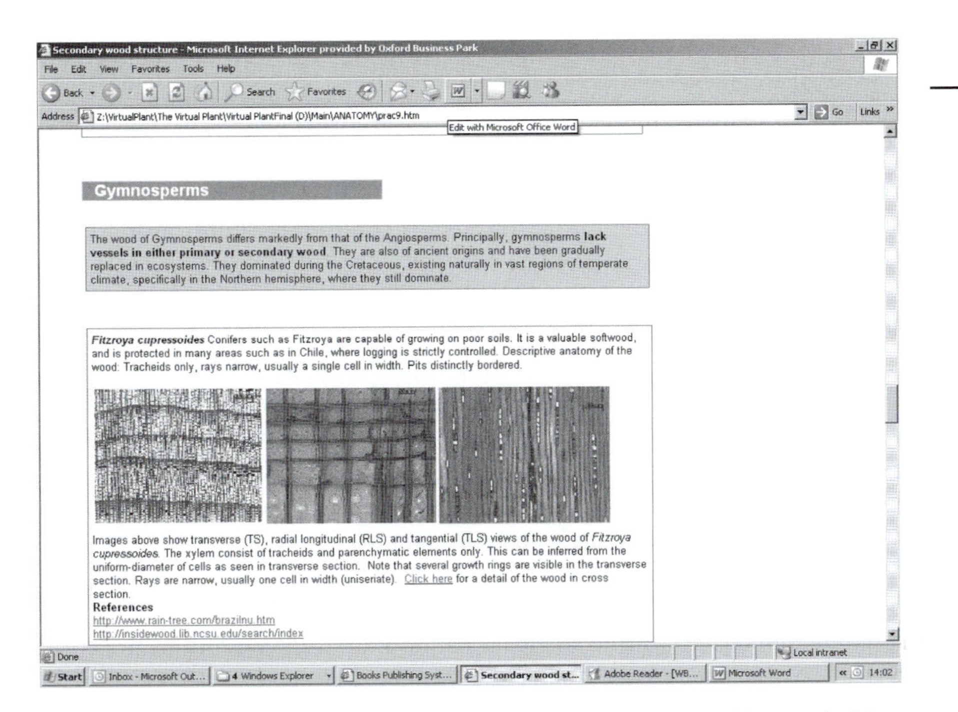

Fig. A2.8 Detail page with information relating to a cross-section of the wood of the conifer *Fitzroya cupressoides*.

not mean you have to try to represent all the cell that you can see, but rather to produce an image that shows the relationship between the three planes of section that we use to understand and explore the structure of the wood.

The exercise cross-references, where appropriate, to the illustrated glossary, dealing with wood-specific definitions. We are particularly grateful to Dr Peter Gasson, from The Royal Botanic Gardens Kew, for kindly and so generously providing the base images that we have used in this exercise.

Exercise 9: Structural adaptations

We have included a section on adaptations to environment, as we believe that the environment that a plant grows in will, with time, force changes in external morphology as well as internal anatomy. It is quite clear that the ability to adapt to environment was a key factor in a plant's survival, as well as the spread of land plants into less-hospitable areas. We explore and highlight some of these factors, and illustrate examples of some simple and complex ways in which plants have successfully adapted to their environments. We have included a number of concept checks, which will point to issues that you should research specifically for issues relating to plant adaptations. Use them to provide a sound guideline for self-study.

There are several important adaptive features that need to be considered here. For the purposes of this discussion, they are divided into categories in this exercise and we focus on:

1 Adapting to immediate environmental conditions (water availability, light intensity and temperature).

2 Competitiveness (climbing, rooting systems and storage).

When looking for adaptations, you will need to have some understanding of what you are looking for. For example, plants growing under dry conditions may well have several structural modifications that are geared primarily to reducing water loss and a reduction of high solar radiation, and the associated temperature effects (Fig. A2.9). Many of these adaptations are visible in leaves, which means one will have to be able to:

1 Distinguish between simple and compound leaves.

2 Determine where and how are leaves produced.

3 Know something about leaf shape and size control.

4 Understand that sun and shade leaves have different anatomies.

5 Realize that shade leaves tend to be thinner than sun leaves.

Figure A2.10 shows part of the apical region of *Oldenburgia grandis*. This plant is endemic to the Western and Eastern Cape regions, where winter

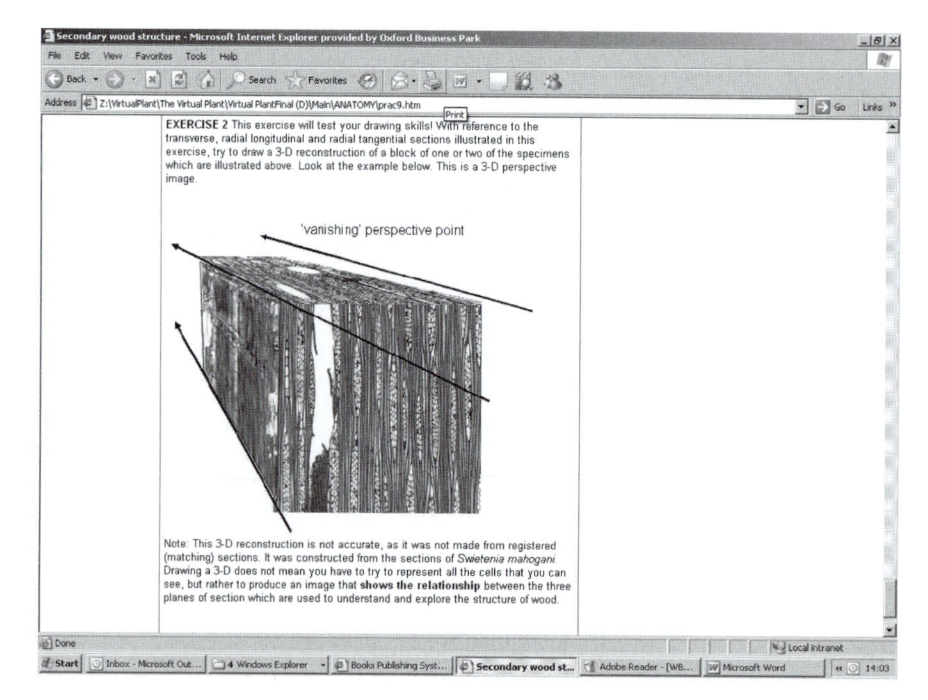

Fig. A2.9 Vanishing-point reconstruction in 3D.

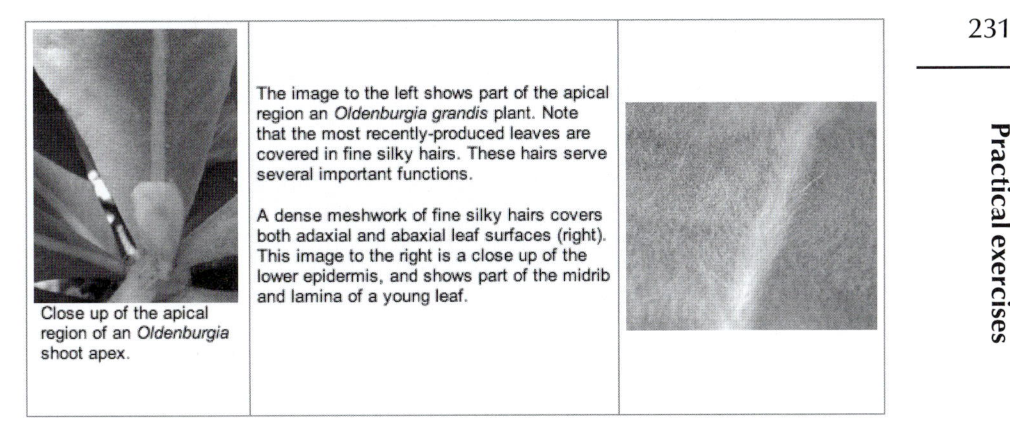

Fig. A2.10 Screenshot of the *Oldenburgia* exercise.

rainfall occurs. In order to prevent desiccation, the most recently produced leaves are covered by a dense meshwork of fine silky hairs which cover both adaxial and abaxial leaf surfaces. This image to the right is a close up of the lower epidermis of a young leaf, and shows part of the midrib and lamina, which are covered in fine silky hairs. These hairs serve several important functions.

1 Reduction in direct and reflected light striking these young leaves.
2 Hairs may reflect some of the light, thereby reducing light stress.
3 Lowering of transpiration rate is associated with hairiness.
4 Hairs may maintain a more ideal leaf temperature.
5 Hairs may induce lower water stress levels.

Basic construction principles vary from plant to plant. All leaves contain photosynthetic tissues, called the mesophyll. The mesophyll is simple in many species, but in the dicotyledons it may be divided into palisade and spongy mesophyll layers. Most leaves are distinctly bifacial, whilst others are unifacial. All leaves are supplied with a vascular system – this may be very varied, but always contain xylem and phloem. An example of an extreme variation can be found in hydrophytes – those plants that live in a watery niche. In the hydrophytes, the major problems that the plants are faced with, dealing with water and sunlight are at a glance, no different from those that terrestrial plants have to contend with, but in the case of hydrophytes, it is simply that they are surrounded and supported by water, and, if they have floating leaves, the leaves need to develop mechanisms to deal with excess light and an efficient transpiration–gas exchange system (Fig. A2.10).

The plant adaptiveness exercise illustrates how plants become modified to adapt to dry high-light environments (Figs. A2.11, A2.12, A2.13).

The micrograph in Fig. A2.11 shows a small vascular bundle in the petiole of *Nymphaea*, the water lily. Notice how the xylem (water conducting

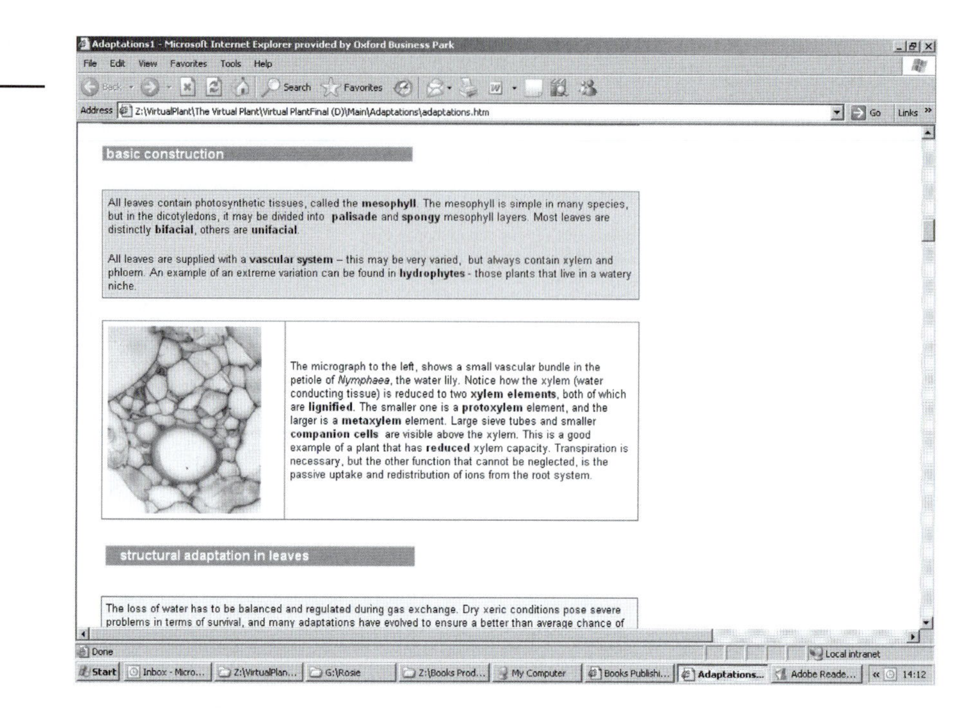

Fig. A2.11 Screenshot of a *Nymphaea* vascular bundle, from plant adaptiveness.

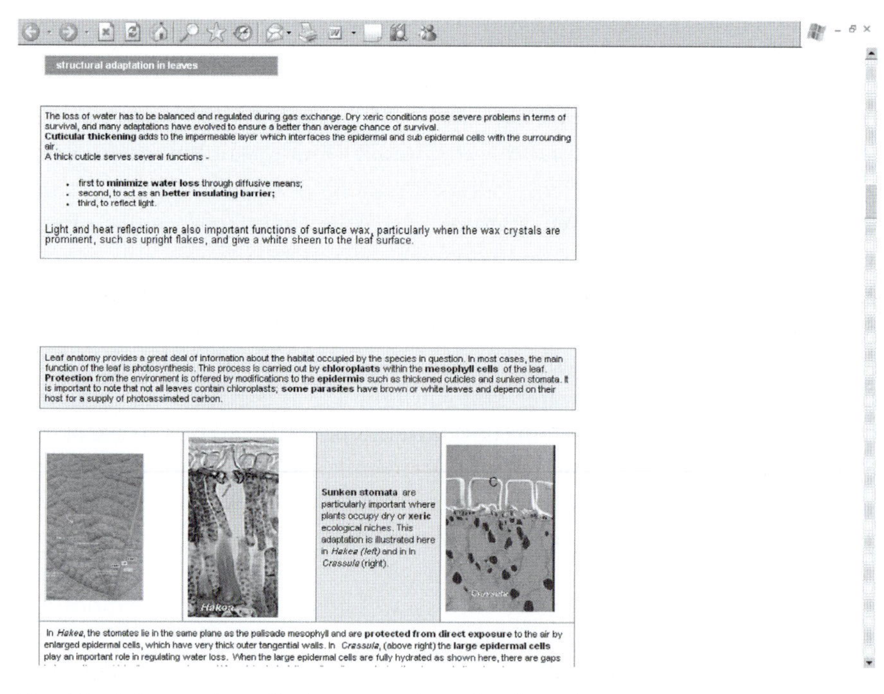

Fig. A2.12 Screenshot of leaf anatomy from the plant adaptiveness exercise.

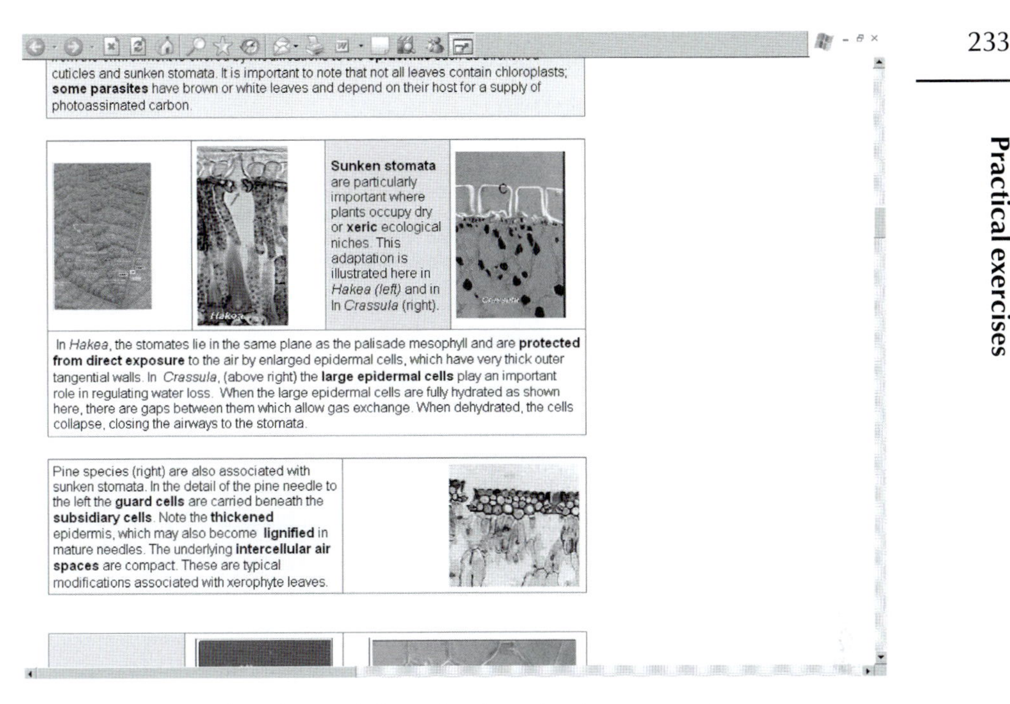

Fig. A2.13 Screenshot of *Crassula* from the plant adaptiveness exercise.

tissue) is reduced to two xylem elements, both of which are lignified. The smaller one is a protoxylem element, and the larger is a metaxylem element. Large sieve tubes and smaller companion cells are visible above the xylem. This is a good example of a plant that has reduced xylem capacity. Transpiration is necessary, but the other function that cannot be neglected, is the passive uptake and redistribution of ions from the root system.

The leaf anatomy shown in Fig. A2.12 provides a great deal of information about the habitat occupied by the species in question. The main function of the leaf is photosynthesis, which is carried out by chloroplasts within the mesophyll cells of the leaf. Protection from the environment is offered by modifications to the epidermis such as thickened cuticles and sunken stomata.

In *Crassula* (Fig. A2.13), when the large epidermal cells are fully hydrated, there are gaps between them, which allow gas exchange. When dehydrated, the cells collapse closing the airways to stomata.

Illustrated glossary

One of the core elements of *The Virtual Plant* is its illustrated hypertext glossary, which for the most part follows the information that has been

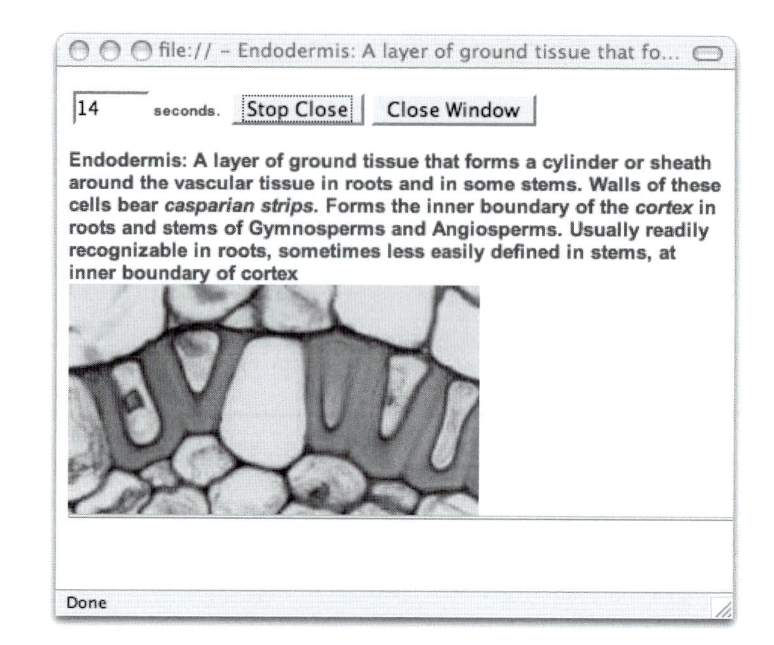

Fig. A2.14 Typical screenshot from the Glossary – the glossary definition for an endodermis.

presented within the text of this book (Fig. A2.14). The glossary contains in excess of 500 pop-up definitions, and these are often associated with a 24-bit colour image, which helps clarify the definition. What makes them unique is that they may be accessed within the exercises, or independently of the exercises.

Teaching material, digital images

The Virtual Plant contains a large collection of images that are being provided for use by instructors and students alike, should they wish to. They are provided free of restrictions, but may not be incorporated into any other publication without the approval of Blackwell Publishing. The images are loosely catalogued under the headings **cells**, **roots**, **stems** and **leaves**. They can be accessed most easily through their own web page.

Clicking on the links will allow the user to navigate to pages illustrating cell type, leaf structure or stem structure (Fig. A2.15). There are about 250 images, >225 Mb.

Clicking on any of the thumbnail links will open the corresponding image at a resolution of 300 pixels per inch, which is high enough to be useful in electronic format (Fig. A2.16). No further information is provided with the images.

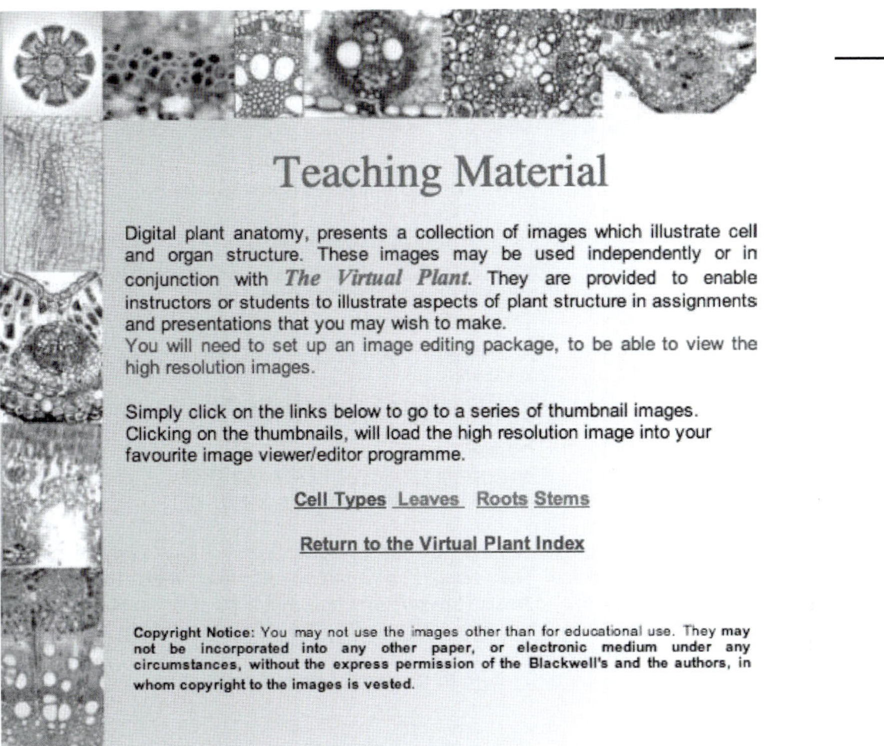

Fig. A2.15 Screenshot of the digital plant anatomy pages.

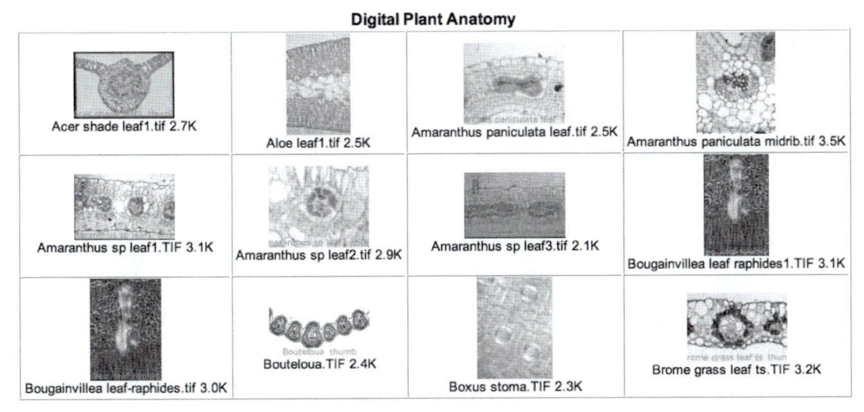

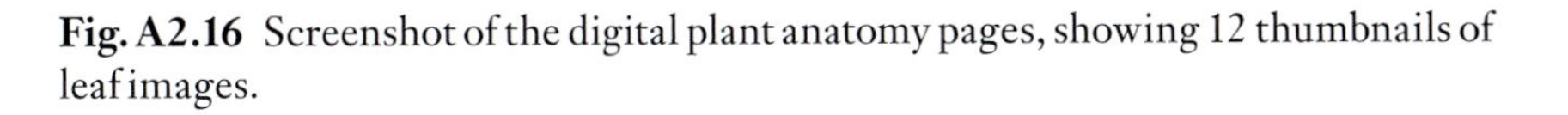

Fig. A2.16 Screenshot of the digital plant anatomy pages, showing 12 thumbnails of leaf images.

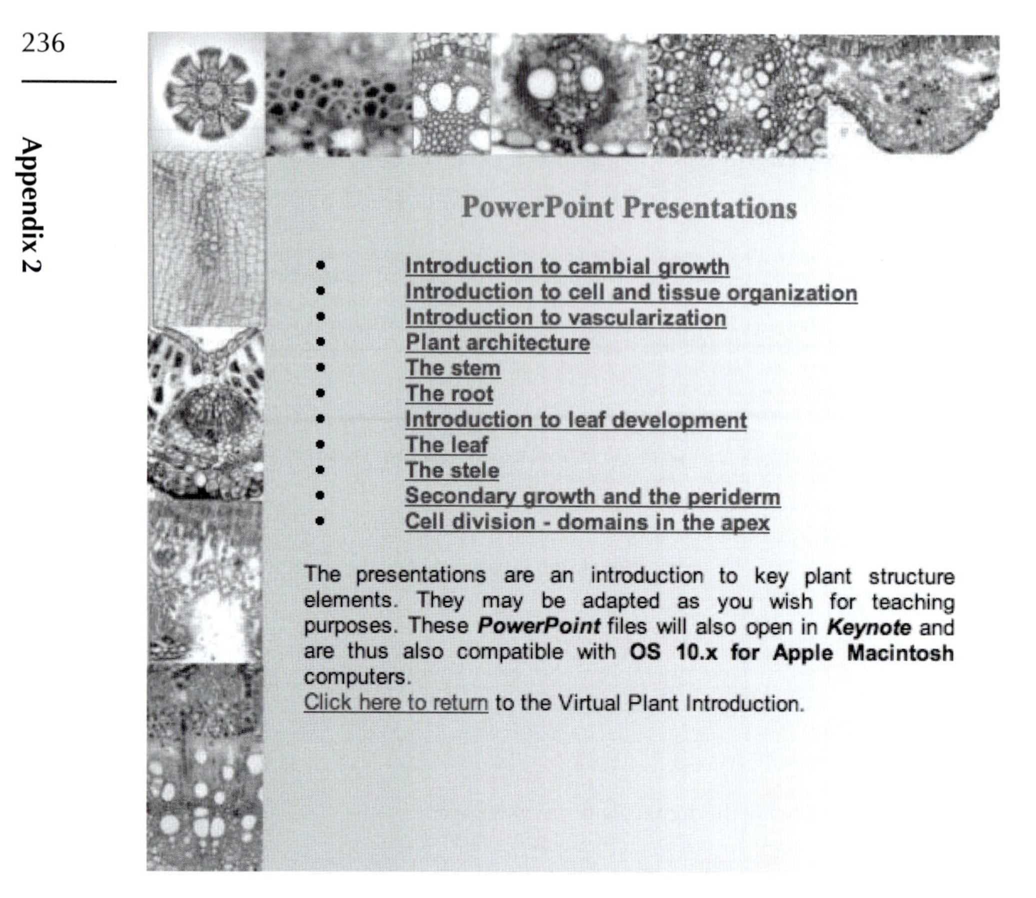

Fig. A2.17 Screen shot of index to PowerPoint Presentations.

Teaching materials, PowerPoint files

We have included a number of PowerPoint Files, which may be of some use to student and tutor alike (Fig. A2.17). The PowerPoint presentations are of variable length, and can of course be modified at will. They have been tested using Keynote and will open on Apple Mac computers, using this presentation package just as easily as they do using PowerPoint. The presentation files have been optimized for projection. Clicking on these files will load them. Mac users who do not have PowerPoint installed will have to associate the files with Keynote.

Factfile Index

The topics presented here cover specific aspects of plant structure-function. They are suitable background reading for advanced plant anatomy courses at the 200 or 300 level. To navigate, just click on the image, or on the text and the relevant material will be displayed for you.

	OVERVIEW: Developmental Plant Anatomy		
	1. Vegetative apical growth	7. Modified and secondary plasmodesmata in higher plants	
	2. The leaf: Functional relationships	8. From single cell to supracell	
	3. The leaf	9. Phloem transport mechanisms	
	4. Morphology and tissue systems	10. Embedding procedures	
	5. The source-sink connection	11. Basic histochemistry	
	6. Plasmodesmata		

Fig. A2.18 Screen shot of the Factfile Index.

Teaching materials: the Factfiles

The 11 Factfiles are introduced briefly below, which we hope gives a clearer understanding of the topics we have covered. With the exception of 10 and 11, which deal with the basics of embedding and histochemistry, the remaining nine deal with development, growth and transport in plants.

1. Cell division in the shoot and root apex

The shoot and root apex are two regions of the plant that are responsible for the production of new vegetative shoot and root growth. Much has been written of late about particular gene systems involved in the division process, and this essay makes no attempts to address these gene systems, but rather takes up points from research carried out by Rinne and van der Schoot in 1998. They reported that the symplasmic fields in the tunica of the shoot apical meristem coordinate morphogenetic events. These authors demonstrated through microinjection how regions within the apical domains could be isolated, which suggests that many of the synchronous and asynchronous cell division events which take place in the shoot apex require plasmodesmatal transport and asynchronous cell divisions within the isolated regions or domains of signals, and at other stages how closure of plasmodesmata would result in isolation of domains. The PowerPoint presentation 'Cell Division – domains in the apex' illustrates a reinterpretation of key data based upon the findings presented by Rinne and van der Schoot.

2. The leaf: functional relationships

No matter whether the leaf is from gymnosperm or angiosperm origin, developmental sequences in these groups appear similar. The differentiation of leaves follows a strict, controlled and sequenced series of events, with marginal, submarginal and procambial initials and derivatives, forming the dermal tissue system, the mesophyll and the vascular system within the leaf. This essay deals with several issues relating to vein maturation within the leaf itself, and the maturation sequence which governs growth phases and the transition from sink to source.

3. The leaf: structure–function relationships

This Factfile deals with the leaf, which is a major lateral appendage of the vegetative shoot. It briefly introduces the concept of heteroblasty and homoblasty, i.e. how sometimes the morphology and anatomy of leaves may change as they mature. It also introduces the concept of micro- and megaphyll to the reader. Photosynthesis, translocation and transpiration are defined and introduced as the three key processes that plant leaves are involved with. The influences that adaptations to different environments and the types of photosynthetic physiology have on the structure of foliage leaves of gymnosperm, dicotyledon and monocotyledonous foliage leaves is introduced, and the reader is linked back to the diagnostic criteria introduced in *The Virtual Plant* that are used to make distinctions between these three very different groups of plants.

4. Morphology and tissue systems: the integrated plant

This Factfile is included to reinforce the basic arrangements of the tissue systems in vascular plants.

Two problems, mechanical support and movement of water and minerals, need solutions in order for plants to successfully compete in their environments. Briefly, these are:

1 Mechanical support, i.e. development of supporting systems to enable exposure of suitable surface area with cells containing chloroplasts to the sunlight to intercept and fix solar energy.

2 Movement of water and minerals from the soil, through the roots to regions where they can be combined with other materials to build the plant body, and movement of synthesized food material from the site of synthesis to places of growth or storage and from the stores to growing cells.

This Factfile illustrates examples of different mechanical support systems and where possible, integrates these with the transport system.

5. The source–sink connection

The source–sink connection in plants relates to the connectivity between the source of assimilated material and the pathway that is followed by this material, to a sink, which can be defined as a (local) region where carbon-based material is metabolized and energy (usually in the form of ATP) is synthesized. Assimilated material is produced through a variety of complex biochemical reactions, and accumulates, usually in sufficient quantities, such that the material will commence movement and follows a diffusive pathway from a region of high concentration to a region of lower concentration, either nearby, or some distance removed from its region of origin (source), or site of production. Diffusion will continue to be the driving force, provided that a concentration gradient is maintained. However, there are few, truly diffusive pathways in plants that are capable of providing or maintaining the necessary diffusion flow rates, to satisfy the demands made by general plant growth and metabolism. It follows that if sustained growth is to be satisfied and achieved, then a better way of mobilization of assimilated material becomes necessary. Clearly, such systems have evolved with time throughout the plant kingdom and the pinnacle of these evolutionary steps lies within the higher vascular plants.

6. Plasmodesmata

Plasmodesmata are narrow strands of cytoplasm that connect neighbouring plant cells across the intervening primary cell wall, lined by the plasmalemma. They thus form interconnecting cytoplasmic bridges, about 50–120 nm in diameter, through which materials can be transported intercellularly. This Factfile deals with formation and function in signals and bulk flow of assimilates and other micro- and macromolecules. It also deals with regulation or 'gating' (opening or closing) of these structures, which can then regulate flow between cells as well as from more distant cells, to the phloem tissue in leaves. The origins, classification and transport functions of plasmodesmata are also discussed. The Factfile is richly illustrated with electron micrographs and diagrams, which help illustrate their unique, variable structure and function.

7. Modified and secondary plasmodesmata

This Factfile deals with plasmodesmatal form and function. We discuss plasmodesmatal formation during cell division by entrapment of endoplasmic reticulum in the cell plate, as well as modifications including branching as a result of increased cell wall thickness. We discuss the differences between primary, secondary and secondarily modified plasmodesmata. A list of useful references is included.

8. From single cell to supracell

This Factfile explores the development of cells from individual cells, the organization into multicellular systems and, finally, what has been accepted now as a 'supracellular' system which is common to higher plants. We summarize the basic systems needed in order to sustain plant life – nucleus, endoplasmic reticulum, dictyosomes, mitochondria, plastids and vacuoles, in autotrophs and heterotrophs.

9. Phloem transport mechanisms: some basic considerations

This Factfile serves as an introduction to phloem transport in plants and the requirements for either active or passive transport within the phloem. We introduce the reader to the two main phloem transport mechanisms – one requiring osmotic potential gradients, the other energy transformation in the movement of assimilate from source to sink. Several key 'must read' articles are included.

10. Embedding procedures

Plant structure studies and the observation of cell relationships require that plant tissues (root, stem, leaf, flower or fruit) need to be fixed and sectioned. This Factfile takes the reader through some basic fixation techniques for the preparation of light micrographs, using tried and trusted formalin acetic acid alcohol techniques, and discusses wax embedding for sectioning. This Factfile is by no means definitive, and the reader is referred to an article by Feder and O'Brien (1968), which describes more advanced techniques using acrolein (difficult to get hold of) and glutaraldehyde, as these yield far superior results.

11. Basic histochemistry

This Factfile has been included to reinforce the need to make use of some basic histochemistry when examining plant cells. The preservation of structural details of cells and tissues is influenced by the condition of the material at the time of collection and by any subsequent preparative steps applied to kill and fix the material. In other words, if you wish to prepare sections in which the structural details are well preserved one should select healthy plant material. Exceptions to this rule apply only when the researcher is interested in observing the effects of disease, fungal infection, insect damage, etc. on the normal structure of the plant material being examined. We strongly advocate the use of living material and freehand sections to discover the structural complexity and variation in wall components. Cellulose, lignin and pectin distribution are very useful in determining structure, and starch is most useful as an indicator of photosynthetic activity and carbohydrate storage for example. Readers are warned that many of the chemicals and procedures referred to here are potentially harmful. Please exercise caution when handling chemicals in the laboratory environment.

Glossary

We have included this glossary as a support and learning aid to those using this text. We have avoided including an exhaustive list of words, terms and phrases, which may be found in most textbooks. Indeed, we have been selective to the point of only including reference material that we feel is relevant to the text and those which may be useful to the reader who wishes to look for the meaning of a word or phrase.

Abaxial: The surface that is directed away from the axis. Normally used when referring to the lower epidermis dorsiventral leaf, but can also be used to describe the location of particular cell types, tissues or structures, for example, abaxial sclerenchyma; abaxial stomata; abaxial trichomes. (Opposite: adaxial.)

Abscission zone: A specialized zone that is formed after physiological and structural changes occur within an organ, which leads to the loss of the organ. For example, the abscission layer may form in petioles (*Gossypium*) or in distal part of a leaf blade (*Streptocarpus*) or in the stalks of fruits or flowers.

Accessory cell: *See* subsidiary cell.

Accessory transfusion tissue: Transfusion tissue in the mesophyll of some of gymnosperm leaves, not related to vascular bundles.

Acicular crystal: needle-shaped crystal.

Acropetal: Term used to describe either the relative position of an organ or structure in relation to the shoot or root apical meristems, or direction of growth. Also: proceeding towards apex (as e.g. in development).

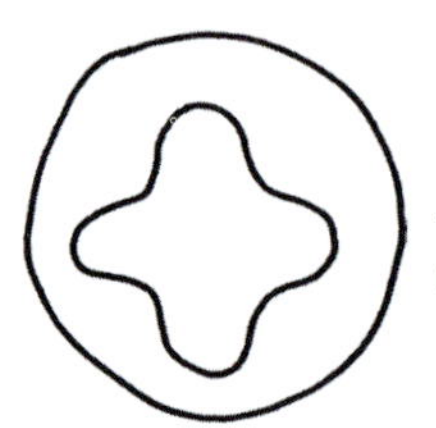

Actinostele: A protostele, with the xylem star-shaped in TS.

Adaxial: The surface that is directed towards the axis in dorsiventral leaves. Used also to describe the location of cells, tissues and structures.

Adaxial meristem: Meristematic tissue present adaxially in leaf. Cell division in this meristem contributes to thickness of the leaf, principally in the petiole and midrib.

Adnation: Concrescence of organs or tissues of a different nature, e.g. a stamen and petal or a leaf and stem.

Adventitious organ: Organ developing an unusual position, e.g. roots at nodes of a stem or buds on root cuttings.

Aerenchyma: Parenchymatous tissue characterized by presence of large intercellular spaces. Principal function is to aid gas exchange in submerged roots and stems and to enhance buoyancy; also found in hydrophytic and in many mesophytic leaves. The cells may have a variety of forms, but are usually characterized by wall evaginations or large intercellular spaces.

Aggregate ray: Groups of small rays in secondary vascular tissue separated by fibres or axial parenchyma, giving the superficial appearance of a far larger ray.

Albuminous cells: Certain cells in phloem rays or phloem parenchyma of gymnosperms, related physiologically and situated adjacent to sieve cells; unlike companion cells, usually originating from different cells than the sieve cells. Also applied to cells in certain seeds containing albumen.

Aleurone grain: Granules of reserve protein, present in many seeds.

Aliform paratracheal parenchyma: *See* Parenchyma.

Glossary

Amphicribal: *See* Vascular bundle.

Amphivasal: *See* Vascular bundle.

Amyloplast, amyloplastid: A leucoplast specialized to store starch.

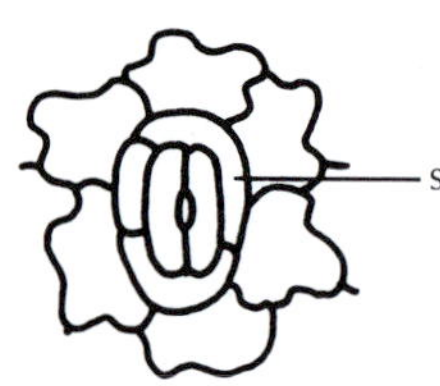

Anisocytic stoma: Stoma in which the guard cell pair has three surrounding subsidiary cells (s) that are of unequal size.

Annular thickening: Secondary wall thickenings deposited as rings on the inner face of primary walls of tracheary elements. Commonly occurs in protoxylem.

Anomalous secondary growth: unusual type of secondary growth in thickness of an organ.

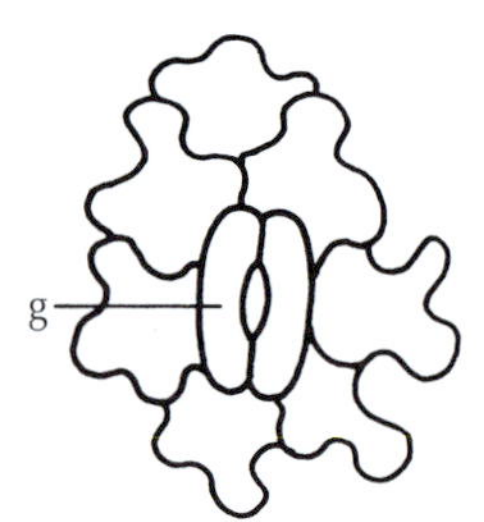

Anomocytic stoma: A stoma in which epidermal cells surrounding the guard cell pair (g) are not morphologically distinct from other epidermal cells.

Anotropal contrast: A form of incident light phase contrast microscopy which gives a dark background.

Anticlinal: A surface, or cell walls, which are perpendicular to the surface of an organ. Also used to describe planes of cell division.

Antipodal cells: Cells of the female gametophyte present at the chalazal end of the embryo sac in angiosperms.

Aperture (of pollen grain): An area of characteristic shape in which exine is completely lacking or in which nexine alone is present; a pollen tube emerges via such an area. (Cf. stoma): the pore between a pair of guard cells.

Apex: Distal portion of organ, i.e. root or shoot (or leaf).

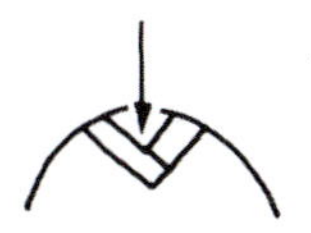

Apical cell or apical initial: A cell that remains in the meristem, perpetuating itself whilst dividing to form new cells that make up the body of the plant (in lower plants).

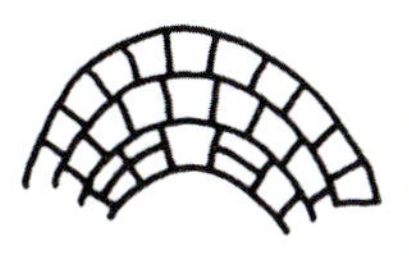

Apical meristem: A single cell or several layers of apical cells which are self-perpetuating and which by division in certain planes produce the precursors of the various tissues of the plant.

Apoplast, apoplastic: A continuum composed of cell walls and intercellular spaces of a plant or an organ; description of movement of substances through cell walls. Includes the lumen of tracheary elements.

Aseptate: Used to describe a cell, which lacks internal septa, e.g. an aseptate fibre.

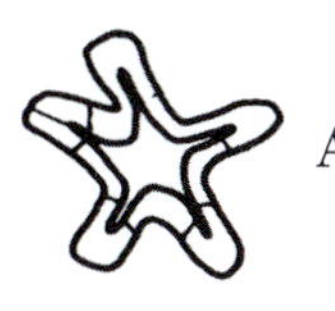

Astrosclereid: A branched sclereid.

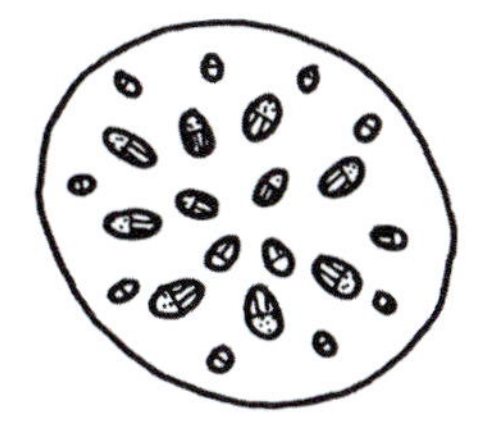

Atactostele: Stele in which, in TS, primary vascular bundles appear scattered throughout the ground tissue in, for example, most monocotyledonous stems. The arrangement is in fact orderly.

Axial parenchyma: *See* Parenchyma, xylem.

Axial system: (i) all cells derived from fusiform cambial initials in secondary vascular tissues; (ii) cells elongated parallel to the long axis of an organ.

Bark: A non-technical term which describes the tissues which occur outside of (exarch to) the vascular cambium in secondarily thickened stems and roots.

Bars of Sanio: *See* Crassulae.

Basal: At, or towards the base of an organ.

Basipetal: Proceeding towards the base (usually of development); towards the base of an organ (i.e. towards the tip of a root).

Bicollateral vascular bundle: A primary vascular bundle, where the phloem occurs on either side of the xylem. Phloem occurs external to the metaxylem, and also internal to the protoxylem elements in this type of bundle in stems. In leaves, the phloem strands occur on either side of the xylem (adaxial and abaxial face of the bundle) in major veins in some families.

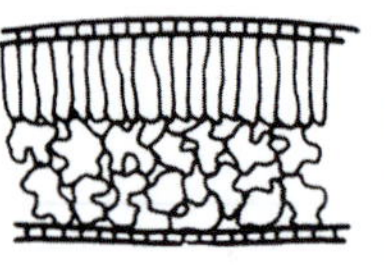

Bifacial: (Dorsiventral) describes a leaf with palisade parenchyma present beneath the adaxial epidermis and spongy mesophyll beneath the other; having distinct dorsal and ventral surfaces.

Blade: That part of a leaf which is distal (away from) the sheath and ligule in monocotyledons; and distal to the petiole or leaf base in dicotyledons.

Body primary: The parts of a plant developing from primary apical and intercalary meristems.

Body, secondary: The parts of a plant made up of secondary vascular tissues and periderm, added to the primary body by the action of the lateral meristems, cambium and phellogen.

Brachysclereid, stone cell: Short, more or less isodiametric, sclereid.

Bract: A reduced leaf in an inflorescence.

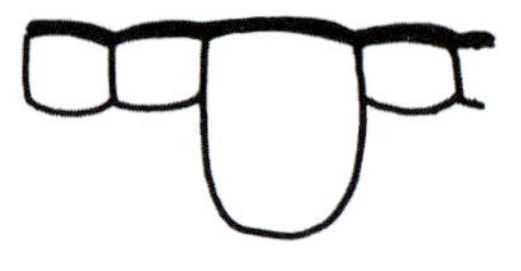

Bulliform cell: Enlarged epidermal cell common in leaves of Gramineae and in more xerophytic monocotyledons (as longitudinal rows of cells); sometimes called 'expansion cells', thought to bring about the unrolling of a developing leaf or 'motor cells' if involved with rolling and unrolling of leaves in response to water status of the leaves. Bulliform cells loose turgor under conditions of water stress, facilitating leaf roll, and consequently, reduction of transpirational water loss via stomata.

Bundle cap: Sclerenchyma or thick-walled parenchyma layer or layers of cells at phloem and/or xylem poles of vascular bundles.

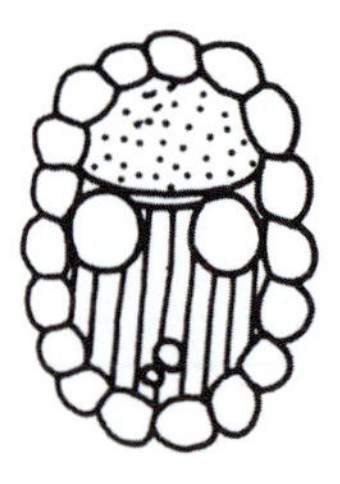

Bundle sheath: A layer or layers of cells surrounding a vascular bundle of leaves and some stems. May have ecophysiological significance in prevention of water loss, if suberized lamellae are present in tangential and/or radial walls of these cells. They act as boundary layer between mesophyll and vascular tissues. May be parenchymatous in younger organs, or lignified in older more mature organs. Often associated with specialized chloroplasts in Kranz (C_4) plants.

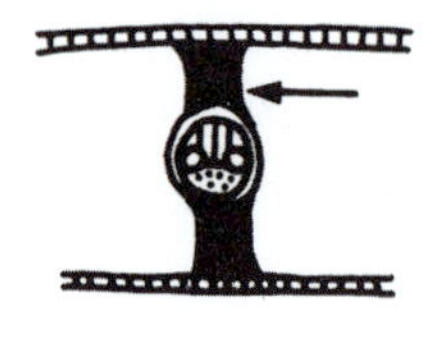

Bundle sheath extension: A strip of ground tissue between vascular bundles and epidermis or hypodermis in leaf adaxially, abaxially or both; consisting of parenchyma or sclerenchyma. Often of characteristic outline in TS for a given genus or species.

Callose: A polysaccharide that is present in sieve areas. The presence of callose is usually indicative of pressure loss within the sieve tube (damage or injury) or in dormant phloem. Also formed as a rejection response when pollen of a different species is received by stigmatic surfaces. On hydrolysis produces glucose.

Callus: Tissue of parenchymatous cells formed as a result of wounding, or a tissue developing in tissue culture.

Calyptrogen: In apical meristem of some roots, meristematic cells giving rise to the root cap; distinct from other apical meristematic cells forming the root itself.

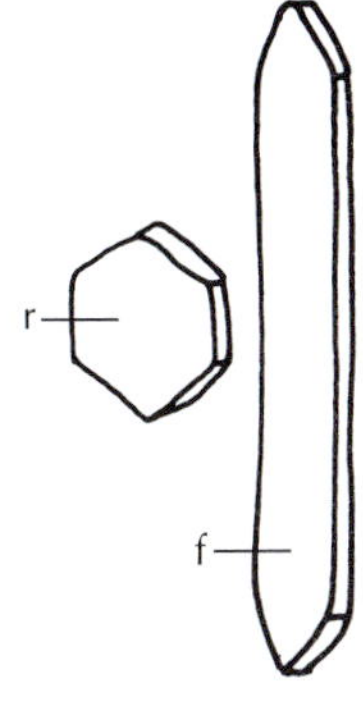

Cambial initials: Self-perpetuating cells in the vascular cambium. Form derivative cells by periclinal division, thus adding to the secondary xylem and secondary phloem or increase in number by anticlinal division, to form additional ray (r) or fusiform (f) cells. *See* Fusiform, Ray initials.

Cambial-like transition zone: A cytohistological zone visible in some shoot apices.

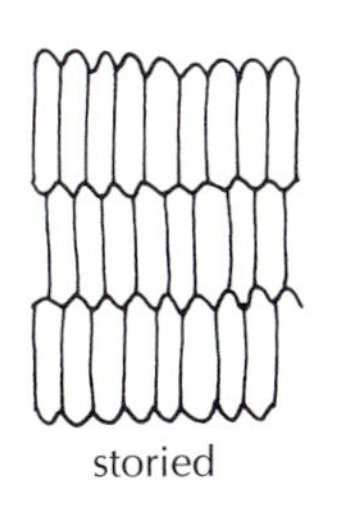
storied

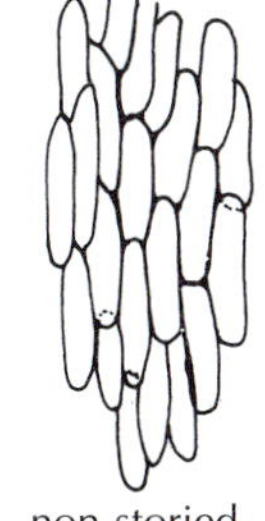
non-storied

Cambium: (relates to vascular cambium) (i) non-storied, composed of fusiform initials which, as seen in TLS, partially overlap one another in a random way and do not form horizontal rows; (ii) storied, composed of fusiform initials, which, as seen in TLS are arranged in horizontal rows. Cambial ray initials can also be non-storied or storied. Resultant rays will appear randomly arranged (non-storied) or in distinct rows (storied) in tangential longitudinal sections of secondary xylem.

Cambium, vascular: A lateral meristem from which the secondary vascular tissues develop: (i) fascicular (f), the cambium forming within the vascular bundle; (ii) interfascicular (i), the cambium between the vascular bundles.

Carousal: Transversely oriented thickenings in tracheid walls of gymnosperms accompanying the pit pairs and formed by the intercellular material of primary wall layers. Also called bars of Sanio.

Caruncle: A fleshy outgrowth of the integuments at the micopylar region of a seed.

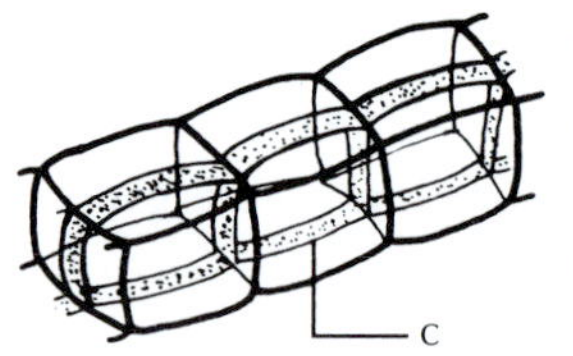

Casparian strip, band: A band-like structure (c) in the primary wall containing lignin and suberin. Particularly characteristic of endodermal cells of roots where the band is present in radial and transverse anticlinal walls. Similar cells are sometimes observed in stems, between cortex and stele, also in exodermis cells of some roots. In ferns, individual vascular bundles can be enclosed by such an endodermis. Hairs on leaf and stem in some xerophytes may have such bands. They are thought to have physiological significance, by controlling apoplastic transport of water and other solutes including ions in solution across the end-odermis, or in reducing water loss through the cell wall in hairs. *See* Passage cell.

Cell: The structural and functional unit of a living organism. In plants, most cells are characterized by the presence of a cell wall.

Cell plate: The part of a cell wall developing between the two daughter nuclei in telophase of cell division.

Cellulose: Carbohydrate consisting of long chain molecules comprising anhydrous glucose residues as basic units; a principle constituent of plant cell walls.

Central mother cells: Cytohistological zone of shoot apex in the region below the surface layers; commonly used in describing gymnosperm apices.

Chalaza: Region in ovule where integuments and nucellus connect with the funiculus.

Chimera: Combination in a single plant organ of cells or tissues of different genetic composition; e.g. yellow margins of *Sanseveria* leaves.

Chlorenchyma: Specialized parenchyma cells, which contain chloroplasts. Chlorenchyma is usually found in the outer cortex of stems, and in the mesophyll of leaves.

Chlorophylls: The green pigments in chloroplasts.

Chloroplast: A specific cell organelle in which photosynthesis takes place; usually disc-shaped.

Chromoplast: Plastid, containing pigment (*see* Plastid).

Cicatrice: Scar left by separation of one part or organ from another; e.g. hair base of deciduous hair.

Coencyte: Group of protoplasmic units, a multinucleate structure. In angiosperms usually refers to multinucleate cells such as the embryo sac.

Coleoptile: Sheath surrounding apical meristem and leaf primordia of grass embryo.

Coleorrhiza: Sheath surrounding the radicle of a grass embryo.

Collateral: *See* Vascular bundle.

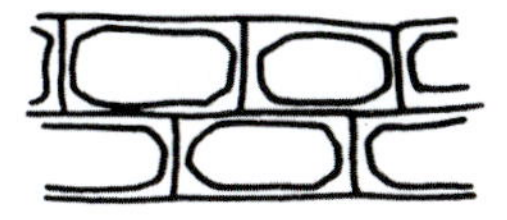

Collenchyma: Supporting or mechanical tissue in young organs and in certain leaves. The walls are mainly cellulosic, with thickening that is (i) even, (ii) angular, (iii) lamellar with wall thickening mainly on anticlinal cell walls or (iv) lacunar. In the latter, the collenchyma tissue is associated with characteristically large intercellular spaces, and the wall thickening tends to be greatest opposite intercellular spaces Walls contain a high proportion of calcium pectate.

Colleter: Multicellular glandular hair with stalk and head having a sticky secretion.

Columella: (i) in some roots central portion of root cap in which cells are arranged in longitudinal files; (ii) in other usage, means a small pillar.

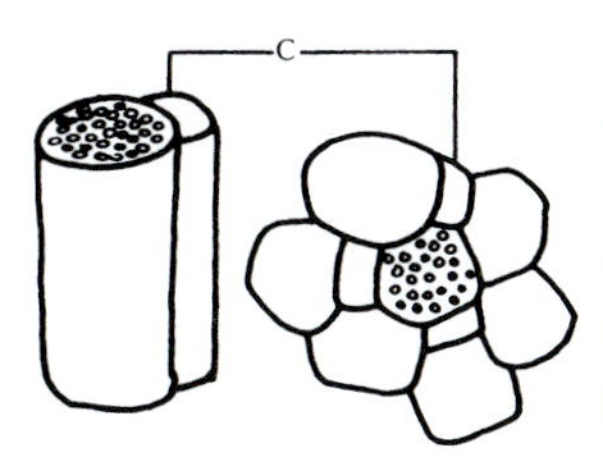

Companion cell: Specialized parenchyma cell (c) associated with, and derived from, the same mother cell as sieve tube member. Companion cells maintain symplastic continuity with sieve tube members via specialized pore-plasmodesmata.

Complementary cells: Loose tissue formed towards periphery by phellogen of the lenticel; cell walls may or may not be suberized.

Compression wood: Reaction wood in conifers formed on abaxial side of branches, etc.; dense in structure with strong lignification of tracheid walls.

Cork: *See* phellem.

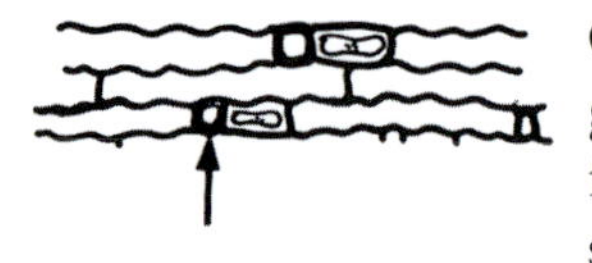

Cork cell: (i) dead cells, arising from the phellogen whose walls are impregnated with suberin; function usually protective; (ii) in the epidermis, a short cell with suberized walls, as in grasses.

Corpus: The cells below the surface layer(s) (tunica) of angiosperm shoot apex in which cell divisions take place in various planes, giving rise to increase in apex volume (tunica-corpus theory).

Cortex: Region of ground tissue between the epidermis or the periderm and the vascular cylinder. The cortex may be entirely parenchymatous, or be composed of chlorenchyma, collenchyma and sclerenchyma.

Costal: *See* Intercostal.

Crassulae: Transversely oriented thickenings in tracheid walls of gymnosperms accompanying the pit pairs. Also called bars of Sanio.

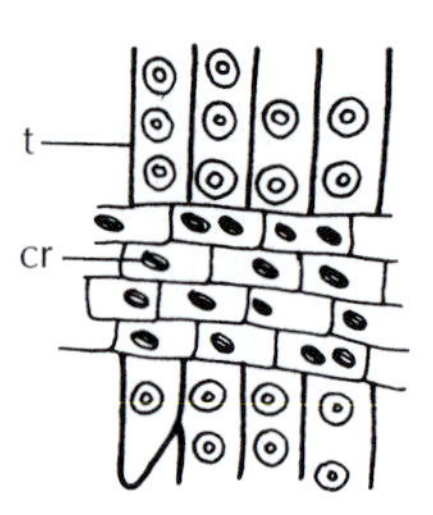

Cross field: Area formed by the walls of a ray cell and an axial tracheid as seen in RLS; mainly used in description of conifer woods; cr = cross field pit, t = tracheid.

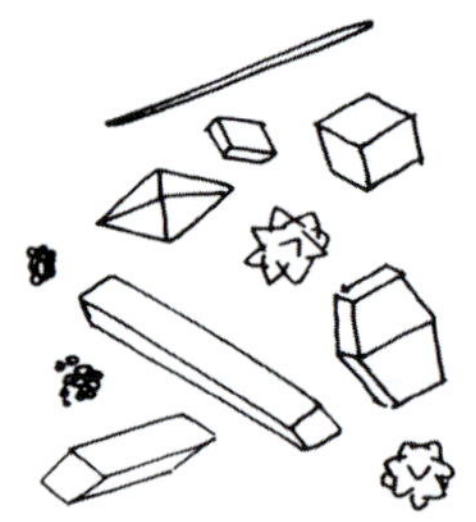

Crystal: Cell inclusion, usually of calcium oxalate, sometimes of calcium carbonate, exhibiting a range of forms; sometimes of taxonomic or diagnostic significance.

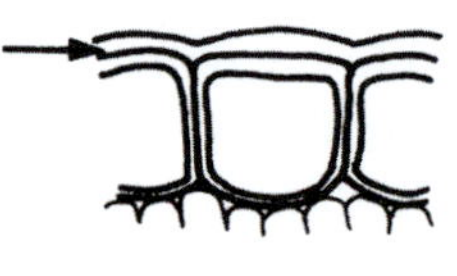

Cuticle: Layer of cutin, a fatty substance that is almost impermeable to water; present on outer walls of epidermal cells, sometimes extending into supra- and sub-stomatal cavities as a very thin lining.

Cutinization: Process of deposition of cutin in cell walls.

Cyclocytic: Stoma in which the guard cell pair is surrounded by one or more rings of subsidiary cells.

Cylinder, central or vascular: That part of the axis of a plant consisting of vascular tissue and the associated ground tissue. Equivalent to term 'stele' but lacking evolutionary implications.

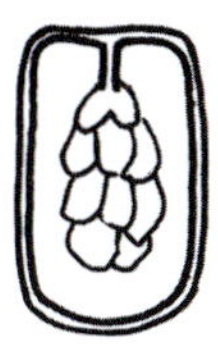

Cystolith: A specific outgrowth of the cell wall on which calcium carbonate is deposited; characteristic of certain plants, e.g., *Ficus*, Moraceae. Cystoliths occur in specialized, enlarged structures (cells), termed lithocysts.

Cytochimera: In a single plant organ, a combination of cells which are of different chromosome number.

Dedifferentiation: A controlled reversal of the differentiation process in which cells (normally parenchyma) may become meristematic.

Dermal tissue: Epidermis or periderm.

Dermatogen: Meristem forming epidermis.

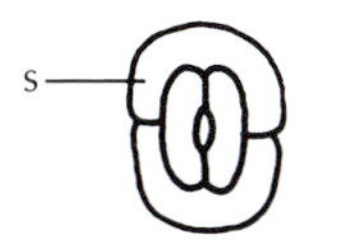

Diacytic stoma: A stoma in which the guard cell pair has one subsidiary (s) cell at either pole, the end walls of which are transverse to the long axis of the stoma.

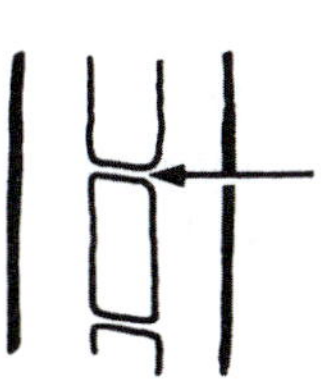

Diaphragm: A partition of cells in an elongated air cavity in an organ; may be transverse or longitudinal.

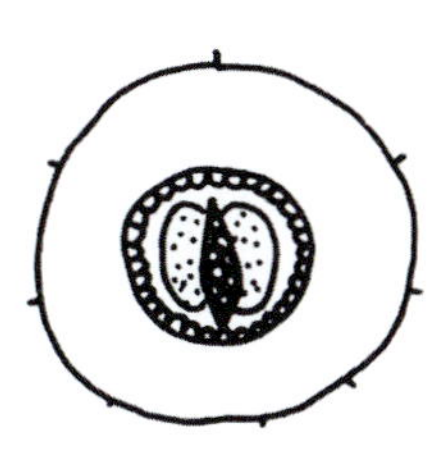

Diarch: Primary root with two protoxylem strands (and poles).

Dictyostele: A siphonostele in which the leaf gaps are large and partly overlap one another and dissect the vascular system into separate bundles, each with the phloem surrounding the xylem. Occurs widely in ferns.

Diffuse porous wood: Secondary xylem in which there is a gradual reduction in diameter between the xylem vessels formed at the beginning and at the end of the season's growth.

Dorsiventral leaf: *See* Bifacial leaf.

Druse: A crystal (compound), which approaches a spherical shape, with component crystals that protrude from the surface.

Duct: A longitudinal space formed schizogenously or lysigenously or schizo-lysigenously. May contain secretions or air.

Ectocarp: The outermost layer of the pericarp (fruit wall). *See* Exocarp.

Elaioplast: An oil-producing and -storing leucoplast.

Elaiosome: An outgrowth on a fruit or seed that contains large oil-storing cells.

Emergence: A projection of the surface of a plant organ consisting of epidermal cells and cells derived from underlying tissues.

Endarch: Towards the centre of the axis.

Endarch xylem: A primary xylem strand in which the first-formed elements are closest to the centre of the axis, as in the shoots of most spermatophyta.

Endocarp: The innermost layer of the pericarp (fruit wall).

Endodermis: A layer of ground tissue that forms a cylinder or sheath around the vascular tissue in roots and in some stems. Walls of these cells bear Casparian strips. Forms the inner boundary of the cortex in roots and stems of gymnosperms and angiosperms. Usually readily recognizable in roots, sometimes less easily defined in stems, at inner boundary of cortex.

Endodermoid layer: Layer of cells surrounding central vascular cylinder of stem, in position of endodermis, but in which Casparian strips are not distinguishable. (The distinction between endodermis and endodermoid layer is not always recognized by some; may be referred to as the starch sheath.)

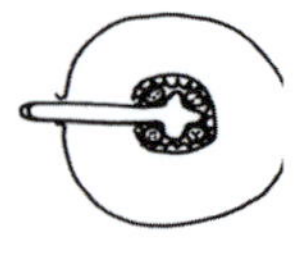

Endogenous: Developing from within, such as a lateral root, or a substance (e.g. a plant growth regulator) or some other naturally synthesized substance produced by a plant.

Endosperm: A nutrient tissue formed within the embryo sac of the angiosperms as a result of double fertilization.

Endothecium: A layer of cells situated below the epidermis in the pollen sac wall having characteristic wall thickenings.

Ephemeral: Of plants having a short life span, usually less than one year, or of plant parts, e.g. hairs, falling soon after development.

Epiblast: A small growth present opposite the scutellum in the embryo of some Gramineae.

Epiblem: The outermost layer (epidermis) of primary roots.

Epicarp: *See* Exocarp.

Epicotyl: The true stem of a developing seedling, which is developed from the shoot apical meristem. It occurs above the cotyledons.

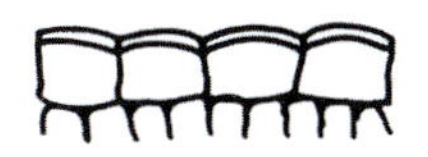

Epidermis: The outermost layer or layers of cells of primary tissues derived from the protoderm. Sometimes comprising more than one layer – multiseriate epidermis or multiple epidermis.

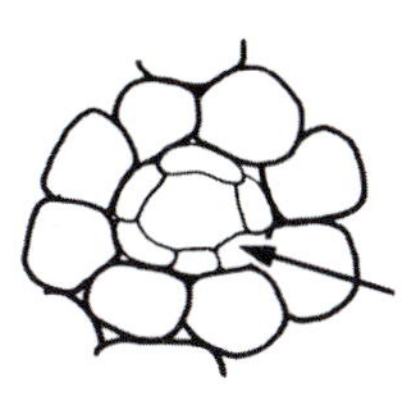

Epithelial cell: Cell lining a cavity or canal; usually with secretory function.

Epithem: The tissue between the vein ending and the secretory pore of a hydathode.

Ergastic: The non-protoplasmic product of the metabolic processes within the cell. Common ergastic substances include starch grains, oil droplets and crystals, which are found in the cytoplasm, vacuoles and cell walls.

Eustele: Considered to be phylogenetically the most advanced type of stele. The vascular tissue forms strands around the pith or (post-secondary growth) a hollow cylinder, composed of collateral or bicollateral vascular bundles.

Exalbuminous seed: A seed lacking endosperm when mature.

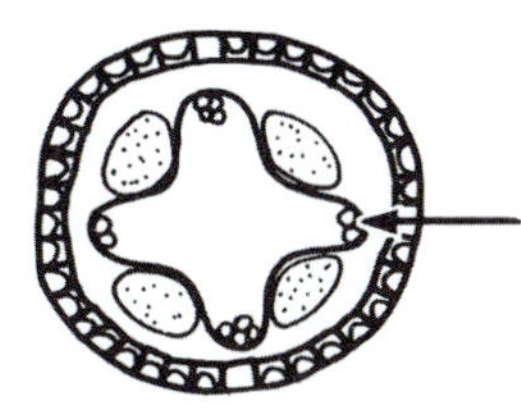

Exarch: The location of cell and tissue types away from the centre of the axis.

Exarch xylem: A strand of primary xylem in which the first-formed elements are furthest from the centre of the axis, as in roots of spermatophyta.

Exine: The outer wall of a mature pollen grain.

Exocarp, epicarp: The outermost layer of the pericarp (fruit wall).

Exodermis: Present in some roots as a modified layer or layers of cells in the outer cortex. Walls of these cells are thickened to a greater or lesser extent and contain suberin lamellae in the periclinal walls.

Exogenous: Developing from outer tissues, e.g. an axillary bud.

Expansion cell: Bulliform cell.

Fascicular: Part of a vascular bundle, or cells which are situated within a vascular bundle, e.g. fascicular cambium.

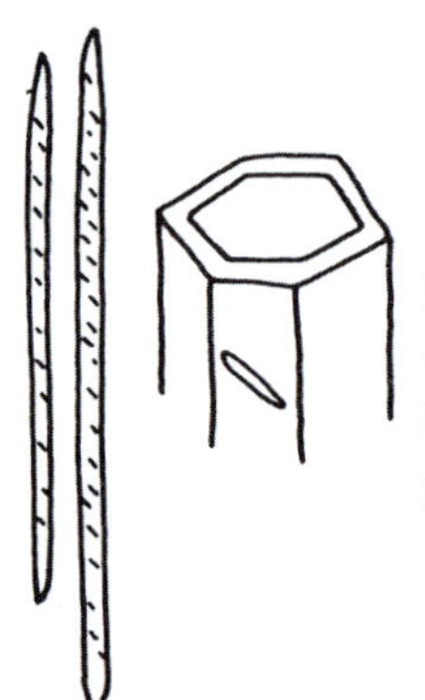

Fibre: An elongate, sclerenchymatous cell with tapered ends. Walls become lignified at maturity. The cells may be with or without cytoplasm at maturity and may or may not have a living protoplast at maturity.

The following are forms of fibre:
—, **gelatinous:** Inner layers of the secondary wall may swell on absorption of water.
—, **libriform:** In secondary xylem with few, simple pits.
—, **pericyclic:** Exarch to the outer regions of the vascular system, or associated with the phloem (then primary/secondary phloem fibre).

—, **septate:** With thin, transverse septa forming after laying down of secondary wall.

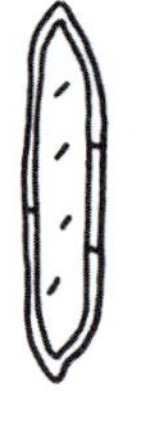

Fibre-sclereid: Cells with wall characteristics intermediate between those associated with a fibre or a sclereid, or if the fibre sclereid develops from parenchyma in non-functional phloem.

Fibre-tracheid: Cell of secondary xylem with characteristics intermediate between a fibre and a tracheid, with pointed ends; bordered pits, with slit-like apertures.

Funiculus: Stalk attaching an ovule to the placenta.

Fusiform: Elongated tapering cell. *See* cambium.

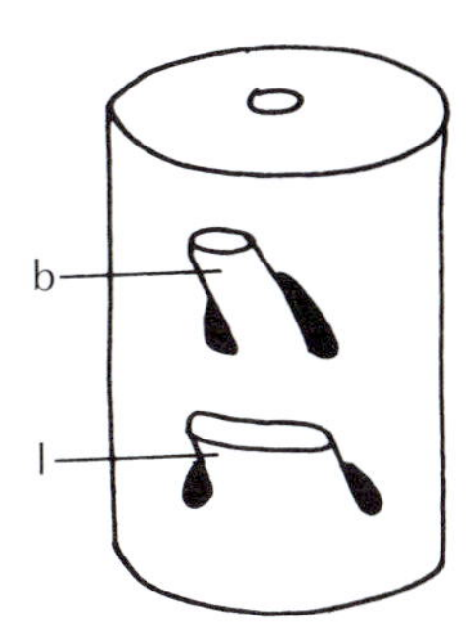

Gap, branch: A parenchymatous region in a siphonostele above the position where a branch trace connects with the vascular cylinder of the stem (b).

Gap, leaf: A parenchymatous region in a siphonostele above the position where a leaf trace connects with the vascular cylinder of the stem (l).

Graft: The physical union of different individuals. *See* Scion.

Ground tissue: Tissues in stem or root, derived from fundamental or ground meristem. Usually composed of parenchyma, collenchyma or sclerenchyma.

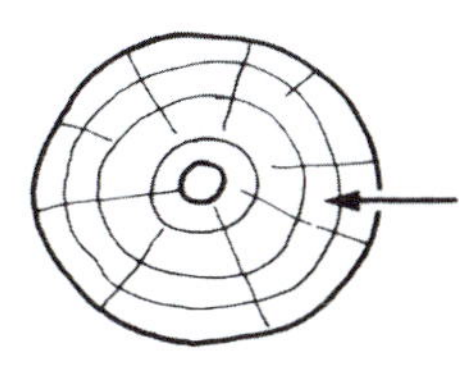

Growth ring: Increment of secondary xylem, usually indicative of seasonal growth patterns, since more than one ring may occur within the space of a year, the term 'annual ring' should be used with caution.

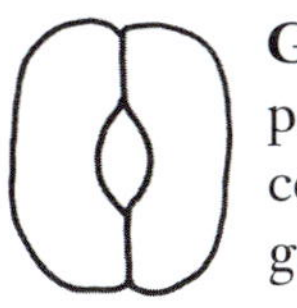

Guard cells: A pair of specialized epidermal cells bordering a pore and constituting a stoma; changes in shape of the guard cells effect the opening or closing of the pore through which gaseous exchange may take place.

Gum: A non-technical term applied to some of the materials arising from breakdown of certain components of plant cells.

Haplostele: A prostele with a more or less circular cross-section to the xylem.

Hardwood: General term for secondary xylem of angiosperms.

Heartwood: Inner part of wood of a trunk or branch which has lost the ability to conduct water; generally darker than sapwood because of the materials deposited in it.

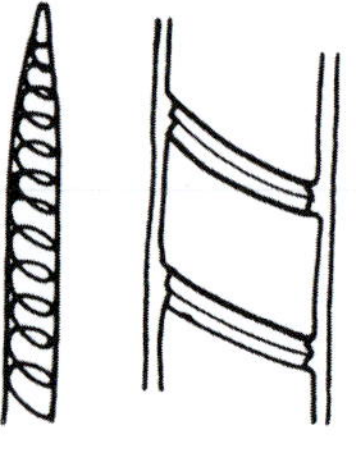

Helical wall thickening, 'spiral' wall thickening: Secondary or tertiary wall material deposited on a primary or secondary wall respectively in certain tracheary elements.

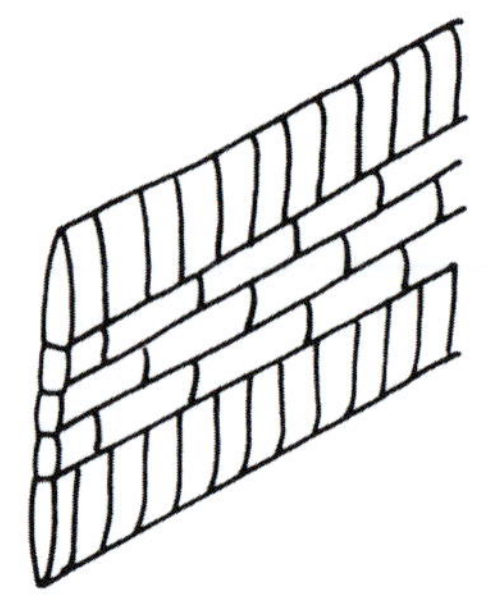

Heterocellular ray: Ray in secondary vascular tissues composed of more than one cell form. In dicotyledons these are all parenchymatous cells; in gymnosperms, tracheids or radial resin canals may be present with the parenchymatous cells; radial canals also occur in some angiosperms.

Heterogeneous rays: Rays of two types present together, e.g. uniseriate and multiseriate, without intermediates. Not to be used to describe heterocellular rays.

Hilum: (i) the funiculus scar in seed; (ii) the portion of a starch grain acting as a nucleus around which the layers are deposited.

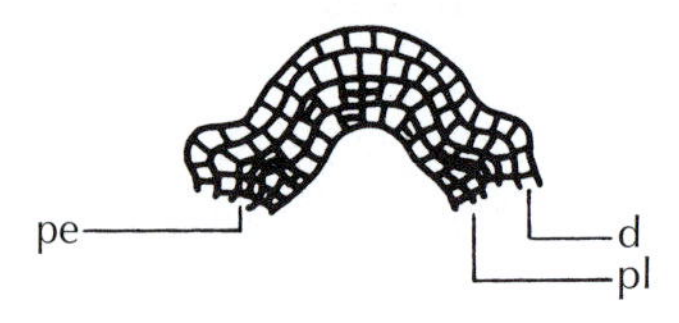

Histogen: In an apical meristem, a layer or layers of cells which develop into one of the three systems of the organ: dermatogen (d) → epidermis, periblem (pe) → cortex, plerome (pl) → vascular system. The number of layers of cells in each can vary from species to species and within a single species, and there may be two or four histogens in some plants.

Homocellular ray: Ray in secondary vascular tissue composed of one (parenchymatous) cell form only.

Homogeneous rays: Rays of one type only present in a wood. Not to be used to describe homocellular rays.

Hydathode: A structural modification of vascular and ground tissues, usually present in leaves, which permits the release of liquid water through a pore in the epidermis. Hydathodes may be secretory in function; thought to be modified stomata.

Hypodermis: Layer or layers of cells immediately below the epidermis, not derived from the same initials as the epidermis (as can be seen by lack of coincidence of anticlinal walls of epidermis and hypodermis), differing in appearance from tissues below them. Some regard any cell layer immediately beneath the epidermis as a hypodermis, but it may be more useful to restrict the term to a cell layer or layers that is/are distinct from the next innermost cortical cells. The root exodermis is a specialized hypodermis.

Idioblast: A cell clearly distinguishable from others in the tissue in which it is embedded, in size, structure or content; e.g. sclereid, or tanniniferous idioblast, or crystal containing cell.

Imperforate: Not perforated, of an intact cell wall; frequently used to describe the pit areas in tracheids.

Initial: A meristematic cell that gives rise to two cells: one of which differentiates into some distinct cell type and the other remaining as the initial within the meristem.

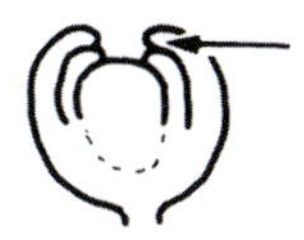

Integument: Enveloping layer surrounding the nucellus.

Intercalary: Meristematic growth that is not associated with apices – usually occurs in internodes and developing leaves.

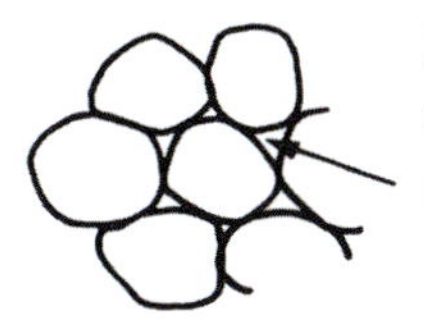

Intercellular space: A space between cells of a tissue; may arise by (a) splitting apart of cells along the middle lamella (schizogenous) or by (b) dissolving cells (Iysigenous) or (c) by tearing apart of cells (rhexigenous).

Intercostal: Tissue between veins in leaves; the tissue above and below veins is termed costal.

Interfascicular: Tissues that occur between vascular bundles in the primary stem (primary medullary rays).

Internode: That part of a stem between two nodes.

Intervascular pitting: Pitting between tracheary elements.

Interxylary: Within or surrounded by xylem, e.g. interxylary cork, cork developing amongst elements of xylem tissue, or interxylary phloem.

Intine: The inner wall of a mature pollen grain.

Intraxylary: On the inner side of the xylem.

Intrusive growth: Specialized growth, usually associated with the apex of cells, where the cell intrudes between adjacent cells, following a line of separation of the middle lamella. A common form of growth of some fibres.

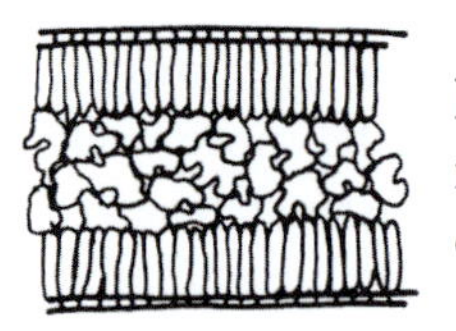

Isobilateral leaf: A leaf in which the palisade parenchyma occurs adaxially and abaxially beneath the epidermis of the leaf.

Lacuna: Space between tissues, usually filled with air. *See* Intercellular space.

Lamina: Blade, or expanded part of leaf.

Laticifer: Cell or series of cells with characteristic latex fluid contents; usually tubular in shape, branched or unbranched.

Laticifer, articulated: Compound laticifer, formed of longitudinal series of cells, with walls between cells entire or perforated.

Laticifer, non-articulated: Single cells which may be coenocytic and branched but are not joined to form long tubes.

Laticiferous cell: Non-articulated or simple laticifer.

Laticiferous vessel: Articulated laticifer with walls between adjacent cells perforated.

Leaf buttress: Initial stage of development of a leaf, from a leaf primordium.

Leaf sheath: The lower part of a leaf, which encases the stem more or less completely.

Lenticel: Part of the periderm, distinguished from the phellem itself, in having intercellular spaces; the tissues may or may not be suberized. Lenticels play a role in gas exchange in stems once the periderm has been formed and the epidermis, which contained stomata, has been lost.

Leucoplast: A colourless plastid.

Lignin: An organic complex of high carbon-content substances, distinct from carbohydrates; present in matrix of cell walls of many cells.

Lignification: Process occurring during secondary growth of plant cell walls. The term refers specifically to cells whose walls have become impregnated with lignin.

Lithocyst: A cell containing a cystolith.

Lumen: The internal space bounded by a cell wall in a plant cell.

Lysigenous: Applied to an intercellular space, which originates by the enzymatic dissolution of cells. *See* Schizogenous.

Maceration: The artificial separation of cells, causing the separation of cells at the middle lamella.

Macrosclereid: Elongated sclereid with unevenly distributed secondary wall thickenings.

Marginal: Located at the margin of a structure – e.g. a marginal meristem is located at the margin of a developing leaf primordium.

Mechanical tissue: Cells with more or less thickened walls, e.g. collenchyma of the primary growth and sclerenchyma of primary and secondary growth. Also called supporting tissue.

Medulla: Pith.

Medullary bundle: A vascular bundle located within the pith or central ground tissue of a stem.

Medullary ray: *See* Ray, pith.

Meristele: One of the bundles of a dictyostele. *See* Vascular bundle.

Meristem: Tissue which by division produces new cells which undergo differentiation to form mature tissue and at the same time frequently perpetuates itself.

Meristem, apical: A meristem at the apex of shoot or root which by division gives rise to cells forming the primary tissues of shoot or root.

Meristem, axillary: Meristem located in leaf axil; capable of giving rise to axillary bud.

Meristem, ground: Meristematic tissue originating in an apical meristem, producing tissues other than epidermis and vascular tissues.

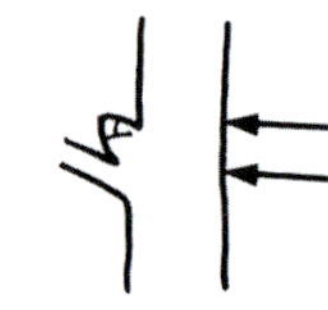

Meristem, intercalary: Meristematic tissue derived from apical meristem which during the course of development of the plant becomes separated from it by regions of more or less mature tissues.

Meristem, lateral: Meristem parallel to the circum-ference of the plant organ in which it occurs, e.g. vascular cambium, phellogen.

Meristem, marginal: Meristem located along the margin of a leaf primordium and forming the leaf blade.

Meristem, mass: Meristematic cells which divide in various planes and contribute to increase in tissue volume.

Meristem, plate: Parallel-layered meristem with planes of cell division in each layer perpendicular to the surface of the organ, which is usually a flat one.

Meristem, primary thickening: Lateral meristem derived from the apical meristem and responsible for the primary increase in width of the shoot axis, commonly found in monocotyledons.

Meristem, rib: (i) one of the regions of the shoot apex; (ii) a meristem composed of parallel series of cells in which transverse divisions characteristically take place.

Meristematic cell: A constituent cell of a meristem; shape, degree of wall thickness and extent of vacuolation varies in cells found in different meristematic regions.

Meristemoid: A cell, or group of cells, constituting an active locus of meristematic activity, usually in somewhat older tissue.

Mesarch xylem: The condition in which protoxylem in a primary xylem strand develops first in the centre of the strand and continues to develop both centrifugally and centripetally, e.g. in shoots of ferns.

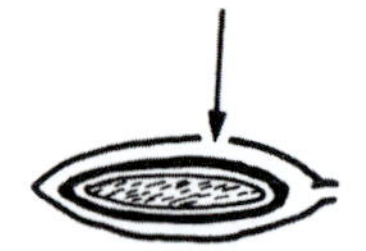

Mesocarp: The central region of a fruit wall (pericarp).

Mesocotyl: Internodal region between scutellar node and coleoptile in the embryo and seedling of a grass.

Mesomorphic: Refers to the structural features normally found in plants adapted for growth in conditions of adequate soil water and a fairly humid atmosphere (mesophytes).

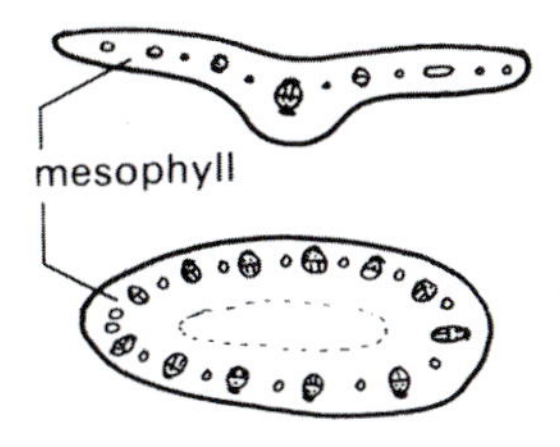

Mesophyll: Chlorenchymatous and other parenchymatous tissues of the leaf blade contained between the epidermal layers.

Mesophyll, plicate: Compact mesophyll cells in which the cell walls have directed projections or folds.

Mesophyte: A plant that requires an environment that is neither too wet, nor too dry.

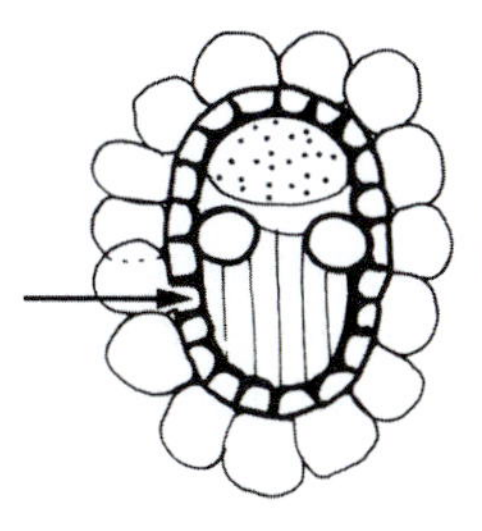

Mestome sheath: A sheath of cells enclosing a vascular bundle, the inner of two sheathes in leaves of the Poaceae (grasses). Often lignified.

Metaphloem: Primary phloem that develops after the formation of protophloem and before secondary phloem (should that also develop).

Metaxylem: Primary xylem which develops after the formation of protoxylem and before secondary xylem (should that also develop).

Microfibril: Submicroscopic, thread-like usually cellulosic components of plant cell walls.

Micropyle: A small opening between the integuments at the free end of an ovule.

Microspore: Male spore from which develops the male gamete.

Microsporocyte: A cell which develops into a microspore.

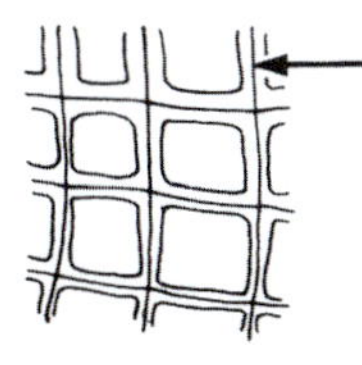

Middle lamella: Layer of intercellular material, mainly pectic substances, the cementing substance between cells.

Mitochondrion: Minute body in cytoplasm, containing respiratory enzymes (also called a chondriosome in old literature).

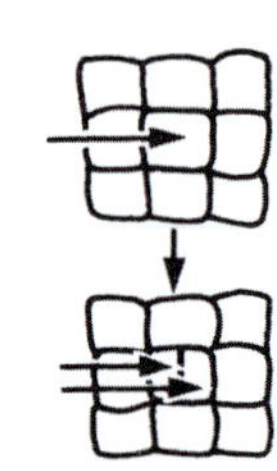

Mother cell: A cell which on division gives rise to other cells and thus loses its identity; e.g. a guard cell mother cell.

Motor cell: *See* Bulliform cell.

Mucilage cell: Cell containing a mucilage, gum or similar carbohydrate material.

Mucilage duct: Duct containing mucilage, gum or similar carbohydrate material.

Multiseriate: Consisting of many layers of cells, e.g. of rays.

Nacreous wall: A cell wall with non-lignified thickening; often associated with sieve elements – may attain considerable thickness – its designation is based on its glistening appearance; collenchyma and some sieve elements in ferns are nacreously thickened.

Nectary: Multicellular, glandular structure capable of secreting sugary solution. Floral nectaries occur in flowers; extrafloral nectaries occur in other plant organs.

Nexine: The inner layer of the exine of a pollen grain.

Node: That part of the stem where one or more leaves are attached. The node is not sharply delimited anatomically.

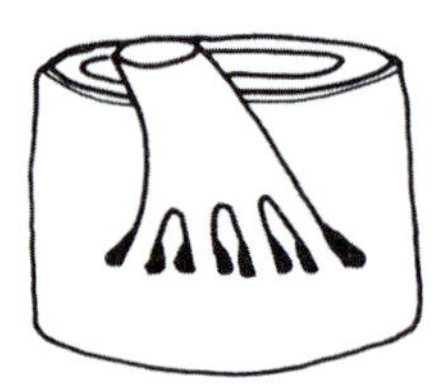

Node, multilacunar: Node with more than one leaf gap in relation to each leaf (usually used when 4 or more gaps are present).

Node, unilacunar: Node with one leaf gap in relation to each leaf.

Non-storied cambium: *See* Cambium.

Nucellus: Tissue within ovule in which the female gametophyte develops.

Ontogeny: The development of a cell, tissue type, or organ from inception to maturity.

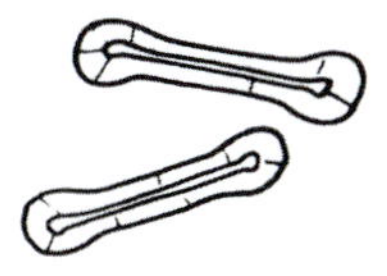

Osteosclereid: A 'bone-shaped' sclereid.

Paracytic stomata: Stoma in which the guard cells have one or more subsidiary cells adjacent and parallel to them on either flank.

Parenchyma: Living cells, frequently thin-walled but sometimes, particularly in the xylem, with lignified, thickened walls variable in shape.

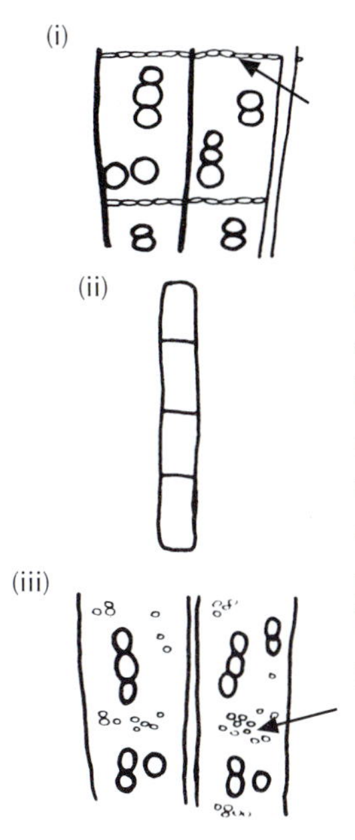

Parenchyma, apotracheal: Axial parenchyma of the secondary xylem, typically not associated with vessels: (i) banded, concentric uni- or multiseriate bands, sometimes complete rings as seen in TS; (ii) diffuse, single cells as seen in TS, distributed irregularly among fibres (usually in axial chains of four cells); (iii) diffuse in aggregates; (iv) initial, bands produced at beginning of growth ring; (v) terminal, bands produced at end of growth. (iv and v not figured)

Parenchyma, palisade: Leaf mesophyll parenchyma (chlorenchyma) characterized by an elongate form of cell, with their long axes arranged perpendicularly to the surface of the leaf.

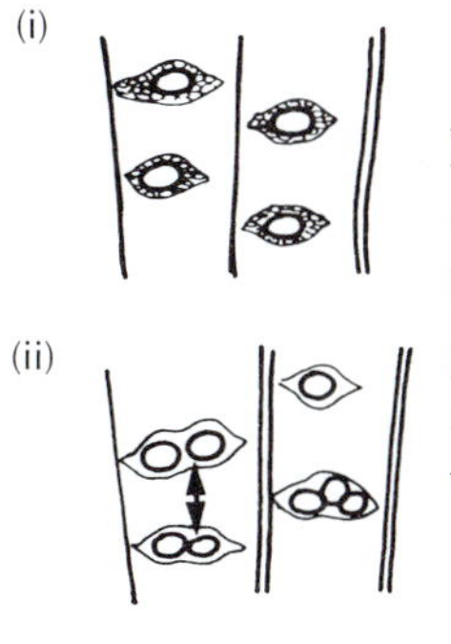

Parenchyma, paratracheal: Axial parenchyma of secondary xylem associated with vessels or tracheids: (i) aliform; (ii) confluent; (iii) scanty incomplete sheath round vessel; (iv) vasicentric complete sheath of variable width round individual or groups of vessels.

Parenchyma, xylem: Parenchyma occurring in secondary xylem: (i) axial; (ii) radial (of rays).

Parthenocarpy: Development of a fruit without fertilization (e.g. cultivated bananas).

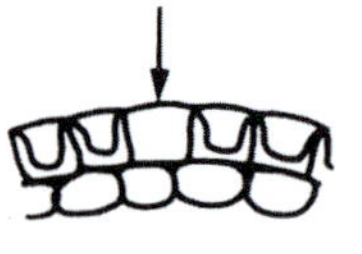

Passage cell: Thin-walled cell in root or stem endodermis or exodermis, conspicuous because of thickened walls of other endodermal cells; Casparian strips present in walls if in an endodermis.

Pectic compounds: Polymers of galacturonic acid and its derivatives; main constituent of middle lamella and intercellular substances, also a component of cell walls.

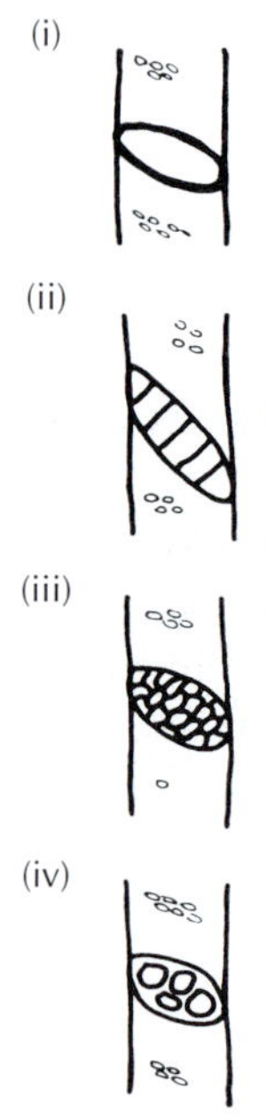

Perforation plate: Perforated end wall of a vessel element: (i) simple, surrounded by rim only; (ii) scalariform, several to numerous elongated pores with bars between them (ladder-like); (iii) reticulate, net-like; (iv) foraminate, numerous more or less circular pores. each with a rim.

Periblem: The meristem forming the cortex, according to Hanstein's system.

Pericarp: Fruit wall developed from the ovary wall.

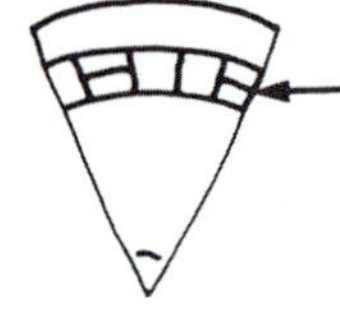

Periclinal: Cell wall or plane of division parallel with the circumference, or the nearest surface of an organ. *See* Anticlinal and tangential.

Pericycle: Part of the ground tissue of the stele located between the phloem and the endodermis. Regularly present in roots, present in few stems.

Periderm: A secondary tissue which replaces the epidermis in stems and in roots which exhibit secondary growth in thickness: consisting of phellogen producing by periclinal division phellem (cork) to the outside and phelloderm to the inside.

Perisperm: Nutrient tissue present in some seeds; originating from the nucellus.

Phellem: Protective, non-living cells with suberized cell walls. Formed centrifugally by the phellogen.

Phelloderm: Layer or layers of parenchymatous cells produced centripetally by the phellogen.

Phellogen: Cork cambium, a secondary lateral meristem, producing phelloderm internally and phellem externally; may be superficial (arising at or close to epidermis) or deep-seated (arising in deeper cortical or phloem layers).

Phloem: Main tissue that translocates assimilated products in vascular plants; composed mainly of sieve elements and companion cells (or albuminous cells), parenchyma, fibres and sclereids. Phloem, included or interxylary; secondary phloem embedded in the secondary xylem of some dicotyledons. Phloem, internal (intraxylary): primary phloem present on the inner side of the primary xylem.

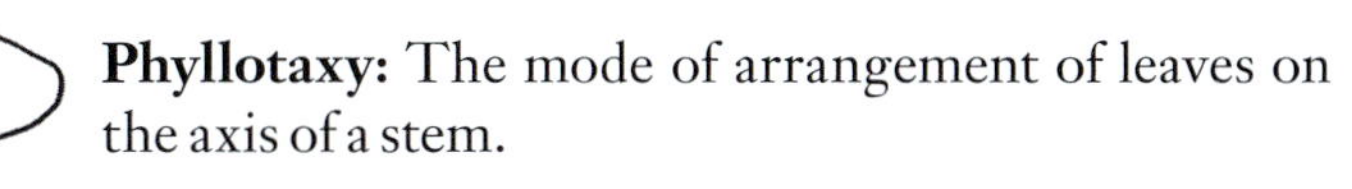

Phyllotaxy: The mode of arrangement of leaves on the axis of a stem.

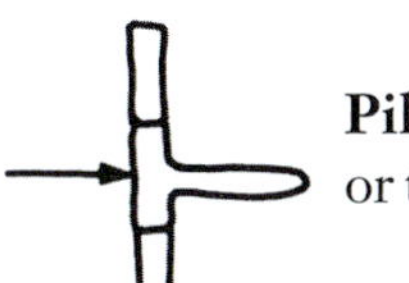

Piliferous cell: Cell, usually of the epidermis, bearing a hair or trichome.

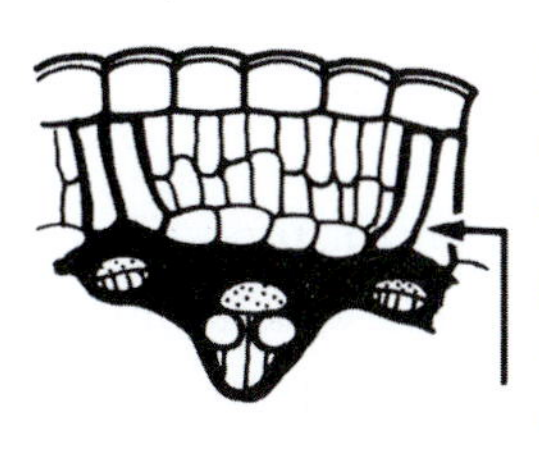

Pillar cell: (i) description of subepidermal sclereids in the seed coat of some Leguminosae; (ii) in Restionaceae, specialized, lignified cells of the stem parenchyma sheath extending to the epidermis and dividing the chlorenchyma into longitudinal channels.

Pit: A thin area of a secondarily thickened cell wall consisting of middle lamella and primary wall only.

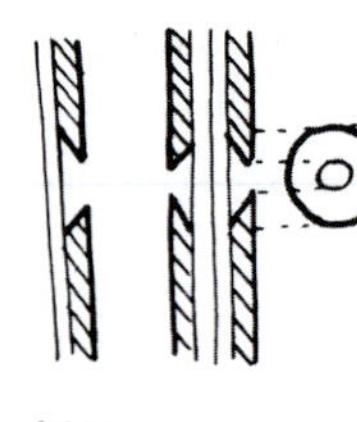

Pit, bordered: a pit in which the aperture is smaller than the pit membrane and in which the secondary wall overarches the pit membrane and pit cavity.

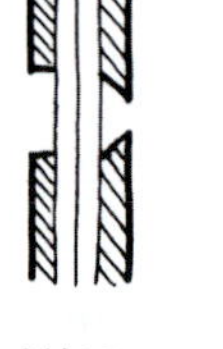

Pit, half bordered: A pit pair in which the aperture is bordered on one side of the middle lamella and not bordered on the other side.

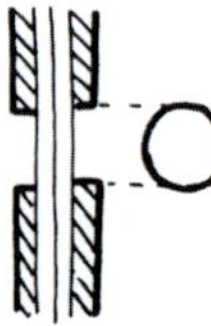

Pit, simple: A pit in which the aperture and pit membrane are similar in size.

Pit, vestured: Bordered pit with projections, either simple or branched, on secondary wall forming border of pit chamber or aperture.

Pith: Central ground tissue of stem and root; often parenchymatous, sometimes sclerotic or containing sclereids or other cell types.

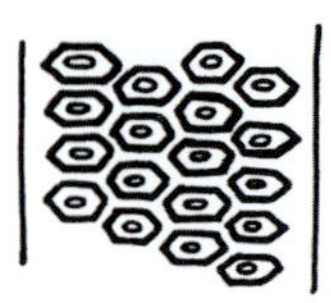

Pits, alternate: In tracheary elements, pits arranged in diagonal rows as seen in TLS and RLS.

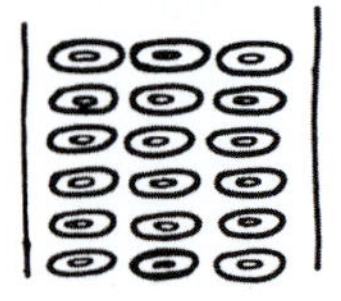

Pits, opposite: In tracheary elements, pits arranged in horizontal pairs or short horizontal rows as seen in TLS and RLS.

Placenta: Region of attachment of the ovules to the carpel.

Placentation: Position of the placenta in the ovary.

Plasmalemma: Membrane delimiting the cytoplasm and occurring next to the cell wall. Also called plasma membrane.

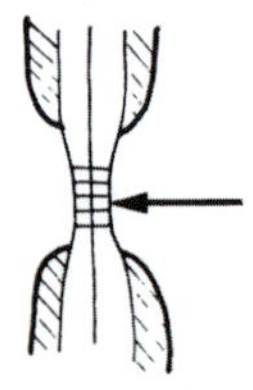

Plasmodesma: (pl. plasmodesmata) Narrow (30–100 nm) diameter, plasmalemma-lined channel in cell walls, which connect protoplasts between adjacent cells. Principal symplastic pathway between living cells.

Plastid: Protoplasmic body separated from the cytoplasm by a double membrane system.

Plastochron: Period of time between the initiation of two successive, repetitive phenomena, e.g. between the initiation of two leaf primordia.

Plectostele: Protostele in which xylem is arranged in longitudinal plates which may be interconnected.

Plerome: The meristematic cells of an apex responsible for the formation of the primary vascular system, its parenchyma and pith (if present), according to Hanstein's theory.

Plicate mesophyll cell: Mesophyll cell with infoldings or ridges of cell wall projecting into the cell.

Plumle: Bud or shoot apex of the embryo.

Pneumatode: A group of cells present in a velamen, with spiral secondary wall thickenings; may also be used for other aerating tissue.

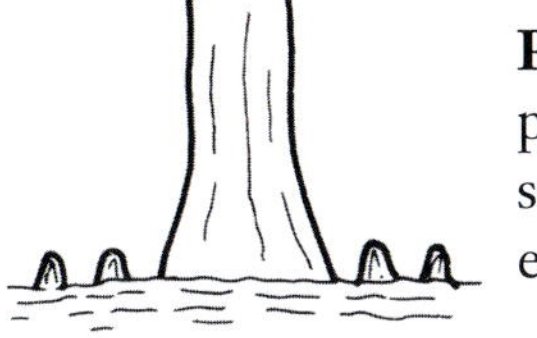

Pneumatophore: Negatively geotropic, aerial root projection formed on certain species growing in swampy ground, e.g. *Taxodium*; serves for gas exchange.

Pollen tube: Projection of vegetative cell of pollen grain, occurring on germination of the grain, covered by intine only.

Pollinium: A mass of pollen grains adhering together and usually dispersed as a unit.

Polocytic: Stoma in which the guard cell pair is situated at the end of a single subsidiary cell.

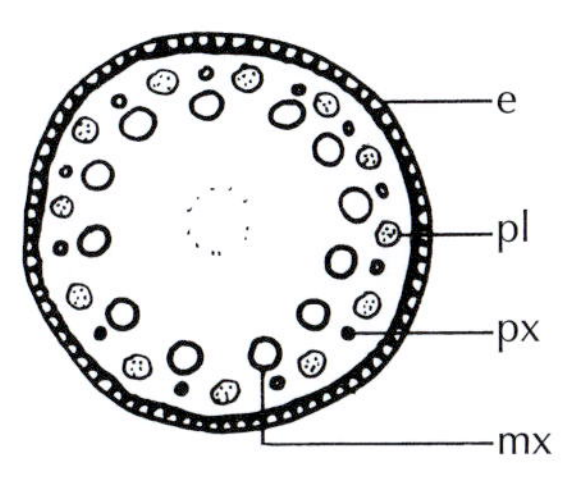

Polyarch: Primary xylem of root with a large number of protoxylem strands; e = endodermis, pl = phloem, px = protoxylem, mx = metaxylem

Polyderm: Protective tissue consisting of alternating bands of endodermis-like cells and non-suberized parenchyma cells.

Pore: In wood, a non-scientific term for the vessel elements as seen in cross-section.

Porous wood: Secondary xylem that contains vessels. (Non-porous wood lacks vessels.)

Primary cell wall: Cell wall that is formed during differentiation of the cell, when cellulose microfibrils have various (random) orientations. In fibres with apical intrusive growth, areas of secondary wall thickening are laid down before the cells have completed their growth in size.

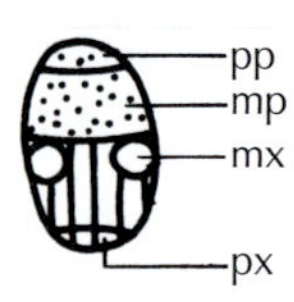

Primary phloem: Phloem tissue that differentiates from the procambium during primary growth and differentiation in a vascular plant. Divided into earlier protophloem (pp) and later metaphloem (mp). Forms the principal axial carbohydrate conducting system in vascular plants. mx, metaxylem; px, protoxylem.

Primary phloem fibres: Fibres (usually lignified, but not always so) located on the outer periphery of the phloem. Originating in the primary phloem, usually amongst the protophloem. Often incorrectly referred to as pericyclic fibres.

Primary pit field: A thin area of the primary wall with a concentration of pores or plasmodesmata.

Primary tissues: Tissues that are derived from the embryo and apical meristems.

Primary xylem: Xylem tissue that differentiates from the procambium during primary growth and of vascular plants, and forms the principal axial water conducting system in monocots and during the primary growth phase in dicots. Protoxylem forms first (px), then metaxylem (mx). Xylem rays are absent from the primary xylem.

Primordium: The earliest stages of differentiation of an organ, group of cells or single cell, e.g. a root primordium.

Procambium: A primary meristem differentiating to form the primary vascular tissues.

Proembryo: An embryo at its earliest stages of development before the start of organ differentiation.

Promeristem: In an apical meristem, the initial cells and their immediate derivatives.

Prop root: A root that is formed and located above soil level. Usually has an adventitious origin.

Prophyll: One of the earliest leaves of a lateral branch.

Proplastid: A plastid in the earliest stages of development.

Prosenchyma: Elongated parenchyma cells with thickened lignified walls; often fibre-like.

Protoderm: A primary meristem or meristematic tissue that gives rise to the epidermis.

Protophloem: The elements of primary phloem which develop first.

Protophloem poles: Term used to describe the location of the phloem elements first to mature in a plant organ.

Protoplast: A living cell unit.

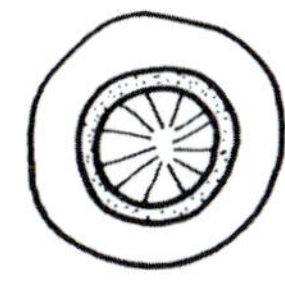

Protostele: The simplest type of stele, in which a solid column of vascular tissue occupies the central region of the organ. Phloem is peripheral to the xylem.

Protoxylem: The elements of primary xylem which develop first.

Protoxylem lacuna: A space surrounded by parenchyma cells in the protoxylem of a vascular bundle. Appears in some plants after the protoxylem has ceased to function, and its constituent elements have been stretched and torn, thus forming the cavity.

Provascular bundle: Procambial strand. *See* Procambium.

Pseudocarp: A false fruit in which floral organs other than carpels participate in forming the fruit wall, e.g. apple.

Pulvinus: The swelling at the base of a leaf petiole or leaflet petiolule.

Radial system: (i) all cells derived from the ray cambial initials in secondary vascular tissues; (ii) cells arranged radially with respect to the long axis.

Radicle: An embryonic root.

Ramiform pit: Simple pit in which canals are coalescent.

Raphe: A ridge on a seed formed by that part of the funiculus which was fused to the ovule.

Raphide: A needle-shaped crystal of calcium oxalate; usually one of a number of crystals arranged parallel to one another in a mucilaginous sac or raphide sac.

Ray, heterocellular: *See* Heterocellular ray.

Ray, homocellular: *See* Homocellular ray.

Ray initial: A meristematic ray-cell producing cell within the vascular cambium.

Ray, multiseriate: Ray of secondary vascular tissue two to many cells wide as seen in TLS.

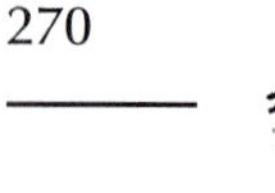

Ray, pith: Parenchymatous interfascicular region of stem.

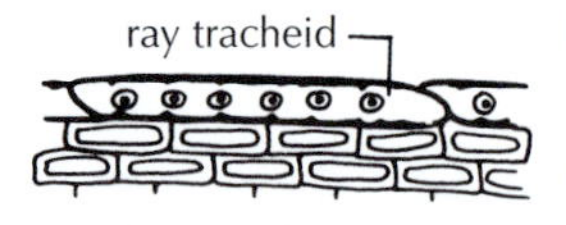

Ray, tracheid: Tracheid occurring in the radial system of wood of some conifers, usually at the ray margins.

Ray, uniseriate: Ray of secondary vascular tissue one cell wide.

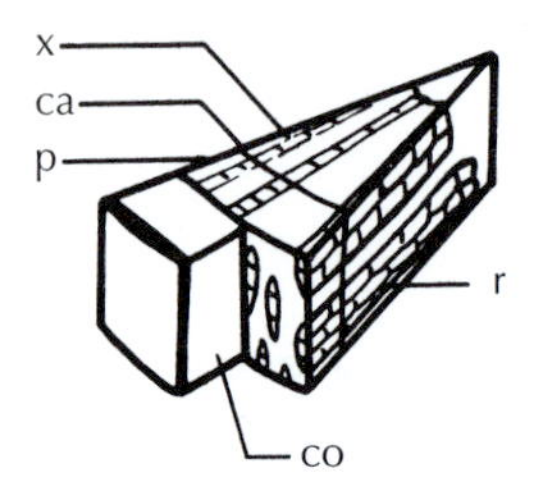

Ray, vascular: A tissue system oriented radially through the secondary xylem (xylem ray) and secondary phloem (phloem ray) and derived from the cambial ray initials; ca = cambium, co = cortex, p = phloem, r = ray, x = xylem.

Reaction wood: Wood with distinctive features forming in leaning or crooked branches and twigs; termed tension wood in angiosperms and compression wood in conifers.

Resin duct or canal: Schizogenous duct containing resin.

Reticulate cell wall thickening: Secondary wall thickening in tracheary elements with a net-like appearance.

Reticulate venation: Leaf blade veins forming an anastamosing, net-like system.

Rhexigenous: Formed by tearing apart of cells. *See* Intercellular space).

Rhizodermis: Name given to the epidermis of a root.

Rhytidome: The outer part of the bark composed of the periderm and all tissues external to it.

Ribosome: Minute protoplasmic organelle containing messenger RNA; concerned with protein synthesis.

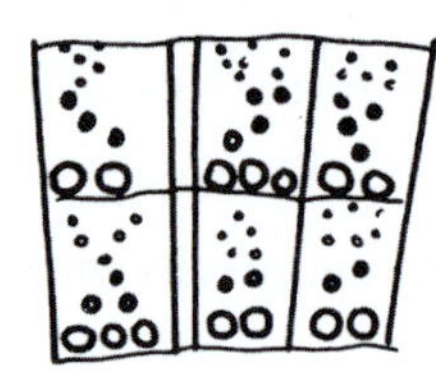

Ring-porous wood: Secondary xylem in which the vessels of the early wood are distinctly wider than those of the late wood and thus form a well-defined zone or ring as seen in a cross-section of the wood.

root cap

main cell ÷

Root cap: Cells cut off by the calyptrogen in the root apical meristem and forming a protective cap cushioning the apex itself.

Root, contractile: Specialized root capable of contraction; helps to maintain a plant or part of a plant at the appropriate depth in the soil.

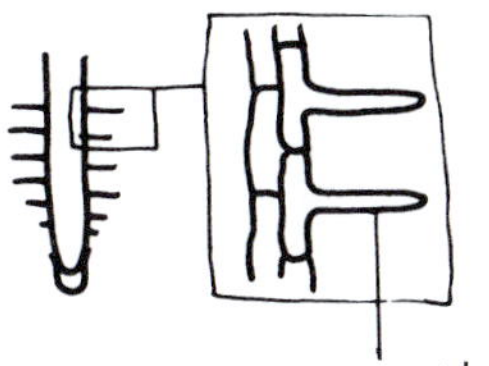

Root hair: A type of unicellular trichome developed from a root epidermal cell; may be short-lived, it absorbs solutions from the soil.

Sapwood: Outer part of xylem of a tree or shrub containing living cells, reserve materials and water.

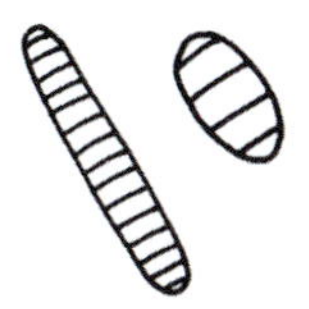

Scalariform: Ladder-like; a closely parallel arrangement of structures on the cell wall of an element, e.g. a type of secondary wall thickening and perforation plate.

Scale: A flattened type of trichome attached along or near to one edge. Lacks vascularization.

Schizogenous: Formed by separation or splitting; usually refers to intercellular spaces which originate by cells parting at the middle lamellae.

Schizo-lysigenous: Applied to an intercellular space, originating by a combination of the two processes, resulting in the separation and degradation of cell walls.

Scion: The part of a plant that is inserted to form a graft with another part of a different plant which forms the stock (the rooted part, rootstock).

Sclereid: A form of sclerenchymatous cell with lignified walls, usually relatively short; a range of types exists:

—, **astrosclereid,** branched or ramified sclereid;
—, **brachysclereid,** or stone cell, short, more or less isodiametric sclereid;
—, **fibre sclereid,** a cell intermediate in length between a fibre and sclereid;
—, **macrosclereid,** elongated sclereid with uneven secondary wall thickening; when present in testa of leguminous seeds also called a malpighian cell;

—, **ostoesclereid,** 'bone-shaped' sclereid;
—, **trichosclereid,** hair-like sclereid.

Sclerenchyma: A mechanical or supporting tissue of cells with lignified walls; made up of fibres, sclereids and fibre-sclereids.

Sclerification: The process of changing into sclerenchyma by the progressive lignification of secondary walls.

Scutellum: Part of the embryo in Gramineae.

Secondary body: That part of the plant that is added to the primary plant body due to the activity of the lateral meristems, the vascular cambium and phellogen.

Secondary cell wall: Cell wall deposited in some cells over the primary wall, after the primary wall has ceased to increase in its surface area. Secondary cell wall shows a definite parallel orientation of cellulose microfibrils in the TEM.

Secondary growth: In gymnosperms and dicotyledons characterized by an increase in the girth or thickness of the stem and root and resulting from the formation of secondary vascular tissues due to activity of the vascular cambium. May be supplemented by activity of the phellogen. In certain monocotyledons, e.g. *Dracaena*, a peripheral meristem leads to increase in stem thickness by divisions to form new ground tissue and vascular bundles.

Secondary phloem: Phloem tissue formed by the activity of the vascular cambium during secondary growth.

Secondary phloem fibre: A fibre located in the axial system of the secondary phloem.

Secondary xylem: Xylem tissue formed by the activity of the vascular cambium during secondary growth.

Secretory cavity: A cavity filled with the breakdown products of cells which formed the cavity.

Secretory cell: A specialized living cell which secretes or excretes substances.

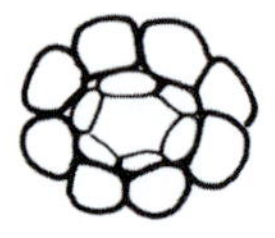

Secretory duct: Duct formed schizogenously, frequently lined by thin-walled secretory epithelial cells which secrete substances into the duct.

Separation layer: The layer or layers of cells which disintegrate in the abscission zone.

Septate fibre: *See* Fibre, septate.

Sexine: The outer layer of the exine of the pollen grain.

Shoot: The stem and its appendages.

Sieve area: An area of the wall of a sieve element which contains a concentration of pores, each callose-lined and encircling a strand of protoplasm which connects the protoplast of one sieve element with that of the next.

Sieve cell: A sieve element with relatively undifferentiated sieve areas; with narrow pores which are uniform in structure on all walls and connecting strands; found in gymnosperms and lower vascular plants.

Sieve element: Cell of the phloem concerned with the longitudinal transport of carbohydrate, classified into sieve cell and sieve tube member.

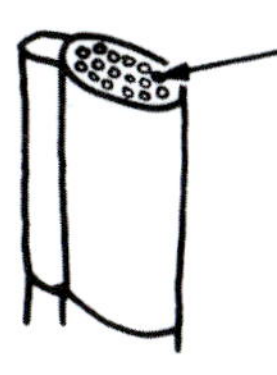

Sieve plate: The part of the common cross wall between concomitant sieve elements bearing one or more highly differentiated sieve areas, typically found in angiosperms.

Sieve tube: A series of sieve tube elements or members joined together end to end.

Sieve tube element or member: One of the series of cellular components in a sieve tube. Shows differentiation of sieve plates and lateral sieve areas.

Silica body: Opaline cell inclusion; the shape of a silica body may be characteristic for a family or group within a family.

Silica cell: (i) a cell containing one or more silica bodies; (ii) an epidermal cell containing a silica body.

Siphonostele: A stele composed of a hollow cylinder of vascular tissue with a central pith: (i) amphiphloic – phloem both to interior and exterior of xylem; (ii) ectophloic – phloem to exterior of xylem cylinder.

Soft wood: Common name for gymnosperm wood, particularly that of the Coniferae. Some gymnosperm wood can, in fact, be very hard.

Solenostele: Amphiphloic siphonostele in which successive leaf gaps are well separated from one another.

Spiral thickening: *See* Helical thickening.

Starch granule: A cell inclusion composed of starch; frequently with a characteristic shape for a particular species or group of species. The radiating chain structure of the crystalline residue produces a characteristic 'Maltese Cross' when the granule is viewed between crossed polars in the microscope.

Starch: An insoluble carbohydrate acting as one of the commonest storage products of plants, composed of anhydrous glucose residues.

Starch sheath: Name given to the innermost layer of the cortex if it is specialized to store starch; probably homologous with the endodermis.

Stele: The part of the plant axis made up of the primary vascular system and its associated ground tissue.

Stele, polycyclic: A stele with two or more concentric circles of vascular tissue.

Stellate: Star-shaped. *See* Aerenchyma and Sclereid.

Stereome: All mechanical tissue of the plant.

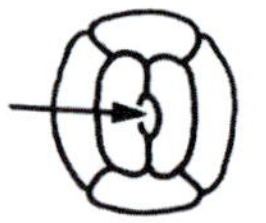

Stoma: A pore in the epidermis encircled by two guard cells; often used to describe both the pore itself and the two guard cells that surround it and regulate its size and thus the rate of gas exchange. Some stomata have additional cells adjacent to the guard cells, which are distinct from other epidermal cells in shape or wall thickness. These are the subsidiary cells. *See* Anisocytic stoma, Paracytic stomata for example.

Stone cell: *See* Brachysclereid.

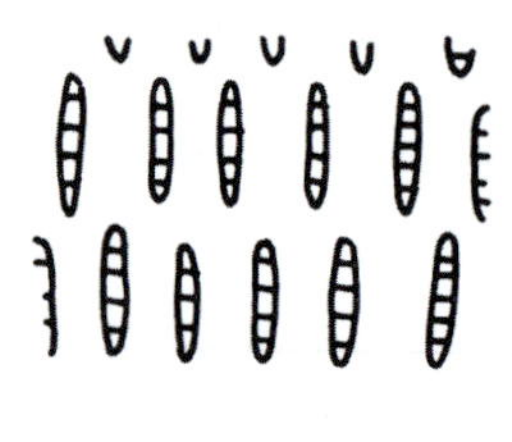

Storied (or storeyed) tissue: Tissue in which the cells are arranged in horizontal series as seen in TLS and RLS: e.g. storied cambium which gives rise to storied xylem and phloem; storied rays may be apparent even when other tissues have lost their regular arrangement during growth adjustments: Produces ripple marks in wood.

Stroma: The structural framework of a plastid.

Styloid: Crystal with elongated prismatic shape; ends flat or pointed.

Suberin: Fatty substance, similar in nature to cutin, in the cell wall of cork tissue, in the Casparian strip of the endodermis and in the inner bundle sheath cells of Poaceae leaves.

Suberization: The process of deposition of suberin in cell walls.

Subsidiary cells: Epidermal cells which together with the guard cells make up the stomatal apparatus. Subsidiary cells are frequently distinguishable from other epidermal cells by their shape or wall thickness. Occasionally they can only be discerned by developmental studies.

Surface layer, surface meristem: A histological zone in the gymnosperm apex.

Suspensor: The connection between the main part of the embryo and the basal cell; may have a function in nutrition of an embryo.

Syncarpy: Fusion of carpels in a flower and ovary.

Synergids: Cells in the mature embryo sac present alongside the egg cell.

Tangential: At right angles to the radius of the stem or roots, or at right angles to the rays in secondary vascular tissue. Tangentially oriented cell walls are frequently also periclinal.

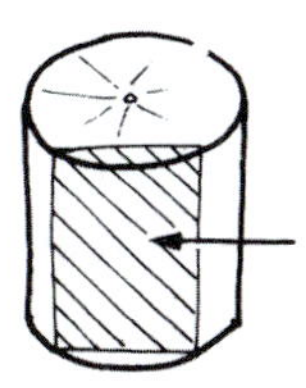

Tangential longitudinal section (TLS): A section made at a tangent to and parallel with the long axis of a cylindrical organ.

Tannin: A collective term used for a range of polyphenolic substances deposited in certain plant cells; common, e.g., in bark, whence it is extracted for tanning.

Tapetum: Innermost layer of cells of pollen sac wall; their contents nourish developing pollen grains and also provide part of the protein involved in recognition systems between pollen and stigma: (i) amoeboid, a tapetum in which the protoplasts of its cells penetrate between the pollen mother cells; (ii) glandular, a tapetum in which the cells disintegrate in their original position.

Taproot: The first or primary root that forms a direct continuation between the radicle of the embryo.

Tendril: A modified leaf or stem, a slender coiling structure, which functions in the support of stems of climbing plants.

Tension wood: Reaction wood formed on the upper side of branches or leaning or bent stems of dicotyledons; fibres characteristically gelatinous and little lignified.

Tepal: A perianth member in flowers that lack distinction between petals and sepals, e.g. *Tulipa.*

Tertiary wall: Wall thickening to inner side of secondary wall.

Testa: The seed coat.

Tetrarch: Primary xylem of roots, having four protoxylem strands or poles.

Tissue, conjunctive: (i) a special type of parenchyma associated with included phloem in dicotyledons having anomalous secondary thickening; (ii) parenchyma present between secondary vascular bundles in monocotyledons having secondary thickening.

Tissue, expansion: An intercalary tissue in the outer part of the inner bark, formed mainly by the phloem rays, which enables the bark to expand without tearing.

Tissue, ground: All tissues of mature plants except the epidermis, periderm and vascular tissues.

Tissue, mechanical: Tissue composed of supporting cells, e.g. sclerenchyma and collenchyma.

Tissue, transfusion: Tissue surrounding or associated with vascular bundles in the leaves of gymnosperms, composed of tracheids and varying amounts of parenchyma cells. Assumed to be involved in the transfer of solutes to and from the vascular tissue.

Tonoplast: Membrane of the cytoplasm where it borders a vacuole.

Torus: Central thickening of pit membrane in a bordered pit of certain gymnosperms and occasional angiosperms; made up of middle lamella and primary wall material.

Trabecula: Bar-like projection of cell wall crossing a cell lumen.

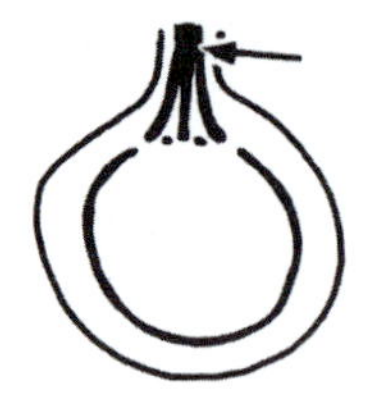

Trace: (i) branch vascular system joining the main stem and vascular supply of a branch; (ii) leaf vascular system joining the main stem to the leaf vascular system.

Tracheary element: Xylem element involved in water transport; includes vessel elements, tracheids and tracheoidal vessel elements. A useful term used to describe water-conducting tissue when the exact cell type has not been ascertained.

Tracheid: An imperforate tracheary element, i.e. with intact pit membrane between it and adjacent elements.

Tracheoidal vessel element: A perforated vessel element which is very elongated, often narrow and looks in all other respects like a tracheid.

Transfer cell: Parenchymatous cell with minute ingrowths of cell wall; concerned with movement of materials, e.g. in seedlings.

Transfusion tissue: *See* Tissue, transfusion.

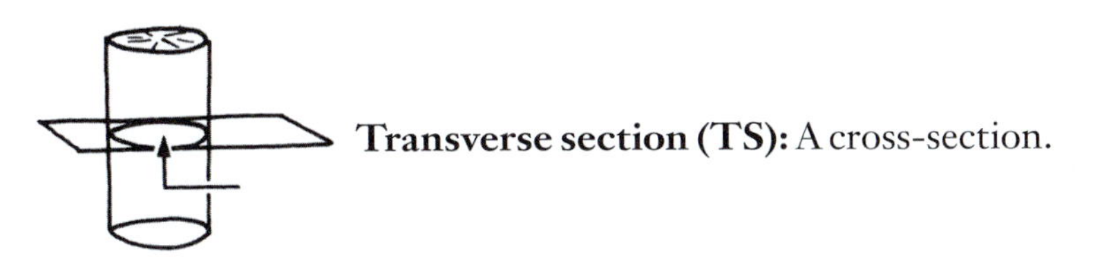

Transverse section (TS): A cross-section.

Traumatic tissue: Wound tissue, e.g. callus or resin-filled cavities of a traumatic resin duct.

Triarch: Primary xylem of root, having three protoxylem poles or strands.

Trichoblast: Specialized cell in root epidermis which gives rise to a root hair.

Trichome: Epidermal appendage, includes hairs, scales and papillae; may be glandular or non-glandular.

Trichosclereid: A type of branched sclereid, with hair-like branches extending into intercellular spaces.

Tunica: Outermost layer or layers of cells in apical meristem of angiosperm shoot in which most cell divisions are anticlinal; in the corpus, cell divisions are anticlinal, periclinal and also in other planes.

Tylosis: Intrusion of a ray or axial parenchyma cell into a vessel element lumen by perforation of a pit membrane; may or may not be lignified; occur rarely in tracheids.

Tylosoid: An epithelial cell which proliferates in an intercellular cavity such as a resin duct.

Unifacial leaf: A leaf that develops from one side of the leaf primordium only and in consequence has only an encircling adaxial or abaxial epidermis (may be secondarily flattened and appear dorsiventral).

Uniseriate: Cells in one layer, e.g. uniseriate ray.

Uniseriate ray: In secondary vascular tissues, a ray that is one cell wide.

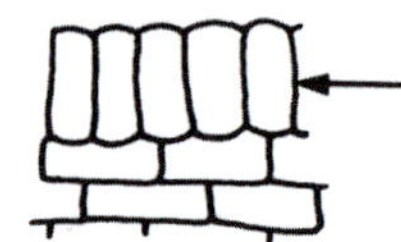

Upright ray cell: Cell of rays of secondary vascular tissues; longer axially than radially.

Vacuolation: Vacuole formation.

Vacuole: A volume enclosed in the cytoplasm separated from it by the tonoplast, containing cell sap.

Vascular: Referring to tissues of the xylem or phloem or both.

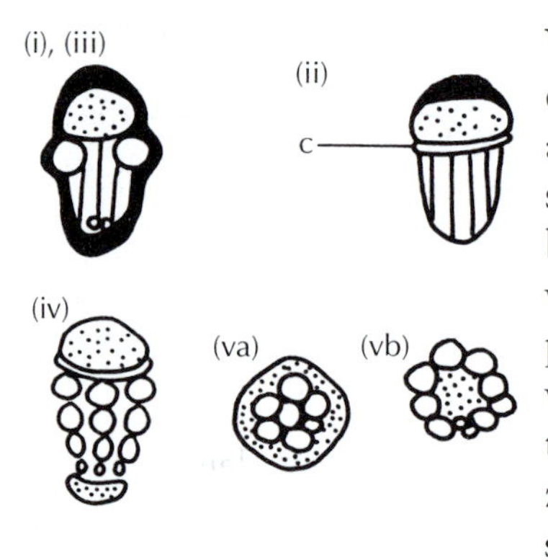

Vascular bundle: An organized strand of conducting tissue composed of xylem and phloem, and in most dicotyledonous stems, cambium: (i) closed lacking cambium, as in monocotyledons; (ii) open with cambium (c); (iii) collateral with one phloem and xylem pole; (iv) bicollateral with two phloem poles, one at either end of the xylem pole, but with only one cambial zone; (va) concentric amphicribal, xylem surrounded by phloem and (vb) concentric amphivasal, phloem surrounded by xylem.

Vascular cambium: A lateral meristem that forms secondary vascular tissues, which are arranged into an axial and a radial system.

Vascular cylinder: The vascular region of the axis. Term used synonymously with stele or central cylinder, and in a more restricted sense, excludes the central pith.

Vein: Vascular bundle or group of closely parallel bundles in a leaf, bract, sepal, petal or flat stem.

Velamen: Multiseriate epidermis, a characteristic tissue of many aerial roots in Orchidaceae and Araceae; may be present in some terrestrial roots.

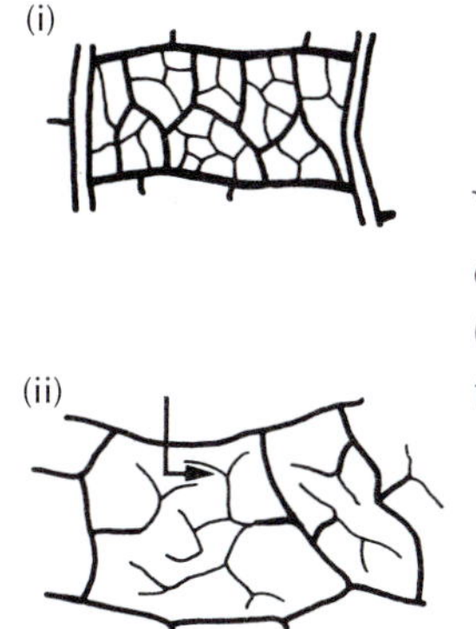

Venation: The arrangement of veins in an organ: (i) closed vein endings which anastome in a leaf blade; (ii) open vein endings which are free, i.e. do not anastome in a leaf blade.

Vessel: A tube-like series of vessel elements or members with perforate common end walls.

Vessel element or member: A tracheary member of a vessel, with perforated end walls.

Wall, tertiary layer: A layer of thickening material to the inner side of a secondary wall, e.g. tertiary spiral wall thickening.

Wart: Fine granular protrusions on inner surface of secondary wall of tracheids, fibres and vessels.

Xeromorphic: Refers to specialized structural adaptations and features of plants adapted to live in dry habitats (xerophytes).

Xerophyte: A plant adapted to dry conditions.

Xylem: The main tissue concerned with water conduction in vascular plants, characterized by the presence of tracheary elements.

Xylem elements: Cells composing the xylem tissue – vessels, tracheids, sclereids and parenchymatous elements.

Xylem fibre: Two types of fibre are recognized in the xylem: (i) fibre-tracheid, which has bordered pits; and (ii) libriform fibre which has pits with no border and a slit-like aperture on the outer face. There are intermediate cell types between the two.

Cited references

Chapter 6

Goggin, F. L., Medville, R. & Turgeon, R., 2001. Phloem loading in the tulip tree. mechanisms and evolutionary implications. *Plant Physiol.* **125**(2), 891–899.

Gottwald, J. R., Krysan, P. J., Young, J. C., Evert, R. F. & Sussman, M. R., 2000. Genetic evidence for the in planta role of phloem-specific plasma membrane sucrose transporters. *Proc. Natl Acad. Sci. USA* **97**, 13979–13984.

Hoffmann-Thoma, G., van Bel, A. J. & Ehlers, K., 2001. Ultrastructure of minor-vein phloem and assimilate export in summer and winter leaves of the symplasmically loading evergreens *Ajuga reptans* L., *Aucuba japonica* Thunb., and *Hedera helix* L. *Planta* **212**(2), 231–242.

Komor, E., Orlich, G., Weig, A. & Kockenberger, W., Phloem loading – not metaphysical, only complex: towards a unified model of phloem loading. *J. Exp. Bot.* **47**, 1155–1164.

Ma, F. & Peterson, C. A., 2001. Frequencies of plasmodesmata in *Allium cepa* L. roots: implications for solute transport pathways. *J. Exp. Bot.* **52**(358), 1051–1061.

Oparka, K. & Turgeon, R, 1999. Sieve elements and companion cells – traffic control centres of the phloem. *Plant Cell* **11**, 739–750.

Patrick, J. W. & Offler, C. E., 2001. Compartmentation of transport and transfer events in developing seeds. *J. Exp. Bot.* **52**(356), 551–564.

Soros, C. & Dengler, N., 1998. Quantitative leaf anatomy of C3 and C4 Cyperaceae and comparisons with the Poaceae. *Int. J. Plant Sci.* **159**, 480–491.

Taiz, L. & Zeiger, E., 2002. *Plant Physiology*, 3rd edn. Sinauer Associates, New York.

Turgeon, R., 2000. Plasmodesmata and solute exchange in the phloem. *Australian J. Plant Physiol.* **27**(6), 521–529.

Chapter 8

Haberlandt, G. 1918. *Physiologische Pflanzenanatomie*, 5th edn. Engelmann, Leipzig.

Chapter 10

Feder, N. & O'Brien, T.P., 1968. Plant microtechnique. Some principles and new methods. *Am. J. Bot.* **55**, 123–142.

Foster, A. S., 1950. *Practical Plant Anatomy*, 2nd edn. Van Nostrand, New York.
Gurr, E., 1965. *The Rational Use of Dyes in Biology*. Leonard Hill, London.

Appendix 2

Feder, N. & O'Brien, T.P., 1968. Plant microtechnique. Some principles and new methods. *Am. J. Bot.* **55**, 123–142.
Rinne, P.L.H. & van der Schoot, C., 1998. Symplasmic fields in the tunica of the shoot apical meristem coordinate morphogenetic events. *Development* **144**, 1477–1485.

Further reading

General

Cutler, D. F. & Gregory, M., 1998. *Anatomy of the Dicotyledons*, vol. IV, *Saxifragales*, 2nd edn. Oxford Scientific Publications.

Evert, R.F., 2006, *Esau's Plant Anatomy*, 3rd edn. *Meristems, cells and tissues of the plant body – their structure function and development*. Wiley-Interscience, New Jersey.

Metcalfe, C. R. & Chalk, L., 1979. *Anatomy of the Dicotyledons*, vol I, *Systematic Anatomy of Leaf and Stem*, 2nd edn. Clarendon Press, Oxford.

Metcalfe, C. R. & Chalk, L., 1983. *Anatomy of the Dicotyledons*, vol II, *Wood Structure and Conclusion of the General Introduction*, 2nd edn. Clarendon Press, Oxford.

Metcalfe, C. R., 1987. *Anatomy of the Dicotyledons*, vol. III, *Magnoliales, Illiciales and Laurales*, 2nd edn. Oxford Science Publications.

Chapter 1

Cutter, E. G., 1969. *Plant Anatomy: experiment and interpretation. Part I Cells and Tissues.* Edward Arnold, London.

Cutter, E. G., 1970. *Plant Anatomy: experiment and interpretation. Part 2 Organs.* Edward Arnold, London.

Dickison, W., 2000. *Integrative Plant Anatomy.* Academic Press, San Diego.

Esau, K., 1965. *Plant Anatomy*, 2nd edn. John Wiley, New York.

Esau, K., 1977. *Anatomy of Seed Plants.* John Wiley, New York.

Fahn, A., 1974. *Plant Anatomy*, 2nd edn. Pergamon Press, London.

Gifford, E. M. & Foster, A.S., 1974. *Morphology and Evolution of Vascular Plants*, 3rd edn. W. H. Freeman, San Francisco.

Meylan, B. A. & Butterfield, B. G., 1972. *Three Dimensional Structure of Wood.* Chapman & Hall, London.

Troughton, J. H. & Donaldson, L. A., 1972. *Probing Plant Structure.* Chapman & Hall, London.

Troughton, J. H. & Sampson, F. B., 1973. *Plants, a scanning electron microscope survey.* John Wiley, Australasia Pty Ltd, Sydney.

Chapter 2

Cutter, E. G., 1965. Recent experimental studies of the shoot apex and shoot morphogenesis. *Botanical Review*, **31**, 71–113.

Rinne, P. L. H. & van der Schoot, C., 1998. Symplastic fields in the tunica of the shoot apical meristem coordinate morphogenetic events. *Development* **144**, 1477–1485.

Steeves, T. & Sussex, I., 1989. *Patterns in Plant Development.* Cambridge University Press.

Wardlaw, C. W., 1968. *Morphogenesis in Plants. A contemporary study.* Methuen, London.

Williams, R. F., 1975. *The Shoot Apex and Leaf Growth.* Cambridge University Press.

Chapter 3

Brazier, J. D. & Franklin, G. L., 1961. *Identification of Hardwoods. A Microscopic Key.* Forest Products Research Bulletin, No. 46. HMSO, London.

British Standards 881 and 589, 1974. *Nomenclature of Commercial Timbers, Including Sources of Supply.* British Standards Institution.

Carlquist, S., 1988. *Comparative Wood Anatomy: systematic, ecological, and evolutionary aspects of dicotyledon wood.* Springer-Verlag, Berlin.

Cutler, D. F., Rudall, P. J., Gasson, P. E. & Gale, R. M. O., 1987. *Root Identification Manual of Trees and Shrubs.* Chapman & Hall, London.

Esau, K., 1969. *The Phloem. Handbuch der Pflanzenanatomie.* Gebruder Borntraeger, Berlin.

Gale, R. & Cutler, D. F. 2000. *Plants in Archaeology: identification manual of vegetative plant materials used in Europe and the Southern Mediterranean to c. 1500.* Royal Botanic Gardens, Kew.

Gregory, M., 1994 . Bibliography of systematic wood anatomy of dicotyledons. *IAWA Journal Supplement* 1, 265 pp. Published for the IAWA by the National Herbarium of the Netherlands, Leiden.

IAWA Committee (Wheeler, E. A., Baas, E. A. & Gasson, P., eds.), 1989. IAWA list of microscopic features for hardwood identification. Repr. from *IAWA Journal* **10**, 219–332. Published for the IAWA by the National Herbarium of the Netherlands, Leiden.

IAWA Committee (Richter, H. G., Grosser, D., Heinz, I. & Gasson, P. E., eds.), 2004. IAWA list of microscopic features for softwood identification. Repr. from *IAWA Journal* **25**, 1–70, illust. Published for the IAWA by the National Herbarium of the Netherlands, Leiden.

Jane, F. W., revised by Wilson, K. & White, D. J. B., 1970. *The Structure of Wood*, 2nd edn, A. & C. Black, London.

Kribs, D. A., 1959. *Commercial Foreign Woods on the American Market.* Pennsylvania State University Press.

Metcalfe, C. R. & Chalk, L., 1983. *Anatomy of the Dicotyledons*, Vol. II. Clarendon Press, Oxford.

Phillips, E. W. J., 1948. *The Identification of Coniferous Woods by their Microscopic Structure.* Forest Products Research Bulletin, No. 22. HMSO, London.

Chapter 4

Carlquist, S., 1961. *Comparative Plant Anatomy.* Holt, Rinehart and Winston, New York.

Metcalfe, C. R. (ed.) *Anatomy of the Monocotyledons – Gramineae*, 1960 (Metcalfe, C. R.); *Palmae*, 1961 (Tomlinson, P. B.); *Commelinales Zingiberales*, 1968 (Tomlinson, P. B.);

Juncales, 1968 (Cutler, D. F.); *Cyperaceae*, 1971 (Metcalfe, C. R.); *Dioscoreaceae* 1971 (Ayensu, E. S.); Alismatidae, 1982 (Tomlinson, P. B.); Fridaceae, 1995 (Rudall, P. J.); Araceae and Acoraceae, 2002 (Keeting, R. C.).

Metcalfe, C. R. & Chalk, L., 1950. *Anatomy of the Dicotyledons*, Vols I & II. Clarendon Press, Oxford.

Chapter 5

Carlquist, S., 1992. Wood anatomy and stem of *Chloranthus* – summary of wood anatomy of Chloranthaceae, with comments on relationships, vessellessness, and the origin of monocotyledons. *IAWA Bull.* **13**, 3–16.

Carlquist, S., Dauer, K. & Nishimura, S. Y., 1995. Wood and stem anatomy of Saururaceae with reference to ecology, phylogeny, and origin of the monocotyledons. *IAWA J.* **16**(2), 133–150.

Fisher, J. B., 1975. Eccentric secondary growth in cordyline and other Agavaceae (Monocotyledonae) and its correlation with auxin distribution. *Amer. J. Bot.* **62**(3), 292–302.

Rudall, P.J., 1991. Lateral meristems and stem thickening growth in monocotyledons. *Bot. Rev.* **57**, 150–163.

Tomlinson, P. B. & Zimmermann, M. H., 1969. Vascular anatomy of monocotyledons with secondary growth – an introduction. *J. Arnold Arbor.* **50**(2), 159–179.

Chapter 6

Hickey, L. J., 1973. Classification of the architecture of dicotyledonous leaves. *Amer. J. Bot.* **60**(1), 17–33.

Öpik, H. & Rolfe, S., 2005. *The Physiology of Flowering Plants*, 4th edn. Cambridge University Press.

Ruiz-Medrano, R., Xoconostle-Cazares, B. & Lucas, W. J., 2001. The phloem as a conduit for inter-organ communication. *Curr. Opin. Plant Biol.* **4**(3), 202–209.

Turgeon, R., Medville, R. & Nixon, K. C., 2001. The evolution of minor vein phloem and phloem loading. *Am. J. Bot.* **88**(8), 1331–1339.

Chapter 7

Baskin, C. & Baskin, J., 1998. *Seeds: ecology, biogeography, and evolution of dormancy and germination*. Academic Press, San Diego.

Corner, E. J. H., 1976. *The Seeds of Dicotyledons*. Cambridge University Press.

Davis, G. L., 1966. *Systematic Embryology of the Angiosperms*. John Wiley, New York.

Endress, P. K., 1996. *Diversity and Evolutionary Biology of Tropical Flowers*. Cambridge University Press, Cambridge.

Erdtman, G. (Gunnar), 1897. *Uniform ti Handbook of Palynology* (*Erdtman's Handbook of Palynology*), 2nd edn (Nilsson, S. & Praglowski, J., eds, with contributions by Arremo, Y. et al.). Imprint Munksgaard, Copenhagen, 1969, 1992.

Jansonius, J. & McGregor, D. C. (eds), 1996. *Palynology, Principles and Applications*. American Association of Stratigraphic Palynologists Foundation, College Station, Texas.

Johri, B. M., 1992. *Comparative Embryology of Angiosperms*, Vol. 2. Springer-Verlag, Berlin.

Martin, A. C. & Barkley, W. D., 1961. *Seed Identification Manual.* University of California Press.
Punt, W., Blackmore, S., Nilsson, S. & Le Thomas, A., et al., 1994. *Glossary of Pollen and Spore Terminology.* LPP Foundation, Laboratory of Palaeobotany and Palynology, University of Utrecht, Utrecht.
Roth, I., 1977. *Fruits of Angiosperms. Handbuch der Pflanzenanatomie*, Bd. 10, T. 1. Gebr. Borntraeger, Berlin.
Rudall, P. J., 1992. *Anatomy of Flowering Plants : an introduction to structure and development*, 2nd edn. Cambridge University Press.
Vaughan, J. G., 1970. *The Structure and Utilization of Oil Seeds.* Chapman & Hall, London.

Chapter 8

Benzing, D., 1980. *The Biology of the Bromeliads.* Mad River Press, Eureka.
Carlquist, S., 1975. *Ecological Strategies of Xylem Evolution.* University of California Press, Berkeley.
Fahn, A. & Cutler, D., 1992. *Xerophytes. Handbuch der Pflanzenanatomie*, Bd. 13(3). Borntraeger, Berlin.
Gibson, A., 1996. *Structure–function Relations of Warm Desert Plants.* Springer, New York.
Gibson, A. & Nobel, P., 1986. *The Cactus Primer.* Harvard University Press, Cambridge.
Haberlandt, G., 1918. *Physiologische Pflanzenanatomie*, 5th edn. Engelman, Leipzig.
Roth, I., 1981. *Structural Patterns of Tropical Barks.* Borntraeger, Berlin.
Sculthorpe, C. D., 1985. *The Biology of Aquatic Vascular Plants.* Koeltz Scientific Books, Königstein.
Tomlinson, P., 1986. *The Botany of Mangroves.* Cambridge University Press.
Tomlinson, P., 1990. *The Structural Biology of Palms.* Oxford University Press.

Chapter 9

Cutler, D. F., Rudall, P. J., Gasson, P. E. & Gale, R. M. O., 1987. *Root Identification Manual of Trees and Shrubs.* Chapman & Hall, London.
Gale, R. & Cutler, D., 2000. *Plants in Archaeology. Identification manual of artefacts of plant origin from Europe and the Mediterranean.* Westbury and Royal Botanic Gardens, Kew.
Hayward, H. E., 1938. *The Structure of Economic Plants.* Macmillan, New York.
Jackson, B. P. & Snowdon, D. W., 1968. *Powdered Vegetable Drugs.* J. & A. Churchill, London.
Parry, J. W., 1962. *Spices. Their morphology, histology and chemistry.* Chemical Publishing Company, New York.
Seiderman, J., 1966. *Starke Atlas.* Paul Pary, Berlin.
Trease, G. E. & Evans, W. C., 1972. *Pharmacognosy*, 10th edn. Bailliere Tindall, London.
Wallace, T. E., 1967. *Textbook of Pharmacognosy*, 5th edn. J. & A. Churchill, London.
Winton, A. L., Moeller, J. & Winton, K. B., 1916. *The Microscopy of Vegetable Foods.* John Wiley, New York.

Bradbury, S., 1973. *Peacock's Elementary Microtechnique*, 4th edn, revised. Edward Arnold, London.
Jensen, W. A., 1962. *Botanical Histochemistry*. W.H. Freeman, San Franciscoo.
Purvis, M. J., Collier, D. C. & Walls, D., 1964. *Laboratory Techniques in Botany*. Butterworths, London.
Ruzin, S. E., 1999. *Plant Microtechniques and Microscopy*. Oxford University Press, New York.

Index

Page numbers in *italics* refer to Figures; those in **bold** to Tables.

Abies *32*
Acacia alata *81*, *85*, *98*
Acer 164
Acer pseudoplatanus 165, *166*
achenes 128
acid bog habitat 152
Acmopyle pancheri *65*
acrolein 172–3
  fixation procedure 174–5
adaptations 6–8, 135–53
  ecological 73, 76, 137–8
  hydrophytes 150–2
  mechanical 135–7
  mesophytes 147–50
  practical aspects 152–3
  xerophytes *see* xerophytes
Aegilops crassa *95*, *99*, 102
aerial roots 49, 149
Aerva lanata *81*
Aesculus hippocastanum *129*
Aesculus pavia *44*
Agave 10, 76
Agave franzonsinii *95*, 102
Agrostis 100, 138
Agrostis stolonifera *99*
Ailanthus 159
air spaces
  hydrophytes 150
  mesophyll 74, 97, 112
    xerophytes 146
Ajuga reptans var. *atropurpurescens* 110
Albuca 73
albuminous cells 42, 44, *65*, 65, 108
alcian blue 182
alcohol-based fixatives 171–2
aleurone grains 102
algae 6
Alismatales 67
Allium 18, *19*, 111
Alnus glutinosa 28, *29*, *37*, 165, *167*
Alnus nepalensis *29*
Aloe 9, 76, 77, *78*, 139
Aloe lateritia var. *kitaliensis* 77, *79*
Aloe somaliensis *140*
aloes 13, 76, *78*, 86, 142, 157
Ammophila 139, 142
Ammophila arenaria (marram grass) 82, 92, *141*
Anacardiaceae 86, 139
Anarthria 156
Anarthriaceae *156*, 156
angiosperms 4, 7, 10
  floral part vascularization 121–3
  phloem 65, 108
    secondary 43–5
  taxonomy 155
  wood (secondary xylem) 31–6, *36*
    axial system 33
    growth rings 33, 35, 41
    rays 35–6
    ring porous 33–4, 41
animal feeds 159–60
animal pests 162–3

Annonaceae 130
annuals 7, 8, 57
Anthemis 128
Anthemis arvenis 128, *130*
Anthemis perigina 128, *130*
Anthobryum triandrum 146
anthocyanins 149
Aphelia cyperoides *88*, 88
Apiaceae 128
apical meristems 11, 14, *15*, 15–20
  cell layers 16
  culture technique 21–2
  quiescent zone 16, 17
Apocynaceae 105
apple 76
aquatic plants *see* hydrophytes
Arabidopsis 111
Araceae 149
Araucaria 73
Arbutus unedo *148*
archaeological specimens
  wood fragments 165, 167
  wood products 167–8
arid habitat adaptations *see* xerophytes
arm cells 93, *94*
aroids 49, 136
Artemesia vulgaris *91*
Arundo donax *80*
Asclepiadaceae 105, 139
ash (*Fraxinus*) 41, 164
Asimina triloba *132*
Asparagales 98, 99
Asteraceae 128
Aucuba japonica Thunb. 110
Avena 159
axillary buds 57, 59
Azorella compacta 146

balsa (*Ochroma lagopus*) 33
bamboo 82, 92, 100
Bambus vulgaris *99*, 100
bark 42, 44–5
  medicines extraction 158
bars of Sanio 31
begonia 149
Betula 43
biennials 7, 8
Bignoniaceae 148
black ironwood (*Krugiodendron ferreum*) 33
Boehmeria 62
Bombax (kapok) 90
Borassus 73, 100, *168*
bordered pit 30, 66
Brassica (cabbage) 76
Brassicaceae 84
bridge grafts 25, *26*
British Pharmaceutical Codex 158
British Pharmacopoeia 158
Briza maxima *113*
Bromeliaceae 149, 150
Bromus unioloides 118
Bryonia 105
bud grafts 24, *25*, 25
bud scales 147
buds 6, 70
  adventitious 6, 12, 14, 27
  axillary 57, 59
  mesophytes 147
bulbs 139
bulliform cells 81, 82, *92*, 92
Bulnesia sarmienti 41
bulrush (*Typha*) 151
bundle sheaths 10, 11, 117–18, 135
  C4 plants 111, 112, *113*, 114, 117
  parenchymatous 99, 100, 116
  sclerenchymatous 116

C3 pathway/plants 94, 96, 112
  Cyperaceae 116
  Pooid grasses 114
C4 pathway/plants 94, 96, 97
  bundle sheaths 111, 112, *113*, 114, 116
  Cyperaceae 116, 117
  Panicoid grasses 114
cabbage (*Brassica*) 76
Cactaceae 67, 123, 139
cacti, ribbed columnar 147
calcium carbonate crystals 98
calcium oxalate crystals 98
callus 14
  culture 22
    cytohybrid plant production 26
  grafting techniques 23–26

calyptra (root cap) 17–18, 54
calyptrogen 17, 18
cambium 12, 14, 20, 28, 41, 63
  fascicular 12
  fusiform initials 20, *21*, 28, 32
  horticultural applications 23–7
  interfascicular 12, 63
  ray initials 28
  see also cork cambium
Camellia 73
Camellia japonica *97*
Camellia sinensis *33*
Canada Balsam 183, 184
Cannabis sativa 168
Cannomois 54
cantilevers 136
carbolic acid solution 180
Carex 59, 84, 100
Cariniana legalis 131, *133*
carnation 22
carpels, vascularization 122
Carpinus betulus *42*
Carya (hickory) 41
Caryophyllaceae 146
Casparian strip 52, 62, 118
Cassia angustifolia *81*
Castanea 35
Cattleya granulosa *50*
Cedrus 31, *32*, 73
cellulose acetate replicas 190
Centrolepidaceae 52, *88*, 88, 101
Centrolepis exserta *88*
Cephaelis acuminata 158
Cephaelis ipecacuanha 158
Cephalotaxus 31
cereals 68, 100, 127
charcoal
  archaeological specimens 165, *167*
  fossilized materials 169
Chenopodiaceae 105
chip bud grafts 24, *25*
chloral hydrate 191
chlorazol black solution 180
chlorenchyma 8, 62, 71, 93
  forest floor plants 149–50
  xerophytes 142
chlorophyll 7, 76, 149
chloroplasts 8, 9, 57, 71, 93, 94
  monocotyledonous mesophyll 112, 116
  stem 62
chlorzinciodine solution (CZI; Schulte's solution) 180
Chondropetalum marlothii *64*
chromo-acetic acid 173
Chrysanthemum leucanthemum *85*
Cicer areitinum *129*
Cinchona 27
Cinnamomum camphora 41
Cistus 160
Cistus salviifolius *83*, *89*
classification 154–5, 156, 157
cleared material 191
clearing procedure 177–9
Clematis 137
climbers 8, 12, 59
  mechanical tissues 137
  vascular bundles 63
Clintonia uniflora *80*, 93, *94*
coconut palm 137
Codonanthe 16, *17*, *150*
Coffea 73
Cola acuminata *129*
Coldenia procumbens *91*
Coleus *18*, 108
Coleus barbatus *131*
collenchyma 7, 9, 135
  leaves 102
  stem 62
Commelinaceae 150
companion cells 44, 47, 65, 66, 108, 109–11, 115
  specialization 109–11
Compositae 67
conifers 73, 147
  resin ducts 119
  wood 30–1
  xeric adaptations 139
Convallariaceae 98–9
Convolvulus 84
Convolvulus arvensis *85*
Convolvulus floridus *91*
coppicing 27
Cordyline 12, 15, 26

cork cambium (phellogen) 14, 20
  commercial applications 22–3, *23*
corms 139
cortex 9, 10
  root 49, *50*, 51
  stem 57, 62
Corylus (hazel) 27, *89*, 159, 162, 167
Corylus avellana *148*
cotton (*Gossypium*) *45*, 90
Couratari asterotrichia 131, *133*
cowitch (*Mucuna*) 89, *90*, 162
Crassula 9, *143*
crassulas 142
cricket bat willow (*Salix alba* var. *caerulea*) 41
Crocus 73, 139
Crocus michelsonii *125*
Crocus vallicola *125*
cross-field pits 31, *32*
crystal violet 187
crystals 51, 67
  mesophyll *98*, 98–100
  stem cortex 62
Cucurbita 12
Cucurbita maxima (water-melon) 24, 43
Cucurbita pepo *64*, *129*
Cucurbitaceae 63, 66, 105, 165
Cupressaceae 31, *32*
cushion habit 146
cuticle 7, 8, 9, 74, 76–82
  hydrophytes 151
  mesophytes 148
  patterns/sculpturing 60, 76, 77, 77, *78*, 104
  surface preparations 190–1
  xerophytes 139
cutin 76
cuttings 22, *23*
cycads 83
Cymophyllus fraseri *99*, *113*
Cyperaceae 100, 103
  endodermoid sheaths 118
  leaf bundles 116–17, *117*
Cyperus diffusus *99*
Cyperus papyrus 167, *168*
cystoliths *98*, 100
Cytisus 25
deciduous trees 147
dedifferentiation 14
dehydration 173
  for permanent mounting 182, **183**
  using tertiary butyl alcohol 173–4, **174**, 184
Delafield's haematoxylin 182
Delphinium staphisagria *129*
desert, leaf adaptations 73
Dianthus 84
diaphragms 59, 67, *68*, 68
diarch root vascular system *53*, 53
dicotyledons 4, *5*, 9
  leaves 14, 74, *75*, 93
    secondary growth 73
    surface features *81*, 81
    vein differentiation 107
    venation 104–5, 135
  mechanical tissues 9, 10
  phloem 43–4
  roots 49, 53, 54
  stems 57, 59, 68
    vascular system 58, 59, 63
  taxonomy 155
  transport system 11, 12
  wood (secondary xylem) 31–6
Dielsia cygnorum *83*
Digitalis lanata 159
Digitalis purpurea 158, 159
Dionaea 152
dormancy 7
Dracaena 12, 15
dried plant material 171
Drimys 39
Drosera 89, 152
drought adaptations 73
  see also xerophytes
drug source identification
  forensic investigations 168–9
  medicinal plants 158–9
druses (cluster crystals) 62

Ecballium 12
Ecdeiocolea 142, *145*, 156
Ecdeiocoleaceae *156*, 156
Echinodorus cordifolius *53*

ecological adaptations 137–8
  hydrophytes 150–2
  leaves 73
    cuticle 76
  mesophytes 147–50
  xerophytes *see* xerophytes
economic aspects 154–69
elderberry (*Sambucus*) 68
electron microscopy 191, 193–4
Elegia 73, 140
Elegia parviflora *83*
Eleocharis 118
Eleutharrhena 157
Eleutharrhena macrocarpa *157*
Elodea 17
embedding oven 173, 174
embryo culture 127
embryology 127
endarch vascular anatomy 53
endocarp 128
  stony 128, *132*
endodermis 18, *51*, 51–2
  leaves 118
  stem 58, 62–3
endodermoid sheath 57, 58, 63
endogenous development 18, 56
endosperm 130, *132*
endotesta 130
ephemerals 7, 57, 138
epidermis 7, 8, 9
  bulliform cells 81, 82, *92*, 92
  development 16
  leaf 70, 74, *75*, 76–93, 104
    dicotyledons *81*, 81
    monocotyledons 79, *80*, 81
  mesophytes 148
  pericarp 128
  roots (rhizodermis) 48–9
  silica bodies *99*, 100–2
  slime cells 128
  stem 57, 60
  tannins 102
  xerophytes 139
epiphytes 8, 142–3, 149, 150, 152
  aerial roots 49
  hairs 91
Equisetum *15*
ergastic substances 97–8
Erica 142
Ericaceae 123
eucalyptus 139
Euparal 183, 184
Euphorbia splendens *33*
Euphorbiaceae 67, 139
European Pharmacopoeia 158
Evandra montana *99*
evergreens 73, 147
  xeric modifications 139
evolutionary aspects
  floral parts 123
    vascularization 121–2
  leaves 71, 72
  phloem 65
    secondary 43
  secondary xylem (wood) 38–9
  vessel elements 54, 66
exarch vascular anatomy 53
exocarp 128, *130*
exodermis 49, 57
exogenous development 18, 57
exotesta 130, 131

Fagaceae *40*
Fagus 35
Fagus sylvatica *129*
Fast Green 182, 183
ferns 7, 92, 155
  leaves 93
  phloem 108
  stomata 85
fertilization 6
fibre tracheids 32
fibres 6, 7, 9, 51
  evolutionary aspects 38, 39
  extraxylary 10
  leaves 135, 136–7
  secondary phloem 42, 44
  secondary xylem (wood) 31, 32, 33, *38*, 41
  stem 58, 62, 66
Ficus 73
Ficus elastica *98*, 100
Fimbristylis 116, 118
fixatives 171–3
  non-coagulating 172–3
flavenoids 76

flax (*Linum*) 44, 62
Flemming's fixative 173
Flemming's Triple Stain 187, **190**
flowers 121–7
  cleared preparations 191
  pollinator adaptations 122
  scanning electron microscopy 123–4
  vascularization 121–3, *123*
    bundle fusions 122
food plants
  adulterants/contaminants identification 159–62, *160*
  seeds 127
forensic applications 162, 168–9
forest floor plants 149
formaldehyde-based fixatives 171–2
formalin-acetic acid-alcohol (FAA) 171–3, **172**, 184
formalin-propionic acid-alcohol (FPA) 173
Fraxinus (ash) 41, 164
fruit 121
  histology 127–34
fuchsin, aniline blue, iodine in lactophenol (FABIL) double stain 181
fume cupboard 170, 171, 172, 173, 174

Gahnia 142
Gaimardia 88
gametes 6–7
Gasteria *5*, 9
Gasteria retata *83*
Gaylussacia frondosa *123*, 123
germination 134
Gesneriaceae 149, 150
girder structures 137
Gladiolus 73
glochids 162
Gloriosa superba *80*, *83*, *113*
glycol methacrylate 172, 173
gorse (*Ulex*) 142
Gossypium (cotton) *45*, 90
grafting 20, 22, 24–6
  bridge 25, *26*
  bud 24, *25*, 25
  interfamily 25
  monocotyledons 26
  root 25
grafting tape 24
grape vine (*Vitis vinifera*) 43, 137, 159
grasses 12, 46, 59, 73, 79, 81, 82, 90, 97, 100
  vascular system 112, 115, 116, 117
    Panicoid 112, *114*, 114
    Pooid 112, *114*, 114
  venation 135–6
growth 14
Guaiacum officinale (lignum vitae) 33, *35*
guard cells 57, 61, 83
Guarea 158
Gunnera 136
gymnosperms 4, 7, 10
  leaves 73, 93, 96
  phloem 65, 108
    secondary 42
  stems 57, 68
    vascular system 58
  taxonomy 155
  wood (secondary xylem) 30–1
    cross-field pits 31, *32*
    radial system 31

Haemanthus 139
hairs 61, 87–9, *88*, 160–2, 163
  antiherbivore function 92
  glandular 89, *90*, 90, 118, 162
    insectivorous plants 152
  hydrophyte 151
  irritant 162
  non-glandular 89, 90, *91*
  suberin bands 91, 92, 142
  water-absorbing 91, 150
  xerophyte 91, 142, 144, 146
  see also trichomes
Hakea 105, 142
Hakea scoparia *144*
halophytes 76, 146–7
Haworthia 139, 142
Haworthia greenii *140*

haworthias 76, *78*
hazel (*Corylus*) 27, *89*, 159, 162, 167
heavy metals accumulation 138
Hedera helix (ivy) 110, *132*, 162
Helianthus (sunflower) 62, 119
Hevea brasiliensis 67
hibernation 73
hickory (*Carya*) 41
Hippophae 36
honey, purity evaluation 124
hydathodes 87, 118
hydrochloric acid 180
hydrofluoric acid 176, 177
hydrophytes 150–2
  acid bog plants 152
  air cavities 150–1
  buoyancy aids 151
  hairs 151
  hibernation periods 73
  leaves 150, 151, 152
  mesophyll 74
  stems 62
  stomata 84
  vascular tissue 151
hypocotyl 6, 12, 53
hypodermis 57, *61*, 61
  leaf 74, 92–3
  pericarp 128
  xerophytes 145
Hypolaena 146

identification of source materials 154, 156, 157
  crops devoured by animal pests 162–3
  forensic applications 168–9
  medicinal plants 158
  woods 163–4
idioblasts 100, 102, 128
  secondary phloem 44
Ilex 73
Ilex aquifolium 74, *75*
infiltration 173
  paraffin wax 184
  paraplast 174–5, 184
  permanent slides preparation 184
  through tertiary butyl alcohol series 173–4, **174**
inflorescence axes, vascular bundles 63
insectivorous plants 152
intercalary meristems 14, 73
  use in horticultural propagation 22
intermediary cells 109
Ionidium 158
Ipecacuanha 158
Iridaceae 99
Iris *55*, 139
ivy (*Hedera helix*) 110, *132*, 162

Juglans (walnut) 68, 165
Juncaceae 98, 99, 101, 140
Juncales 100
Juncus 12, *13*, 62, 68, 84, 139
Juncus acutiflorus *50*, *53*
Juncus acutus *64*
Juniperus glauca 24
Juniperus virginiana 24
Justicia 84, *90*, *91*
Justicia cydonifolia *85*

kapok (*Bombax*) 90
Klattia 15
Kniphofia macowanii *80*
Kranz mesophyll 112, 116, 118
Krugiodendron ferreum (black ironwood) 33

Labiateae 59
Laburnum 25
lacuna 49, 66, 112
Lamiaceae 128
lamina 10, 11
Laminaria 6
Landolphia 67
Larix *32*
lateral meristems 14, 20
  horticultural applications 22–7
lateral roots 54, 56
latex 119
latex cells 67, *98*
Lathyrus 59
laticifers 44, 67, 119

Laurus nobilis *43*
Lavendula spica *131*
Laxmannia 119, 146
leaf buttresses 16, 17, *18*, 70
leaf gaps 59
leaf initials 70
leaf traces 11, 13
leaves 6, 7, 70–120
  bulliform cells 81, 82, *92*, 92
  cleared preparations 191
  cuticle 76–7
  development 70, 71
  ecological adaptations 73
  endodermis 118
  epidermis 70, 76–93, *80*, *81*, *82*, *83*, 104
  evolutionary aspects 71, *72*
  food adulterant/contaminant identification 159, 160
  forest floor plants 149
  growth 14–15
  hydrophytes 150, 152
  hypodermis 74, 92–3
  insectivorous plants 152
  light capture adaptations 7–8, 149
  maturation (sink to source transition) 70
  mechanical support tissues 10, *95*, 102–3, 135–7
  medicines extraction 158
  mesophyll 71, 93–102
  mesophytes 147, *148*
  midrib 70
  monocotyledons 79, *80*, 81, 111–18
  movements 82, 92, 139
  procambium 70
  reduction 73, 146
  secretory structures 118–19
  sectioning process 177, *179*
  shedding/life-span 73, 147
  silica bodies *99*, 100–2
  standard examination levels 191, *192*
  stomata 83–7
  structure 74, *75*
  surface preparations *188*, 189
  surface sculpturing 77, 79
  tannins 102
  trichomes 87–92
  tropical rain forest trees 148
  vascular system 46, 71, *72*, 74, *75*, 103–11
    development 107
    phloem 108–11
  veins 70, 135–6
  venation 104–5
    classification system *106*
  xerophytes 139, 142, 143–5
Lecythidaceae 131, 133
Leguminosae 36, 39, 99
lenticels 20, 46
Leptocarpus 13, 88, 142
Leptocarpus tenax *88*, *144*
Lepyrodia scariosa 61–2
lianes 43
light capture adaptations 149
lignum vitae (*Guaiacum officinale*) 33, *35*
Liliaceae *98*
lime (*Tilia*) 33, 41, 44, *45*
Limnophyton obtusifolium *151*
Limonium vulgare *87*, 87
Linum (flax) 44, 62
Liriodendron tulipifera *33*, 110
Liriope 120
Lithocarpus 35
Lithocarpus conocarpa *40*
Lithops 9, 102, 142
low light environment adaptations 149
Loxocarya pubescens *88*

Magnoliales 39
Malpighiaceae 90
Malvaceae 44
Malvaviscus arboreus *45*
mango 128
Marantaceae 150
marjoram (*Origanum vulgare*) *89*, 160
marram grass (*Ammophila arenaria*) 82, 92, *141*
Matricaria lamellata *130*

mechanical support tissues 7, 8, 9–11
adaptative features 135–7
climbers 137
leaves *95*, 102–3, 135–7
cuticle 76
pericarp 128
petioles 136
stems 58, 63–4, 137
xerophytes 142
Medicago 159
medicinal plants 158–9
Meeboldina 88
Melastomataceae 104
Mentha (mint) 87, 159, 160
Mentha spicata *89*
meristems 6, 14–27
apical 14, *15*, 15–20, 21–2
horticultural applications 20–7
intercalary 14, 22, 73
lateral 14, 20, 22–7
leaf 70
secondary thickening 15
mesembryanthemum 142
mesocarp 128
mesophyll 71, 74, *75*, 93–102
crystals *98*, 98–100
ergastic substances 97–8
intercellular spaces 74, 97, 112
Kranz 112, 116, 118
mesophytes 148
palisade layer 71, 74, 93, 148
paraveinal 96, 107
plicate cells 96
sclereids 96, *97*
silica bodies *99*, 100–2
spongy layer 74, 93, 148
water storage 97, 142, *143*
xerophytes 142, *143*, 146
mesophytes 147–50
leaves 147, *148*
mesophyll 74
mesotesta 130, 131
mestome sheath 114, 116, 117, 118
metaphloem 11, 47, 66, 107, 112, 115, 116
metaxylem 11, 66, 107, 112
roots 54
methylene blue 180
microhairs 90
micropapillae 60
microscope attachments 194
microscopy 191, 193–4
optical techniques 194
microtome 175, 184
operating procedure **185**, 185
midrib 9, 11, 70, 103, 107, 135, 136
vascular bundle 74, *75*
mint (*Mentha*) 87, 159, 160
monocotyledons 4, *5*, 9
grafting 26–7
leaves 73, 93, 96, 97, 111–18
blade bundle anatomy 112, *114*, 114
bundle sheaths 116, 117–18
epidermis 79, *80*, 81
growth 14–15
phloem 115
venation 105, 112, 114, *115*, 135, 136
mechanical support tissues 9, 10, 63
meristems 15, 20, 26
roots 48, 49, 53, 54
secondary thickening 20, 26
stems 59, 68
vascular system 58, 63, 64
taxonomy 155
vascular bundles 11, 12–13, *13*, 65
xylem, evolutionary aspects 39
montane environment 146
mosses 7
motor cells 81, 82, *92*, 92
insectivorous plants 152
mountain plants
cushion life forms 146
hairs 142
ultraviolet light protection 145
mounting media 184
mucilage cells 51
Mucuna (cowitch) 89, *90*, 162
Musanga cecropioides *37*
Myristicaceae 130
Myrtaceae 139

Narcissus 73, 139
nectaries 123
  extra-floral 118
Nerium oleander 84, 93
Nestronia umbellulata *123*, 123
nettle (*Urtica*) 89, *90*
Nicotiana (tobacco) 162
nitric acid 191
Nivenia 15
nocturnal folding 82, 92, 136
nodes 12, 13, 57, 59
non-coagulating fixatives 172–3
Nothofagus solandri *40*
Nymphaea 73, 84
Nymphoides 47, 65, 108

oak (*Quercus*) 35, 41, 43, 73, 162, 164, 167
Ochroma lagopus (balsa) 33
Ochroma pyramidalis 33, *34*
oil seeds 127
Olea europaea *97*, *148*
Olivacea radiata *97*
Ophiopogon 120
Opuntia 162
Orchidaceae 149
orchids 18, 49
Origanum vulgare (marjoram) *89*, 160
Orobanche 62
Oscularia deltoides *98*
osmium tetroxide 172
Oxalis exigua 146

paleobotany 169
palisade cells 71, 74, 93, 148
palisade ratio 93–4
palms 13, 15, 59, 65, *99*, 100, 136–7
palynology 124–7
Pandanaceae 65
Pandanus 18
papillae 61, 144, 146
  see also trichomes
paraffin wax infiltration 173, 184
paraplast infiltration 174–5, 184
parenchyma 9
  bundle sheath 99, 100, 116
  central ground tissue (pith) 54, 67
  mesophyll 93
  pulvinus 136
  root cortex 49, 51
  secondary phloem 42, 44
  secondary xylem 33, 34–5
  stem 57, 58, 59
Pariana bicolor *92*
passage cells 52
Passiflora foetida *98*
Pattersonia 15
peach 128
pear (*Pyrus*) 41
Pelargonium 10, 58
perennials 7, 8, 147
  xerophytes 139
pericarp 128, *129*, *130*, *131*
pericycle 18, 52, 54, 60, 111
permanent slides preparation 183–9
  attaching sections to slide 185–6
  infiltration procedures 184
  staining **186**, 186–7, **187**
petals, vascularization 122
petiole 9, 10
  mechanical tissues 136
  standard examination levels 191, *192*
  trace 10, 11
Phalaris canariensis 80, *95*, 102, 117
phelem 20
phelloderm 20
phellogen *see* cork cambium
phloem 6, 8, 10, 11, 28
  angiosperms 43–5, *45*, 65, 108
  development 11, 16, 20
  evolutionary aspects 43, 65
  ferns 108
  gymnosperms 42, 65, 108
  leaf 103, 104, 105, 108–11, *109*
    midrib 70
    monocotyledons 115
  loading 65, 108
    companion cell
      function 109–11

root 53, 54
secondary 12, 41–5, *45*
rays 42, 44
stem 58, 66–7
structure–function relationships 45–7
transport 64–6
unloading 111
phloroglucin/hydrochloric acid 180
Phoenix 73
photosynthesis 7, 71, 76
phyllotaxis 6, 16, 70
phylogeny
secondary xylem (wood) 39
see also evolutionary aspects
Phytolacca americana (pokeweed) 68
Picea (spruce) 31, *32*, 41
Pinaceae 31
Pinguicula 89, 152
Pinus 30, 31, *32*, 73, 146, 159
Pinus ponderosa *83*, 96
Pinus sylvestris *30*
Piper nigrum *64*
Piperaceae 59, 64
Pistia (water lettuce) 151
pith 54, 67–8
plant materials 170–1
Plant Micromorphology Bibliographic Database 155
plasmodesmata 11, 64, 94
monocotyledon leaf blade bundles 114
pericycle 111
phloem
loading 109, 110
unloading 111
Platanus 27
Platymitra siamensis *42*
plum 76
Plumbago 84
Plumbago zeylanicum *81*, *85*, 105, *107*
Poaceae 94, 102, 103
pokeweed (*Phytolacca americana*) 68
pollarding 27
pollen 7, 124–7, *125*
culture 127
stigma interactions 124, *126*, 126–7
polyarch root vascular system *53*, 53
polyester wax infiltration 173
polyphenols 145
Pomoideae 164
Populus 164
Posidonia 152
Potamogeton 73
potato 22, 105
practical microtechnique 170–94
materials 170–1
safety considerations 170
tissue processing *see* tissue processing
preservation 171
prickle hairs 90
primula, pin-eyed/thrum-eyed 127
procambium
leaf 70
strands 11, 16, 107
prosenchyma 9
protandry 124
protective clothing 170
gloves 172, 173, 174
protoderm 70
protogyny 126
protophloem 11, 66, 107, 112
stem 57
protoplast fusion 26
protoxylem 11, 66, 107, 112
root 53
stem 57
Prunoideae 164
Prunus 25
Pseudolarix 31
Pseudotsuga 31
pulvinus 82, 92, 136
Pycnarrhena macrocarpa 157
Pycnarrhena pleniflora *157*
Pycnophyllum micronatum 146
Pycnophyllum molle 146
Pycreus 118
Pyrus (pear) 41

Quercus (oak) 35, 41, 43, 73, 162, 164, 167

Quercus brandisiana *40*
Quercus robur *36*, *40*
Quercus suber 22–3
quiescent zone 16, 17
Quillaja 27

Ranunculacae 84
Ranunculus *55*
Ranunculus acris 53
raphides 62
raspberry 22
rays 46
  angiosperm wood 35–6, *36*, *37*
  evolutionary aspects 39
  gymnosperm wood 31
  initials 20, *21*
  secondary phloem 42, 44
  tracheids 31
red clover (*Trifolium*) *60*
reproduction 6–7, 8
resin ducts 31, 119, 146
Restio 13
Restionaceae 13, *88*, 88, *99*, 101, 139, 140, *156*, 156
Rhapis 13
Rheum officinale 158
Rheum rhaponticum 158
rhizodermis 48–9
rhizomes
  food (herbs) adulterants/ contaminants 159
  medicines extraction 158
  vascular bundles 63
  xerophytes 139
Rhododendron *15*, 73
rhododendrons 90
rhubarb 136
rhytidome (bark) 42, 44–5
Ribes nigrum *23*
ring porous wood 33–4, 41
root cap (calyptra) 17–18, 54
root hairs 48
root stock 24, 25, 26
roots 4, 6, 7, 13, 48–56, *50*, *55*
  adventitious 6, 13, 14, 22
  aerial 49, 149
  apical meristem 17–18, *19*, 20
  cortex 49, *50*, 51
  damage to buildings 164–5
  ecological adaptations 142
  endodermis *51*, 51–2
  epidermis (rhizodermis) 48–9
  food (herbs) adulterants/ contaminants 159
  lateral 54, 56
    development 18, *19*, 52
  mechanical strengthening tissues 10–11
  medicines extraction 158
  pericycle 52
  standard examination levels 191, *192*
  vascular system 46, *53*, 53–4, *55*, *56*
Rosaceae 123, 164
rose 25
ruminate endosperm 130, *132*
ruthenium red 181

Saccharum officinarum 114, *115*, 118
safety considerations 170
safranin 182, 184, 191
Safranin Fast Green staining chart **187**, **189**
Salicaicae 164
Salicornia 9
saline environments 146–7
  leaf adaptations 76
Salix 27, 164
Salix alba var. *caerulea* (cricket bat willow) 41
Salix babylonica L. 110
salt glands *87*, 87
Salvadora persica *61*
Salvia officinalis *90*, *91*
Sambucus (elderberry) 68
Sambucus nigra *33*, *37*
Santalaceae 123
sap, cold habitat adaptations 147
saxifrages 87
scales 61, 92
  water-absorbing 150
  see also trichomes

scanning electron microscopy 191, 193
archaeological wood specimens 165
flowers 123–4
leaf surface features 157
seed coats 131
Schouwia 73
Schulte's solution (chlorzinciodine; CZI) 180
Scilla 139
scion 24, 25, 26
Scirpodendron chaeri *33*
sclereids 7, 44, 51, 54, 58, 59, 62, 66, 67, 128, 148
leaves 96, *97*
sclerenchyma 7, 63
leaves 102, 112
pericarp 128
secondary growth 8, 12, 14, 68
leaves 73
stems 57
secondary plant products 118, 119
secretory canals 33
secretory structures
external 118–19
internal 119
leaves 118–19
sectioning 175–7, *178*
cork supporting material 176–7, *178*, *179*
free-hand 175
leaves 177, *179*
stems 176
twigs 176
wax-embedded specimens 175, 184, 185
wood 175–6, *176*
sedges 46, 59, 100, 112, 115, 116
seed coat *129*, 130
scanning electron microscopy 131
winged seeds 131, *133*
seeds 121, 127–34
germination 134
histology 127–34
nutrient delivery to storage cells 111
winged 131, *133*
xerophytes 139
Selaginella 149
Senecio 9
Senecio scaposus *143*
sepals, vascularization 122
Sequoia 31
sieve areas 43, 44, 65
sieve cells 42, 43, 46, 65, 108
sieve elements 46, 108, 109, 115
symplastic/apoplastic loading 109
sieve plates 43, *44*, 44, 65, 66
sieve tubes 43, 44, 46, 65–6, 115, 116
silica bodies 61, *99*, 100–2, *101*, 162, 163, 180
tissue sectioning 177
silver-leaved composites 89
sledge microtome 175
slime cells 128
slime trichomes 128, *130*
Smilax 105
Smilax hispida *80*
Solanaceae 105, 162
Solanum 25
Sorghum 84
spines 147
spongy mesophyll 74, 93, 148
spruce (*Picea*) 31, *32*, 41
staining 179–83
paraffin sections **186**, 186–7, **187**
stains
permanent 182–3
temporary 180–1
stamens, vascularization 122
standard examination levels 191, *192*
starch grains 11, 97, 158, 159, *161*
starch sheath 57, 58, 62
stegmata 100, 101
stele 58
stellate cells 49, 59, 67, 151

stems 7, 8, 57–69, *60*
 apical meristem 16–17, *17*, *18*
 central ground tissue (pith) 67–8
 cleared preparations 191
 cortex 62
 endodermis 62–3
 epidermis 60
 hydrophytes 150
 hypodermis *61*, 61
 mechanical support tissues 10, 58, 63–4, 137
 secondary thickening 57
 sectioning process 176–7, *178*
 standard examination levels 191, *192*
 stomata 61
 surface preparations 190
 transport tissue 11–12, 66–7
  phloem 64–6
 vascular system 11–12, 57, *58*, 58–9, 63–4, *64*
 xerophytes 139, 140, 143
stigma–pollen interactions 124, *126*, 126–7
stomata 7, 8, 9, 46, 74, 83–7, 104, 162
 actinocytic 84
 amphistomatic 83
 anisocytic 84, *85*
 anomocytic 84, *85*
 cyclocytic 84
 development *86*, 86
 diacytic 84, *85*
 hydrophytes 151
 hyperstomatic 83
 hypostomatic 83
 mesocytic 85
 mesophytes 148
 paracytic 84, *85*, *86*, 86
 polocytic 85
 stem 61
 subsidiary cells 84, 86
 sunken 83, 84
 tetracytic 84
 types 84–5, *85*
 water droplet exudation 86–7
 wax embellishment 77
 xerophytes 139, 140, 143, 146
stomatal crypts 84
Stratiotes 49, *50*, 73
striae 60
suberin 20
 bands 91, 92, 142
 lamellae 114, 117, 118
succulents 142, 146
 halophytes 146–7
 isobilateral palisade tissue 74
sucrose transporters 111
sudan IV 180
sunflower (*Helianthus*) 62, 119
surface preparations 189–91
 cuticle 190–1
 leaf *188*, 189
 replicas 190
 stem 190
Swietenia mahagoni *37*
Syringa *15*

tannin cells 51, 67, *98*
tannins 62, 102
Taraxacum 67
Taxodiaceae *32*
taxonomy 155–7
Taxus 31, *32*
teak (*Tectona grandis*) 41
tertiary butyl alcohol dehydration procedure 173–4, **174**, 184
tetrarch root vascular system *53*, 53
Thamnochortus argenteus *88*
Thamnochortus scabridus *83*
Theaceae 102
Thurnia sphaerocephala 71, 72
Tilia (lime) 33, 41, 44, *45*
Tilia europaea *34*
Tillandsia 91
tissue processing
 clearing 177–9, 191
 dehydration 173, 182, **183**
 fixation 171–3
  procedure for acrolein 174–5
 infiltration 173, 184
  paraplast 174–5

through tertiary butyl alcohol series 173–4, **174**, 184
killing tissues 171
mounting 182, *183*, 185
media 184
permanent slides preparation 182, **183**, 183–9
preservation 171
sectioning 175–7, *178*, 184
staining 179–83, **186**, 186–7, **187**
surface preparations *188*, 189–91
tissue systems 4–13
tobacco (*Nicotiana*) 162
Torreya 31
torsional forces, petiole adaptations 136
torus 30–1
tracheids 10, 30–1, 32, 33, *38*, 41, 66
evolutionary aspects 38, 39
radial (ray) 31
roots 53, 54
Trade Descriptions Act 159, 163
Tradescantia 84
Tradescantia pallida 126
transfer cells 12, 109
translocation 8, 43, 44, 46, 47, 71, 103
sinks 103, 108
sources 103, 108
transmission electron microscopy 191, 193
transpiration 8, 104
leaves 71
with xeric modifications 139, 140
traumatic ducts 31
tree root damage 164–5
triarch root vascular system *53*, 53, 56
trichomes 61, 87–92, 118
slime 128, *130*
Trifolium (red clover) *60*
Trigonobalanus verticillata *91*
Triticum 22
tropical rain forest 148–9
Tulipa 73, 139
tyloses 41
Typha (bulrush) 151

Ulex (gorse) 142
Ulmus 35
ultraviolet light protection 7, 145
reflective surfaces 146, 147
Urtica (nettle) 89, *90*

vascular bundles 10, 11–12, 46, 63
amphicribal 63
amphivasal 63, *64*
bicollateral 63, *64*, 105
collateral 63, *64*, 105
development 11, 16
leaves 71, *72*, 103, 104, 105
midrib 74, *75*
monocotyledons 63
grasses 112
pathways in floral parts 121–3, *123*
fusions 122
stem 63–4, *65*
vascular cambium *see* cambium
vascular system
hydrophytes 151
leaf 74, *75*, 103–11
stem 57, *58*, 58–9
vegetative propagation 20
callus culture 22
embryo culture 127
meristem culture 21–2
veins 10, 11, 103, 104, *107*, 135–6
major 105, 107
minor 105, 107–8
phloem loading 110
velamen 49, 149
venation 104–5, 135
classification system *106*
dicotyledons 104–5
monocotyledons 105, 112, 114, *115*
Verbascum bombiciforme *91*
vessel elements 10, 32, 33–4, *34*, *38*, 39, 53, 54, 66
evolutionary aspects 38–9, 54, 66
perforation plates 32, *33*, 38, 66
wall pitting 32, *33*, 38, 66

Victoria 84
virus-free stock, meristem culture 22
Vitis vinifera (grape vine) 43, 137, 159

walnut (*Juglans*) 68, 165
waste disposal 170, 174
water lettuce (*Pistia*) 151
water loss regulation 7, 8, 9, 104
cuticle 76
xerophyte adaptations 139, 140, 142
water storage cells 97, 142, *143*
water-melon (*Cucurbita maxima*) 24, 43
Watsonia 73, 139
wax 76, 77, *79*, 79, 139, 143
wilting 9, 83, 136
wind
leaf/petiole adaptations 136
xeric adaptations 140, 142, 145
'window' plants 142
Winteraceae 39
Witsenia 15
wood
angiosperms 31–6, *36*, 41
gymnosperms 30–1
identification of source materials 163–4
archaeological samples 165, 167–8
damaging tree roots 164–5
modern building materials 168
practical aspects 163–8
preservation 164
sectioning process 175–6, *176*
see also xylem, secondary
wound healing 20, 22
forestry practices 23, *24*
grafting techniques 24

xerophytes 73, 139–47, 152
boundary layer 143, 144
cold habitat adaptations 146, 147
drought escapers 139
drought resisters 139
hairs 91, 142, 144, 146
hypodermal layers 92–3, 145
leaf adaptations 76, 144–5
mesophyll 74, 97, 146
saline habitat adaptations 146–7
sclerotic 142, 146
stomata 84, 143, 146
surface roughness 143–4
water storage (succulents) 142, 146
waxy cuticle 77, 143
'window' plants 142
xylem 6, 10, 11, 28
apoplastic transport processes 103, 104
development 11, 16, 20
dwarfing root stock 25
leaves 103, 104, 105, 135
midrib 70
roots 53, 54
secondary 12, 28–41, *40*, *42*, *43*
angiosperms 31–6, *36*
axial system 28–9, *29*, *30*, 33
evolution 38–41
gymnosperms 30–1
rays (radial system) 28–9, *29*, *30*, 31, 35–6, *37*
uses 36–8, 41
stem 66–7
structure–function relationships 45–7
xylopodia 139

Zea 49, 100
Zea mays *56*, *58*, *99*, 117
Zostera 152